Functional Metagenomics: Tools and Applications

Trevor C. Charles • Mark R. Liles
Angela Sessitsch
Editors

Functional Metagenomics: Tools and Applications

Springer

Editors
Trevor C. Charles
Department of Biology
University of Waterloo
Waterloo, Ontario
Canada

Mark R. Liles
Department of Biological Sciences
Auburn University
Auburn, AL
USA

Angela Sessitsch
Biorescources
AIT Austrian Institute of Technology
Biorescources
Tulln
Austria

ISBN 978-3-319-61508-0 ISBN 978-3-319-61510-3 (eBook)
DOI 10.1007/978-3-319-61510-3

Library of Congress Control Number: 2017955378

Printed on acid-free paper

This Springer imprint is published by Springer Nature
The registered company is Springer International Publishing AG
The registered company address is: Gewerbestrasse 11, 6330 Cham, Switzerland

Preface

Microbiologists have long understood that most of the microbial world is hidden from view. This has been dramatically emphasized as a result of the application of DNA sequencing to the investigation of microbial communities. The advent of high-throughput DNA sequencing, coupled with accessible public sequence databases, has provided ample fodder for computational analyses of the genomes of uncultivated microbes. It is in this context that we address a major challenge in microbiology—the need for a more complete understanding of gene function that will support models and predictions of cell behavior and community dynamics in particular habitats. Metagenomics, and in particular functional metagenomics, provides a framework within which to address this challenge.

Metagenomics consists of a set of enabling technologies rooted in genomics, microbial genetics, microbial ecology, and bioinformatics, applied to the study of microbial genetic material recovered directly from environmental samples. It provides a deep window into the diversity of life on Earth, which is dominated by microbes. This rapidly emerging field is based on the application of DNA sequencing technology and microbial genetics to the investigation of microbes that to a large extent have not been grown in culture. Microbial communities perform critical services in the environment and are central to processes such as wastewater treatment, bioremediation, food microbiology, and the processes that are critical to the basic understanding of Earth's ecosystems.

Functional metagenomics, which seeks to determine not only what microbes are present but also what they are doing, facilitates the discovery and study of new enzymes or biosynthetic gene clusters without relying on prior cultivation of microbes whose genomes express these enzymes. This is a breakthrough technology whose value cannot be overestimated. It allows access to gene products without having to work with the original microbe that produced the enzyme or metabolite and that may not be able to, or is difficult to, culture. A much better understanding of enzyme function is essential to optimize the processes that occur in microbial communities that are providing essential services. Many enzymes are actually developed as commercial products or used to produce commercial products in industries such as textile, food, brewing, paper, biofuel, detergent, animal feed, bioremediation, green chemistry, and many more.

It is through this lens that this book was envisioned. A broad set of experimental and computational approaches are being developed to provide traction in the quest toward greater understanding of gene function in microbial communities. Initial chapters provide overview and examples of the application of high-throughput sequencing, bioinformatics tools, and different strategies for production of metagenomic libraries and library screening approaches. This is followed by examples of the application of functional metagenomics to microbial communities from different habitats and ecosystems.

This book is intended as a collection of representative studies and views that will provide the reader with a sense of some of the exciting work currently being done in functional metagenomics. We hope that it contributes to further advances in the field.

Waterloo, ON — Trevor C. Charles
Auburn, AL — Mark R. Liles
Tulln, Austria — Angela Sessitsch

Contents

1 **Metagenomic Cosmid Libraries Suitable for Functional Screening in Proteobacteria** 1
Jiujun Cheng, Kathy N. Lam, Katja Engel, Michael Hall, Josh D. Neufeld, and Trevor C. Charles

2 **Expression Platforms for Functional Metagenomics: Emerging Technology Options Beyond *Escherichia coli*** 13
Anna Lewin, Rahmi Lale, and Alexander Wentzel

3 **Engineering of *E. coli* for Heterologous Expression of Secondary Metabolite Biosynthesis Pathways Recovered from Metagenomics Libraries** 45
Lei Fang, Guojian Zhang, and Blaine A. Pfeifer

4 **Functional Analysis in Metagenomics Using MEGAN 6** 65
Sina Beier, Rewati Tappu, and Daniel H. Huson

5 **Enhancing Metagenomic Approaches Through Synthetic Biology** . . . 75
Luana de Fátima Alves, Rafael Silva-Rocha, and María-Eugenia Guazzaroni

6 **Metagenomics for the Discovery of Novel Biosurfactants** 95
Wesley Williams and Marla Trindade

7 **Challenges and Opportunities in Discovery of Secondary Metabolites Using a Functional Metagenomic Approach** 119
Alinne L. R. Santana-Pereira and Mark R. Liles

8 **Enhancing Functional Metagenomics of Complex Microbial Communities Using Stable Isotopes** 139
Marcela Hernández, Josh D. Neufeld, and Marc G. Dumont

9 **Metagenome Assembly and Functional Annotation** 151
Adina Howe, Fan Yang, and Qingpeng Zhang

10 Human Gut Metagenomics: Success and Limits of the Activity-Based Approaches ... 161
Alexandra S. Tauzin, Elisabeth Laville, Davide Cecchini, Hervé M. Blottière, Marion Leclerc, Joël Doré, and Gabrielle Potocki-Veronese

11 Metagenomics of Plant Microbiomes ... 179
G. Brader, E. Corretto, and A. Sessitsch

12 Metagenome Analyses of Multispecies Microbial Biofilms: First Steps Toward Understanding Diverse Microbial Systems on Surfaces ... 201
Christel Schmeisser, Ines Krohn-Molt, and Wolfgang R. Streit

13 Functional Metagenomics of a Replicase from a Novel Hyperthermophilic Aquificales Virus ... 217
David A. Mead, Scott Monsma, Baigen Mei, Krishne Gowda, Michael Lodes, and Thomas W. Schoenfeld

14 Functional Metagenomics and Antimicrobial Resistance ... 243
Fiona Walsh

Chapter 1
Metagenomic Cosmid Libraries Suitable for Functional Screening in Proteobacteria

Jiujun Cheng, Kathy N. Lam, Katja Engel, Michael Hall, Josh D. Neufeld, and Trevor C. Charles

Abstract Functional metagenomics, based on screening/selection of clones from metagenomic libraries, has the potential to make major contributions to our understanding of gene function and the development of biotechnology solutions. However, there are challenges and limitations that must be overcome if that potential is to be realized. These include cloning bias in library construction, host-dependence of gene expression, and library vector host range restrictions. In this chapter, we discuss some of our efforts to improve the quality and availability of metagenomic libraries through the production of a series of metagenomic cosmid libraries from diverse Canadian soils. Although these libraries are suitable for screening in a range of bacteria, they are currently limited to the Proteobacteria. To better capture genes from throughout the diversity of microbial life, it will be desirable to construct and make available metagenomic libraries that are able to support phenotypic screening in correspondingly suitable taxonomic backgrounds. Ongoing work is directed at achieving this important goal.

1.1 Introduction

The depth and breadth of known microbial diversity have been expanded greatly by the application of ever more powerful sequencing technology (Schloss et al. 2016; Hug et al. 2016). Nonetheless, the enormous benefit of accessing data from uncultivated microorganisms is tempered by the acknowledgement that the functions encoded by much of these newly determined DNA sequences cannot be reliably assessed and evaluated. Although the absence of most microbial diversity from pure culture represents a major limitation for gene discovery, functional metagenomics based on phenotypic screening may be the ideal methodological approach for circumventing this limitation.

J. Cheng • K.N. Lam • K. Engel • M. Hall • J.D. Neufeld • T.C. Charles (✉)
Department of Biology, University of Waterloo, Waterloo, ON, Canada, N2L 3G1
e-mail: trevor.charles@uwaterloo.ca

T.C. Charles et al. (eds.), *Functional Metagenomics: Tools and Applications*,
DOI 10.1007/978-3-319-61510-3_1

In general, the use of the term "functional metagenomics" implies a very specific function-based "wet-lab" methodology. Although the term is occasionally co-opted to mean something different, such as sequence-based metagenomics with a focus on gene function (Dinsdale et al. 2008; Roller et al. 2013), or even completely redefined to mean the study of functional members of the microbiota that influence human health (Li et al. 2008), such uses are rare in the scientific literature; the "wet-lab" distinction continues to hold sway. Functional metagenomics, in the traditional sense, involves using DNA that has been isolated from microbial communities to study the functions of proteins and other gene products encoded by that DNA. The process usually consists of extracting DNA directly from environmental samples, cloning that DNA in libraries, introducing those libraries into surrogate hosts where they can be expressed, and selecting or screening for functions of interest. This approach can facilitate the discovery of novel gene products such as enzymes for which DNA sequence is not predictive of function. The information from these analyses can contribute to the annotation of genome and metagenome sequences. In this way, functional metagenomics complements sequence-based metagenomics, similar to the way that molecular genetics of model organisms has provided knowledge of gene function that has been widely applicable in pure culture genomics and systems biology.

1.2 *Escherichia coli* as the Host for Metagenomic Libraries

Traditionally, functional metagenomics has been performed in *Escherichia coli*. This is directly related to the overwhelming dominance of *E. coli* as a model organism and its foundational role as an integral part of the molecular biology toolbox. Most functional metagenomic libraries are constructed in vectors that replicate in *E. coli*. These range from small-insert plasmid libraries, through to medium-insert cosmid and fosmid libraries, and large-insert BAC libraries (Kakirde et al. 2010). Each of these types of libraries has their advantages and disadvantages, but biases in library construction and library maintenance and host dependence of gene expression may have major impacts on the experimental outcomes of functional metagenomics.

Because metagenomic libraries are almost always constructed and maintained in *E. coli* host strains, and this is not likely to change, screening often also occurs in an *E. coli* background. For example, to isolate clones conferring antibiotic resistance, recombinant host cells can be applied directly to selective media-containing antibiotics. This example, although simple, has been useful for exploring the antibiotic resistance gene reservoir harboured in the human gut microbiota (Sommer et al. 2009). However, screening solely in *E. coli* strains may limit success due to possible incompatibilities that prevent expression in a given background. Depending on the target activity, functional screens can exhibit a low hit rate (Uchiyama and Miyazaki 2009), the reasons for which might include barriers at the levels of both transcription and translation. For example, promoters, codon usage, and regulator elements are all host-dependent factors that influence gene expression. Strategies to improve screening efficiency in *E. coli* have included the introduction of heterologous sigma factors to direct transcription initiation (Gaida et al. 2015), using T7 RNA polymerase (Terrón-González et al. 2013) and employing hybrid ribosomes (Kitahara et al. 2012). Despite these efforts, it will be necessary to continue the development of different screening hosts, especially for complementation of functions that are not available in *E. coli*. Fortunately, this is an area of investigation that has not been neglected.

1.3 Alternate Hosts for Screening Metagenomic Libraries

We have established that screening in surrogate hosts other than *E. coli* may provide additional success from functional screening due to the variation in regulatory and structural elements required for gene expression between the original organisms and *E. coli*. Though it is arguably difficult to quantify, one estimate of how much of the metagenome is accessible by screening in *E. coli* is ~40%, based on analysis of 32 genomes from different bacteria and archaea, counting ORFs with ribosome-binding sites and promoters that are predicted to be recognized in *E. coli* (Gabor et al. 2004). The fraction of "inaccessible" genes depends, of course, on the particular environmental sample DNA and its underlying microbial community composition. Regardless, to address this limitation, metagenomic libraries can be transferred from *E. coli* to other surrogate hosts that may be more suitable for screening. This may be done efficiently using conjugation or, if the recipient species is amenable, transformation or electroporation. The transferred clones must be able to replicate in the recipient host, either autonomously or after integration into the genome. Also, the issue of possible barriers to transcription and translation in *E. coli* is a particularly important methodological limitation.

Alternative expression hosts that have been used in functional metagenomics include *Agrobacterium tumefaciens*, *Caulobacter vibrioides*, *Rhizobium leguminosarum*, *Ralstonia metallidurans*, *Pseudomonas fluorescens*, *Pseudomonas putida*, *Xanthomonas campestris*, *Burkholderia graminis*, *Sinorhizobium meliloti*, *Bacillus subtilis*, *Thermus thermophilus*, and *Streptomyces albus* (Li et al. 2005; Aakvik et al. 2009; Uchiyama and Miyazaki 2009; Craig et al. 2010; Taupp et al. 2011; Schallmey et al. 2011; Liebl et al. 2014; Leis et al. 2015; Cheng and Charles 2016; Cheng et al. 2017; Iqbal et al. 2016). The vectors used for these metagenomic libraries contain single broad host range *oriV*, multiple *oriV* to support replication in *E. coli* as well as the screening hosts, or recombinase-based systems that facilitate integration into the chromosome of the screening host. Not only do these vectors allow screening in different host backgrounds, but they also make it possible to take advantage of sophisticated genetic analysis, in many cases using specific mutants and strain constructs. Despite what may at first glance appear to be a large number of possible screening hosts for the existing vectors and libraries, they actually only represent a very small proportion of phylogenetic diversity. Hence, there is a need for further expansion of the hosts for metagenomic screening to better represent the entirety of microbial diversity.

1.4 Considerations for Metagenomic Library Construction

There are important considerations for undertaking functional metagenomic approaches. First, consideration must be given to choosing an appropriate environment for the desired target genes. For instance, a rumen sample from a grass-fed cow may be ideal for generating a metagenomic library that is enriched with genes encoding enzymes for cellulose degradation (Gong et al. 2012), whereas a sample from a hot spring site would be more suited to isolation of genes encoding thermostable enzymes (Leis et al. 2015). Second, an appropriate vector must be selected for the library backbone. The choice depends on various factors, such as whether a small-insert or large-insert library is desired and, in the former case, whether expression vectors would be advantageous to help drive gene expression in *E. coli* (Kaddurah-Daouk et al. 2011). Third, surrogate host(s) other than *E. coli* may be considered, as indicated above, for either an attempt to increase the hit rate (Tebbe and Vahjen 1993; Ufarté et al. 2015) or for complementation of specific phenotypes (Wang et al. 2013). Finally, other logistics in the screening strategy have to be considered, such as whether to pool clones for screening or to instead keep clones arrayed and carry out individual clone screening. In the latter case, the achievable throughput must be very carefully considered because, depending on the particular screen, clone-by-clone screening may not be a feasible strategy, although the design of automated microfluidic screening strategies is an exciting area of development (Colin et al. 2015).

Although there are limitations and biases with existing approaches for constructing metagenomic libraries (Ekkers et al. 2012), as there are with all methods, functional metagenomics remains a powerful experimental strategy that has the potential to help improve our understanding of the mechanisms that underlie biological phenomena as well as aid in the functional annotation of the ever increasing number of metagenomes.

1.5 Enrichment of Desired Sequences

Not only can functional selections discover novel gene products, they can also greatly reduce the sheer quantity of genetic material to be sequenced. For example, a high-throughput functional metagenomic approach was used to find enzymes in the human gut involved in dietary fibre catabolism, reducing the amount of metagenomic DNA to be sequenced from 5.4×10^9 bp to 8.4×10^5 bp, a reduction of almost four orders of magnitude, simply by selecting for the growth of library clones on different polysaccharides (Tasse et al. 2010). Using this approach, the authors identified 73 carbohydrate-active enzymes, corresponding to a five-fold enrichment in the target-gene identification over shotgun metagenomic sequencing. If enrichment can be performed prior to sequencing, a great deal of time and resources can be saved, not to mention the value of having experimental data regarding function.

Enrichment can also be effective prior to library construction. This can involve subjecting environmental samples to traditional enrichment culture before DNA is extracted. More recently, it has become possible to enrich for DNA from members of the community that are performing certain metabolic tasks using techniques based on stable-isotope probing (Neufeld et al. 2007). For example, an enriched metagenomic library constructed from multiple displacement amplification products of DNA pooled from ^{13}C-cellulose incubations of soil microbial communities was used to isolate clones expressing cellulose-degrading ability at a higher rate than previously reported for non-enriched libraries (Verastegui et al. 2014).

1.6 Library Construction

The first step of functional genomics is the construction of metagenomic libraries. Many investigators prefer cosmid- or fosmid-based libraries because of their high cloning efficiency and large insert size. DNA fragments are first extracted from the environmental sample of choice, such as soil, faeces, or water. These fragments are typically enriched for those of high molecular weight by size selection using pulsed-field gel electrophoresis (PFGE). Following end repair and ligation to a linearized, blunt-ended cos-based vector, λ phage heads are used to package the resulting ligation mixture through recognition of the vector *cos* site. Transduction of *E. coli* with this packaged ligation mixture results in libraries that typically contain inserts of 25–40 kb, depending on the size of the vector backbone. Among the many advantages of using cos-based vectors and phage transduction to construct clone libraries, two important considerations are the high efficiency of transduction and the reduced likelihood of insert concatemers.

The degree to which the content of metagenomic libraries is an accurate reflection of the content of the source samples is often overlooked. When this is evaluated, biases are often observed. It was first hypothesized that these biases were linked to DNA GC content perhaps due to uneven DNA fragmentation (Temperton et al. 2009; Ghai et al. 2010; Danhorn et al. 2012). However, we recently demonstrated that fragmentation was not a cause of bias for a human gut metagenomic library and that the dominant reason for library cloning bias may be strong spurious transcription from σ^{70}-like sequences, which causes vector instability in *E. coli* (Lam and Charles 2015). Incorporation of transcriptional termination sequences adjacent to the fragment insertion site on the vector may reduce this source of cloning bias.

1.7 An Example of a Collection of Broad Host Range Metagenomic Libraries

In an effort to produce a functional metagenomics resource that could be freely shared with other researchers, we developed the concept of open resource metagenomics (Neufeld et al. 2011). Here, we describe the development of a collection of metagenomic libraries to be made available under this aegis. We collected representative Canadian soils encompassing vast taxonomic diversity and used these samples to construct metagenomic cosmid libraries. We present these publicly available libraries, in addition to the key methodology used for their construction, as part of the Canadian MetaMicroBiome Library (CM^2BL) project (http://cm2bl.org).

We initiated CM^2BL with the goal of establishing a permanent functional metagenomics resource. Surface soil samples (0–10 cm depth) were collected from 14 locations across Canada spanning multiple biomes and ecozones, including Arctic tundra, oil sands, forest, peatlands, agriculture soils, and municipal compost (Table 1.1). Each sample was collected at three sites that were approximately 5 m apart, combined and sent by courier on ice, or hand delivered, to University of Waterloo. Physical and chemical measurements of the soils were performed at the University of Guelph Agricultural and Food Laboratory (Guelph, Ontario, Canada).

Table 1.1 Canadian MetaMicroBiome soil samples and metagenomic libraries (for details see http://www.cm2bl.org/samples.html)

		Soil characteristics				Metagenomic library		
Sample ID	Habitat	Bulk density	Total carbon (% dry)	Total nitrogen (% dry)	pH	Number of clones	Insert size (kb)	Coverage (×)
1AT	Arctic tundra	0.21	46.9	1.42	3.9	178,100	27.1	1026
2ATN	Arctic tundra	1.05	3.7	0.25	6.7	62,260	31.1	412
4TS	Oil sand 1	1.23	2.1	0.11	7.6	73,000	37.4	581
5BF	Boreal coniferous forest	1.16	1.1	0.08	4.6	56,370	29.7	356
6TD	Temperate deciduous forest	1.10	3.6	0.26	6.4	2,306,580	40.2	19,728
7TR	Temperate rainforest	0.62	10.8	0.35	4.9	68,200	33.7	469
8NP	Northern peatlands	0.38	27.2	1.22	5.5	NA	NA	NA
9WLM	Wetland soil	0.26	43.3	2.21	5.0	64,470	19.7	270
10AS	Agricultural—Soybean	1.10	2.4	0.22	7.6	760,000	37.5	6064
11AW[a]	Agricultural—Wheat	1.10	1.9	0.19	7.4	8,806,400	41.2	77,196
12AC[a]	Agricultural—Corn	1.67	NA	NA	7.8	79,060	33.4	561
13CO	Compost	0.86	11.7	0.92	8.0	42,000	34.2	305
19TS	Oil sand 2	1.12	2.8	0.07	6.0	149,880	33.8	1078
20CG	Community garden	0.87	10.2	0.63	7.6	118,300	36.9	929

The coverage of bacterial genomes was calculated based on the average size of 4.7 Mb microbial genome (Raes et al. 2007)
NA not available
[a]Previously published (Cheng et al. 2014)

The Arctic tundra soil sample (1AT) had the highest content of carbon and nitrogen but the lowest pH of 3.9. The soils from boreal coniferous forest (5BF), temperate rain forest (7TR), northern peatlands (8NP), wetland (9WLM), and oil sands 2 (19TS) were also acidic (pH < 6.0). Compost soil (13CO) had the highest pH of 8.0.

Prior to library construction, metagenomic DNA from the soil samples (http://www.cm2bl.org/samples.html) was isolated using a method described previously (Cheng et al. 2014). Co-extracted humic acids were removed using synchronous coefficient of drag alteration electrophoresis (Engel et al. 2012) or PFGE (Cheng et al. 2014). To explore the diversity of microbes in CM^{2}BL soil samples, the V3 regions of 16S rRNA genes were sequenced as previously described (Bartram et al. 2011), and the 16S rRNA reads were deposited in the Sequence Read Archive of European Bioinformatics Institute (PRJEB9449).

Taxonomic affiliations of 16S rRNA sequences were assigned through an AXIOME2 (https://github.com/neufeld/AXIOME2) pipeline, including paired-end assembly with PANDAseq v2.5 with a quality threshold of 0.9 (Masella et al. 2012), and sequence clustering at 97% sequence identity with UPARSE using USEARCH v7.0.1090 (Edgar 2013). Taxonomic classifications were predicted by RDP v2.2 (Wang et al. 2007) using the Greengenes 13_8 reference set (McDonald et al. 2012). The OTU table was generated by QIIME 1.8.0 (Caporaso et al. 2010), and taxonomy plots were prepared with ggplot2 (Wickham 2016). The 16S rRNA gene sequence analysis confirmed considerable genetic diversity of the microbial communities, dominated by *Proteobacteria*, *Acidobacteria*, *Actinobacteria*, and *Bacteroides* (Fig. 1.1).

In order to perform phenotypic screening in a broad range of surrogate hosts, we constructed metagenomic libraries based on the low-copy number IncP Gateway entry cosmid pJC8 (Cheng et al. 2014). Depending on the size distribution of isolated soil DNA, fragment ranges of 25–50 kb or 30–75 kb were excised from PFGE and recovered by electroelution. These fragments were not generated by restriction enzyme digestion to circumvent bias that might arise from sequence-dependent digestion. Following end repair, purified DNA was ligated to cosmid pJC8 that was cut with Eco72I and dephosphorylated. The ligated product was packaged in vitro with Gigapack III XL packaging extracts (Agilent Technologies) and then transduced into *E. coli* HB101. Recombinant cosmid clones were selected on LB agar plates with tetracycline (15 μg/ml), pooled, then saved in 1-ml aliquots at −70 °C in a final concentration of 7% DMSO. *E. coli* library clones were selected randomly for analysis of cosmid DNA. The average sizes of cloned metagenomic DNA and coverage of bacterial genomes were calculated based on average insert sizes of HindIII-EcoRI-BamHI or EcoRI-HindIII fragments and the total number of recombinant library clones.

A total of thirteen metagenomic cosmid libraries were constructed and maintained in *E. coli* HB101 (Table 1.1), including two libraries reported previously (Cheng et al. 2014). Clones recovered for each library ranged from 4.2×10^3 (compost, 13CO) to 2.9×10^6 (agricultural wheat soil, 11AW). Analysis of randomly selected cosmids with restriction enzyme digestion indicated that the average sizes of cloned metagenomic DNA were 20 kb (wetland soil, 9WLM) to 40 kb (temperate deciduous forest, 6TD) (Table 1.1). The generated soil DNA libraries contained

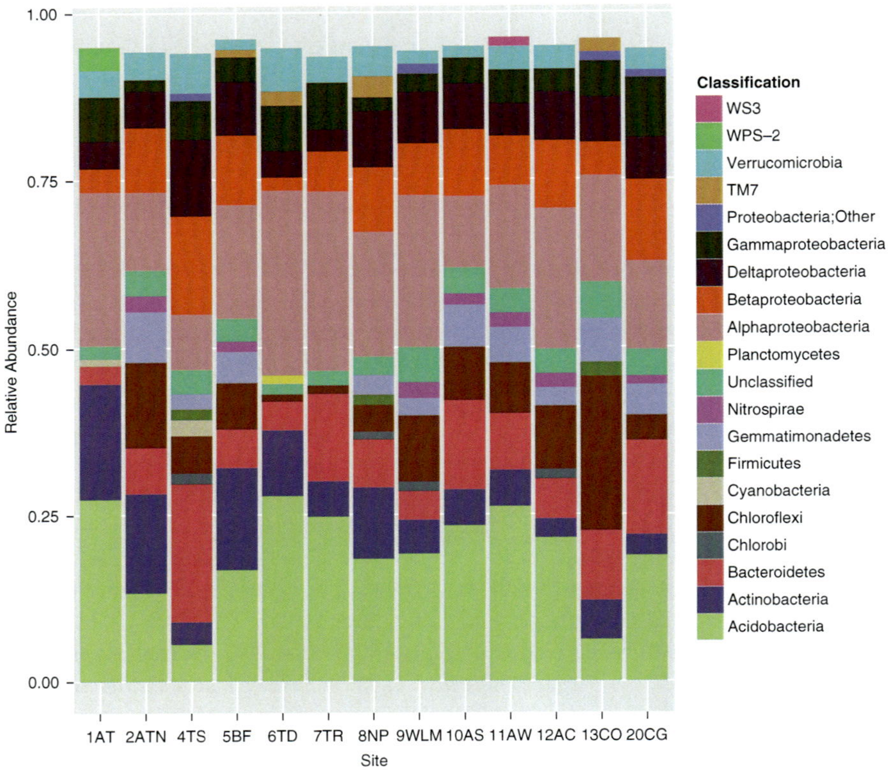

Fig. 1.1 Bacterial community composition of CM²BL samples. Taxonomic profiles of soil samples were analysed based on the sequences of V3 regions of bacterial 16S rRNA genes that were PCR amplified from metagenomic DNA. Any phyla (or class, for *Proteobacteria*) that did not have at least 1% relative abundance in that sample is not shown

1.3–92.7 Gb of metagenomic DNA, which represents approximately 305–19,728 bacterial genomes, assuming an average genome size of 4.7 Mb in a soil community bacteria (Raes et al. 2007). Detailed information describing each metagenomic library and its availability can be accessed through the CM²BL website (http://cm2bl.org/samples.html). This resource thus represents a collection of high-quality metagenomic libraries that are freely available for functional metagenomics in a wide range of Proteobacteria.

1.8 Concluding Statements

The future of functional metagenomics will likely see the development of a greater variety of alternative hosts for functional screening, which will not only lead to an increase in the aggregate hit rates of functional screens but also make available a broader range of phenotypes for functional complementation. We encourage efforts geared to advance the development of surrogate hosts that better represent the whole of microbial diversity and continue to expand the construction of metagenomic libraries that are suitable for screening in these hosts. This will be necessary if functional metagenomics is to continue its contributions to knowledge of microbial gene function.

Acknowledgements We wish to thank Drs. Kenneth J. Reimer, Paul Grogan, Richard A. Frank, Sylvie A. Quideau, Richard S. Winder, Roland I. Hall, Tim R. Moore, Kari E. Dunfield, and Clark Reichert for collecting the soil samples. This work was partially supported by a Strategic Projects grant to TCC and Discovery Grants to JDN and TCC, both from the Natural Sciences and Engineering Research Council of Canada (NSERC).

References

Aakvik T, Degnes KF, Dahlsrud R et al (2009) A plasmid RK2-based broad-host-range cloning vector useful for transfer of metagenomic libraries to a variety of bacterial species. FEMS Microbiol Lett 296:149–158. doi:10.1111/j.1574-6968.2009.01639.x

Bartram AK, Lynch MDJ, Stearns JC et al (2011) Generation of multimillion-sequence 16S rRNA gene libraries from complex microbial communities by assembling paired-end Illumina reads. Appl Environ Microbiol 77:3846–3852. doi:10.1128/AEM.02772-10

Caporaso JG, Kuczynski J, Stombaugh J et al (2010) QIIME allows analysis of high-throughput community sequencing data. Nat Methods 7:335–336. doi:10.1038/nmeth.f.303

Cheng J, Romantsov T, Engel K, Doxey AC, Rose DR, Neufeld JD, Charles TC (2017) Functional metagenomics reveals novel ß-galactosidases not predictable from gene sequences. PLOS ONE 12(3):e0172545

Cheng J, Charles TC (2016) Novel polyhydroxyalkanoate copolymers produced in *Pseudomonas putida* by metagenomic polyhydroxyalkanoate synthases. Appl Microbiol Biotechnol 100(17):7611–7627. doi:10.1007/s00253-016-7666-6

Cheng J, Pinnell L, Engel K et al (2014) Versatile broad-host-range cosmids for construction of high quality metagenomic libraries. J Microbiol Methods 99:27–34. doi:10.1016/j.mimet.2014.01.015

Colin P-Y, Kintses B, Gielen F et al (2015) Ultrahigh-throughput discovery of promiscuous enzymes by picodroplet functional metagenomics. Nat Commun 6:10008. doi:10.1038/ncomms10008

Craig JW, Chang F-Y, Kim JH et al (2010) Expanding small-molecule functional metagenomics through parallel screening of broad-host-range cosmid environmental DNA libraries in diverse proteobacteria. Appl Environ Microbiol 76:1633–1641. doi:10.1128/AEM.02169-09

Danhorn T, Young CR, DeLong EF (2012) Comparison of large-insert, small-insert and pyrosequencing libraries for metagenomic analysis. ISME J 6:2056–2066. doi:10.1038/ismej.2012.35

Dinsdale EA, Edwards RA, Hall D et al (2008) Functional metagenomic profiling of nine biomes. Nature 452:629–632. doi:10.1038/nature06810

Edgar RC (2013) UPARSE: highly accurate OTU sequences from microbial amplicon reads. Nat Methods 10:996–998. doi:10.1038/nmeth.2604

Ekkers DM, Cretoiu MS, Kielak AM, Elsas JDV (2012) The great screen anomaly—a new frontier in product discovery through functional metagenomics. Appl Microbiol Biotechnol 93:1005–1020. doi:10.1007/s00253-011-3804-3

Engel K, Pinnell L, Cheng J et al (2012) Nonlinear electrophoresis for purification of soil DNA for metagenomics. J Microbiol Methods 88:35–40. doi:10.1016/j.mimet.2011.10.007

Gabor EM, Alkema WBL, Janssen DB (2004) Quantifying the accessibility of the metagenome by random expression cloning techniques. Environ Microbiol 6:879–886. doi:10.1111/j.1462-2920.2004.00640.x

Gaida SM, Sandoval NR, Nicolaou SA et al (2015) Expression of heterologous sigma factors enables functional screening of metagenomic and heterologous genomic libraries. Nat Commun 6:7045. doi:10.1038/ncomms8045

Ghai R, Martin-Cuadrado A-B, Molto AG et al (2010) Metagenome of the Mediterranean deep chlorophyll maximum studied by direct and fosmid library 454 pyrosequencing. ISME J 4:1154–1166. doi:10.1038/ismej.2010.44

Gong X, Gruninger RJ, Qi M et al (2012) Cloning and identification of novel hydrolase genes from a dairy cow rumen metagenomic library and characterization of a cellulase gene. BMC Res Notes 5:566. doi:10.1186/1756-0500-5-566

Hug LA, Baker BJ, Anantharaman K et al (2016) A new view of the tree of life. Nat Microbiol 1:16048. doi:10.1038/nmicrobiol.2016.48

Iqbal HA, Low-Beinart L, Obiajulu JU, Brady SF (2016) Natural product discovery through improved functional metagenomics in *Streptomyces*. J Am Chem Soc 138(30):9341–9344. doi:10.1021/jacs.6b02921

Kaddurah-Daouk R, Baillie RA, Zhu H et al (2011) Enteric microbiome metabolites correlate with response to simvastatin treatment. PLoS One 6:e25482–e25410. doi:10.1371/journal.pone.0025482

Kakirde KS, Parsley LC, Liles MR (2010) Size does matter: application-driven approaches for soil metagenomics. Soil Biol Biochem 42:1911–1923. doi:10.1016/j.soilbio.2010.07.021

Kitahara K, Yasutake Y, Miyazaki K (2012) Mutational robustness of 16S ribosomal RNA, shown by experimental horizontal gene transfer in *Escherichia coli*. Proc Natl Acad Sci U S A 109:19220–19225. doi:10.1073/pnas.1213609109

Lam KN, Charles TC (2015) Strong spurious transcription likely contributes to DNA insert bias in typical metagenomic clone libraries. Microbiome 3:22. doi:10.1186/s40168-015-0086-5

Leis B, Angelov A, Mientus M et al (2015) Identification of novel esterase-active enzymes from hot environments by use of the host bacterium *Thermus thermophilus*. Front Microbiol 6:275. doi:10.3389/fmicb.2015.00275

Li Y, Wexler M, Richardson DJ et al (2005) Screening a wide host-range, waste-water metagenomic library in tryptophan auxotrophs of *Rhizobium leguminosarum* and of *Escherichia coli* reveals different classes of cloned *trp* genes. Environ Microbiol 7:1927–1936. doi:10.1111/j.1462-2920.2005.00853.x

Li M, Wang B, Zhang M et al (2008) Symbiotic gut microbes modulate human metabolic phenotypes. Proc Natl Acad Sci U S A 105:2117–2122. doi:10.1073/pnas.0712038105

Liebl W, Angelov A, Juergensen J et al (2014) Alternative hosts for functional (meta)genome analysis. Appl Microbiol Biotechnol 98:8099–8109. doi:10.1007/s00253-014-5961-7

Masella AP, Bartram AK, Truszkowski JM, Brown DG, Neufeld JD (2012) PANDAseq: PAired-enD Assembler for Illumina sequences. BMC Bioinformatics 13:31

McDonald D, Price MN, Goodrich J et al (2012) An improved Greengenes taxonomy with explicit ranks for ecological and evolutionary analyses of bacteria and archaea. ISME J 6:610–618. doi:10.1038/ismej.2011.139

Neufeld JD, Vohra J, Dumont MG et al (2007) DNA stable-isotope probing. Nat Protoc 2:860–866. doi:10.1038/nprot.2007.109

Neufeld JD, Engel K, Cheng J et al (2011) Open resource metagenomics: a model for sharing metagenomic libraries. Stand Genomic Sci 5:203–210. doi:10.4056/sigs.1974654

Raes J, Korbel JO, Lercher MJ et al (2007) Prediction of effective genome size in metagenomic samples. Genome Biol 8:R10. doi:10.1186/gb-2007-8-1-r10

Roller M, Lucić V, Nagy I et al (2013) Environmental shaping of codon usage and functional adaptation across microbial communities. Nucleic Acids Res 41:8842–8852. doi:10.1093/nar/gkt673

Schallmey M, Ly A, Wang C et al (2011) Harvesting of novel polyhydroxyalkanaote (PHA) synthase encoding genes from a soil metagenome library using phenotypic screening. FEMS Microbiol Lett 321:150–156. doi:10.1111/j.1574-6968.2011.02324.x

Schloss PD, Girard RA, Martin T et al (2016) Status of the Archaeal and Bacterial Census: an update. mBio 7:e00201–e00216. doi:10.1128/mBio.00201-16

Sommer M, Dantas G, Church GM (2009) Functional characterization of the antibiotic resistance reservoir in the human microflora. Science 325:1128–1131. doi:10.1126/science.1176950

Tasse L, Bercovici J, Pizzut-Serin S et al (2010) Functional metagenomics to mine the human gut microbiome for dietary fiber catabolic enzymes. Genome Res 20:1605–1612. doi:10.1101/gr.108332.110

Taupp M, Mewis K, Hallam SJ (2011) The art and design of functional metagenomic screens. Curr Opin Biotechnol 22:465–472. doi:10.1016/j.copbio.2011.02.010

Tebbe CC, Vahjen W (1993) Interference of humic acids and DNA extracted directly from soil in detection and transformation of recombinant DNA from bacteria and a yeast. Appl Environ Microbiol 59:2657–2665

Temperton B, Field D, Oliver A et al (2009) Bias in assessments of marine microbial biodiversity in fosmid libraries as evaluated by pyrosequencing. ISME J 3:792–796. doi:10.1038/ismej.2009.32

Terrón-González L, Medina C, Limón-Mortés MC, Santero E (2013) Heterologous viral expression systems in fosmid vectors increase the functional analysis potential of metagenomic libraries. Sci Rep 3:1107. doi:10.1038/srep01107

Uchiyama T, Miyazaki K (2009) Functional metagenomics for enzyme discovery: challenges to efficient screening. Curr Opin Biotechnol 20:616–622. doi:10.1016/j.copbio.2009.09.010

Ufarté L, Potocki-Veronese G, Laville É (2015) Discovery of new protein families and functions: new challenges in functional metagenomics for biotechnologies and microbial ecology. Front Microbiol 6:563. doi:10.3389/fmicb.2015.00563

Verastegui Y, Cheng J, Engel K et al (2014) Multisubstrate isotope labeling and metagenomic analysis of active soil bacterial communities. mBio 5:e01157–14. doi:10.1128/mBio.01157-14

Wang Q, Garrity GM, Tiedje JM, Cole JR (2007) Naive Bayesian classifier for rapid assignment of rRNA sequences into the new bacterial taxonomy. Appl Environ Microbiol 73:5261–5267. doi:10.1128/AEM.00062-07

Wang L, Hatem A, Catalyurek UV et al (2013) Metagenomic insights into the carbohydrate-active enzymes carried by the microorganisms adhering to solid digesta in the rumen of cows. PLoS One 8:e78507. doi:10.1371/journal.pone.0078507

Wickham H (2016) ggplot2: Elegant graphics for data analysis. Springer-Verlag, New York

Chapter 2
Expression Platforms for Functional Metagenomics: Emerging Technology Options Beyond *Escherichia coli*

Anna Lewin, Rahmi Lale, and Alexander Wentzel

Abstract *Escherichia coli* is the prime workhorse for various metagenomic applications due to the multitude of efficient tools available for genetic manipulation and controlled heterologous gene expression. However, metagenome-based bioprospecting efforts continuously target a wider spectrum of ecological niches in order to harvest new enzymes and bioactive compounds for industrial and medical applications from the enormous pool of natural microbial diversity. Consequently, the development of robust and flexible screening platforms that allow functional evaluation of an expanded fraction of the highly diverse metagenomic information is widely addressed in Functional Metagenomics research. The heterologous recognition of transcriptional regulators and promotors, diverse codon usages among environmental microorganisms, and sufficient supply of precursors for secondary metabolite formation are major challenges that are addressed by an increasing spectrum of alternative expression and host systems. This includes optimized broad host-range transfer and expression vectors, screening hosts for improved gene expression and metabolite formation, as well as cell-free expression systems to cover proteins that due to toxicity are inaccessible by in vivo screening methods. In this chapter, we provide a current overview of the state of the art of selected expression systems and host organisms useful for functional metagenome screening for new enzymes and bioactive metabolites, as emerging options beyond what is currently available in and for *E. coli*.

A. Lewin • A. Wentzel (✉)
Department of Biotechnology and Nanomedicine, SINTEF Materials and Chemistry, Trondheim, Norway
e-mail: Alexander.Wentzel@sintef.no

R. Lale
Department of Biotechnology and Food Science, PhotoSynLab, Norwegian University of Science and Technology, Trondheim, Norway

T.C. Charles et al. (eds.), *Functional Metagenomics: Tools and Applications*,
DOI 10.1007/978-3-319-61510-3_2

2.1 Introduction

Metagenomics has since its introduction in the late 1990s (Handelsman et al. 1998) proven to be a powerful tool for describing microbial communities and their metabolic potentials irrespectively of cultivability. Over the years, both sequence- and function-based screening approaches have led to the discovery of numerous new enzymes and metabolites fulfilling various academic and industrial needs (Ferrer et al. 2015; Fernandez-Arrojo et al. 2010; Novakova and Farkasovsky 2013). The pipeline for Functional Metagenomics spans from sampling, isolation of high-quality environmental DNA (eDNA), and its cloning (including vector design) to metagenomic library construction (including host transformation and transfer), heterologous gene expression, and production of functional molecules in amounts sufficient for detection in high throughput screening (Fig. 2.1). The function-based screening route of metagenome-based bioprospecting therewith complements the sequence-based route, in which eDNA is sequenced using next-generation sequencing methods and resulting sequence datasets mined bioinformatically for genes of interest (Lewin et al. 2013).

Irrespective of the chosen screening route, successful bioprospecting of a metagenomic library starts with the isolation of the eDNA. Its quality and quantity are of major importance for the achievable number of clones of the constructed library and consequently the representation of biodiversity in an environmental sample (Zhou et al. 1996). In order to capture as much of the biodiversity as possible, the applied DNA isolation procedures need to be highly effective in sampling from the diverse microorganisms inhabiting the selected environment (Kakirde et al. 2010). In addition, isolated DNA needs to have a high degree of purity and be free of contaminating substances, such as humic acids that are often present in soil and hamper efficient library construction (Tebbe and Vahjen 1993). Several studies document eDNA isolation procedures that resulted in contamination-free high molecular weight (HMW) DNA (Zhou et al. 1996; Brady 2007; Liles et al. 2008; Pel et al. 2009; Cheng et al. 2014). Contaminating compounds co-isolated with the eDNA can also be successfully removed by gel electrophoretic methods, including conventional (Craig et al. 2010), pulse-field (Cheng et al. 2014), or nonlinear electrophoresis (Pel et al. 2009), followed by size selection of the random fragmented DNA, prior to cloning.

New and improved enzyme discovery is currently the largest field of application for Functional Metagenomics tools. Aside from the catalytic function itself, beneficial properties like robustness under harsh conditions or high activity at low temperatures are often required in industrial applications. Consequently, dependent on the aims of a bioprospecting approach, different environments might serve as eDNA sources (Taupp et al. 2011). The microbial habitat to be sampled usually reflects the desired properties, i.e., subjecting a metagenomic library originating from a thermal vent or a hot deep subsurface oil reservoir to thermostable enzyme screening is likely to have a higher success rate compared to subjecting a

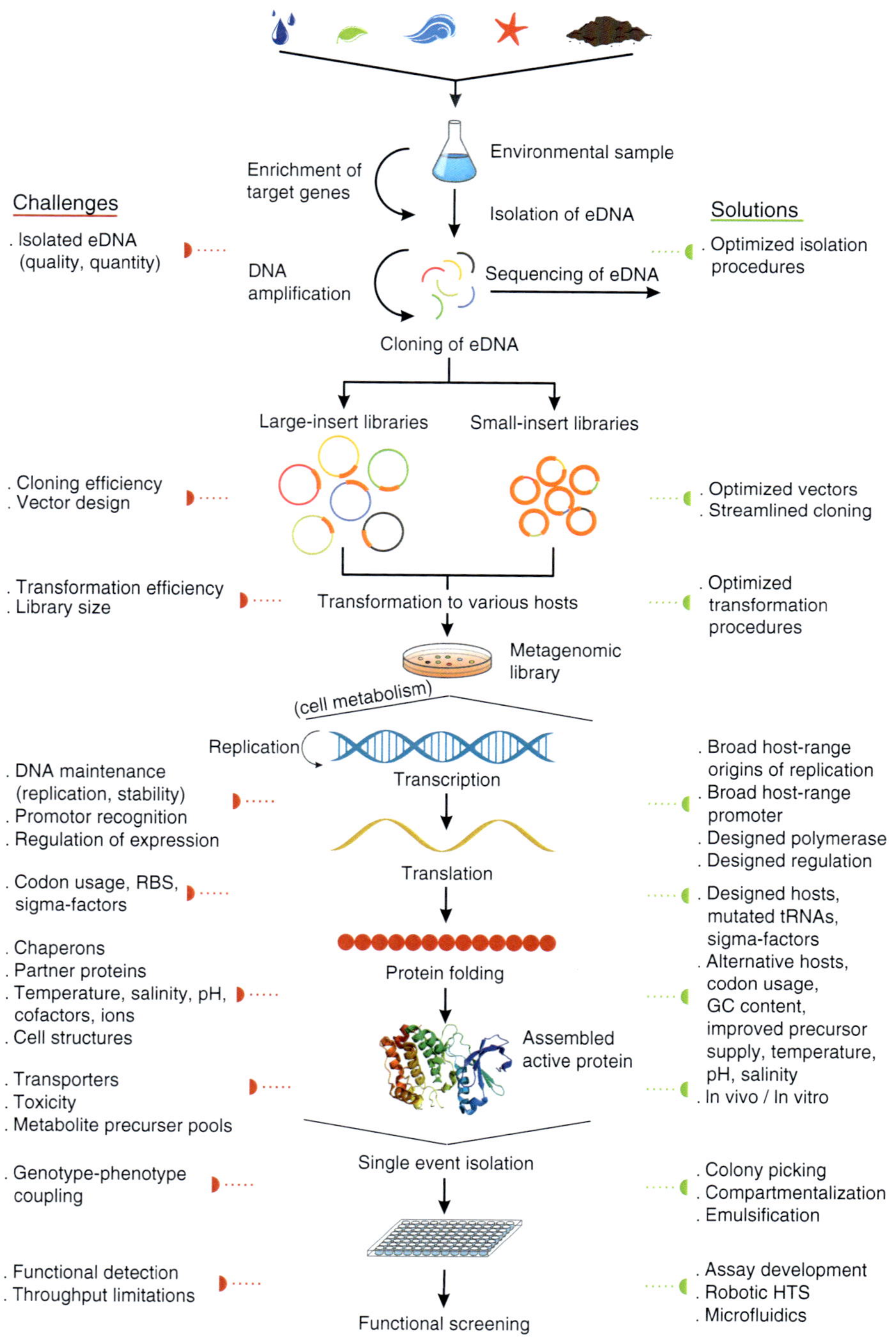

Fig. 2.1 Graphical representation of the Functional Metagenomics biodiscovery pipeline with its key challenges and potential solutions

glacier or permafrost soil-originating library to the same screening. In many examples such directed metagenomics sampling strategies aiming to increase probability of finding the desired properties have proven successful (Vester et al. 2015; Taupp et al. 2011). Selected examples are, among many others, a cold-adapted esterase enzyme from Antarctic desert soil (Hu et al. 2012), hydrolytic enzymes from cow rumen metagenome (Ferrer et al. 2007), and thermostable lipolytic enzymes from water, sediment, and biofilm samples from the Azores, Portugal (Leis et al. 2015a). However, due to often lower microbial density within some, particularly extreme environments, sufficient DNA yields may not be readily obtainable (Kennedy et al. 2008; Vester et al. 2015; Kotlar et al. 2011). In such cases, isolated metagenomic DNA can be subjected to isothermal amplification (like Phi29 whole genome amplification, WGA) in order to increase DNA yields prior to cloning (Rodrigue et al. 2009; Zhang et al. 2006). However, the challenge of this technology with respect to the formation of amplification artifacts, like chimeras, duplications, and inversions, needs to be considered. Therefore it is well suited for small-insert libraries for the purpose of enzyme discovery, but less suitable for large-insert library cloning where intact biosynthetic gene clusters are targeted.

Following sampling and successful isolation, the eDNA is usually either sequenced directly or cloned in suitable vectors for functional screening approaches (Sect. 2.2). The choice of the vector usually depends on the envisioned eDNA insert sizes, as well as the screening targets and methodology. However, for successful expression of genetic information contained in metagenomic DNA libraries, several additional factors need to be taken into account (Fig. 2.1). Suitable vector systems need to carry host-compatible selection markers, replicate stably and autonomously (ideally in combination with the possibility to control the copy number), may contain functional gene regulatory elements like inducible promotors for high level expression, and preferably enable vector transfer to other host organisms. Suitable host organisms in turn need to provide functionality of the vector elements involved in the production of functional products and allow efficient transcription and translation (Sect. 2.3). In addition, proper folding, possible cofactor supply, sufficient precursor availability for metabolite product formation, as well as means for nontoxic product localization, like secretion mechanisms, are needed. In order to meet the different demands for functional expression, such as codon usage, different assay temperatures, precursor requirements, etc. (Lam and Charles 2015; Uchiyama and Miyazaki 2009), different approaches can be applied in order to maximize the probability of successful expression (Fig. 2.1). *E. coli* systems designed and optimized for this purpose have so far been most widely used and extensively covered elsewhere (Guazzaroni et al. 2015). The scope of this chapter is therefore to summarize developments of various hosts and heterologous expression systems for functional metagenome screening beyond the common systems available for *E. coli* only.

2.2 Cloning and Expression Vectors for Environmental DNA

Selection of a suitable vector system for random metagenomic library construction will largely be guided by (1) the expected DNA size encoding the targeted compound of interest, (2) the envisioned subsequent screening approach involving one or more expression hosts, (3) the desired design of the library to be established, as well as in some occasions (4) the quantity of DNA available (Sect. 2.2.1). For approaches to identify new enzymes, small-insert libraries with eDNA sizes of 5–10 kb will in most cases be sufficient to obtain a sufficiently large number of complete gene sequences. Isolation of DNA for small-insert libraries is normally straightforward, since DNA shearing is not a major concern. However, it needs to be considered that a library with an average insert size of 10 kb will require 3–20 times more clones compared to a library with inserts of 30–40 kb to cover the same amount of genetic potential (Sabree et al. 2009). Hence comparably larger amounts of DNA are needed. To identify encoded functions that rely on single genes or small gene loci (e.g., enzyme function or genetic determinants of antibiotic resistance (Riesenfeld et al. 2004)), small-insert libraries are normally sufficient (Kakirde et al. 2010; Sabree et al. 2009). However, in cases where a desired function depends on multiple gene products, libraries harboring larger inserts are needed. These are normally constructed as cosmid, fosmid (30–40 kb), or bacterial artificial chromosome (BAC) libraries (up to ≥100 kb). The construction of comprehensive large-insert libraries can be very laborious, both with respect to the isolation of HMW DNA and successful cloning and transformation of the host. In addition, the lower stability of large inserts in the generated library needs to be considered. Also, the aspect of a higher degree of degradation of low guanine + cytosine (G + C) content DNA and some DNA modifications, which can impair cloning of HMW DNA, can result in a bias within large-insert libraries (Danhorn et al. 2012).

The choice of suitable vector systems is usually also related to the available expression host organism for subsequent screening experiments (Sect. 2.3). Moreover, for some targets, screening in multiple hosts can increase the hit rates (Mullany 2014). Hence library transfer and broad host-range capabilities of an expression vector (Sect. 2.2.2) can be desired characteristics (Craig et al. 2010; Aakvik et al. 2009; Kakirde et al. 2010).

2.2.1 *Small- and Large-Insert Random Cloning Vectors*

Cloning vectors useful for small-insert metagenomic library construction usually contain a defined promoter for transcription of the inserted DNA sequence. In some cases they are even equipped with two promoters (dual promotor vectors), flanking both sides of the cloning site in order to achieve gene expression regardless of insert orientation (Lammle et al. 2007). The promoters can have different, independent induction mechanisms in order to achieve expression in only one direction at a time to prevent potential mRNA duplex formation that may result in lower protein production (Lale et al. unpublished). For cloning and construction of small-insert metagenomic libraries in *Escherichia coli* as primary host organisms, standard cloning vectors, such as pUC derivatives, pBluescript SK(+), and pTOPO, or their derivatives (Mullany 2014; Sabree et al. 2009) are frequently used.

In order to allow metagenomic library clones to cover entire biosynthetic pathways, like secondary metabolite clusters, large-insert libraries are required. Such libraries can be generated as cosmids or fosmids based on phage packaging of the eDNA ligated to a respective vector fragment or for very large inserts (up to 100 kb or more) as BACs (Kakirde et al. 2010; Danhorn et al. 2012). Fosmid and cosmid cloning vectors carry inserts of 30–40 kb, and both approaches utilize phage-based transfer of the cloned DNA into the host, usually *E. coli*. Consequently, the resulting library clones carry inserts within a narrow size range, determined by the packing capacity of the phage particle, and generally rely on gene expression from promoters included in the cloned insert. Cosmids are hybrid plasmids containing *cos* sequences from the λ phage, whereas fosmids are based on the F-factor replicon from *E. coli*. Compared to cosmids, fosmids are more tightly regulated with respect to copy number and are hence more stable (Kim et al. 1992; Kakirde et al. 2010). Both cosmids and fosmids are designed to carry antibiotic resistance markers and have broad host-range capabilities (Craig et al. 2010; Cheng et al. 2014; Aakvik et al. 2009; Wexler et al. 2005). Due to the frequent use of both cosmid and fosmid systems for metagenomic library construction, several variants (including commercial ones) are available (Lam et al. 2015; Mullany 2014; Kim et al. 1992; Parks and Graham 1997; Li et al. 2011; Terron-Gonzalez et al. 2013).

For random cloning of very large inserts, 40–100 kb and above, BACs are normally used, relying on the F-factor replicon (Danhorn et al. 2012; Shizuya et al. 1992). BAC vectors have been used in several metagenomic studies (Brady 2007) using, e.g., soil samples (Rondon et al. 2000) and murine bowel microbiota (Yoon et al. 2013). Similar to fosmids and cosmids, there are different BAC systems available, with some of them allowing inducible high copy numbers (Mullany 2014; Warburton et al. 2009; Wild et al. 2002) and/or having broad host-range capability (Mullany 2014; Aakvik et al. 2009; Kakirde et al. 2010). The US-based company Lucigen Corp. (Madison, WI; www.lucigen.com) has developed dedicated broad host-range vector systems for use in Functional Metagenomics. The pBAC-SBO and pSMART-BAC-S vectors both attribute efficient library construction in *E. coli* and are transferable to both Gram-positive and Gram-negative hosts. They have features allowing selection in several host organisms and gene expression from both insert-flanking regions, and are inducible in copy number (see Chap. 1). pSMART-BAC-S vector provides integration in the host genome only, whereas the pBAC-SBO vector allows both chromosomal integration, as well as extrachromosomal propagation in the recipient.

For DNA experiencing superhelical stress due to, e.g., regions dense in tandem and/or inverted repeats, cloning into circular plasmids can be challenging. In such cases, linear plasmids, such as the pJAZZ vector series (Lucigen), have been designed which can carry large DNA inserts and contain features like transcriptional terminators flanking the cloning site to hinder vector-insert transcriptional interference (Godiska et al. 2010).

2.2.2 Broad Host-Range Expression Vectors

Depending on the desired activity, functional screening in different (or several) hosts can be of high value. As mentioned, *E. coli* is the most commonly used host both for library construction and functional screening. However, for certain screening activities, such as thermostable enzymes, or for bioactive secondary metabolite production, hosts like *Thermus thermophilus* (Angelov et al. 2009) and *Streptomyces* (or other *Actinobacteria*), respectively, might be beneficial due to their inherent features (Kakirde et al. 2010; Martinez et al. 2004) (see Sects. 2.3.1 and 2.3.2). Metagenomic libraries can be constructed directly in the host where they will be screened. However, the number of transformants obtained is often much lower in such hosts compared to the number of clones that can be obtained in *E. coli*. Thus, the common method is to utilize shuttle and/or broad host-range vectors for library construction in *E. coli*, which allows library transfer and screening in the host organism of choice. There are various such vectors available, both for small and large inserts. *E. coli–Bacillus subtilis* shuttle systems (plasmid and BAC) have been used for screening soil metagenomes for antimicrobial activities (Biver et al. 2013), and the pMDB14 vector (McMahon et al. 2012) can be shuttled between *E. coli*, *Pseudomonas putida*, and *Streptomyces lividans*, allowing gene expression in different hosts, similar to other systems reported (Sosio et al. 2000; Martinez et al. 2004). For development of psychrophilic expression systems, *E. coli* shuttle vectors such as a pGEM derivative and a pJRD215 derivative have been constructed, allowing the transfer of constructed libraries from *E. coli* to, e.g., *Psychrobacter* sp. and *Shewanella livingstonensis* (Cavicchioli et al. 2011; Miyake et al. 2007; Tutino et al. 2001). Also, *E. coli–T. thermophilus* shuttle systems have been designed (Angelov et al. 2009; Leis et al. 2015b). Apart from these, several other broad host-range systems have been developed. The pUvBBAC system supports replication in both Gram-positive and Gram-negative bacteria and allows functional screening in *Listeria* hosts (Hain et al. 2008). pGNS-BAC-1 presents opportunities for a copy induction in *E. coli*, as well as replication and functional screening in a broad spectrum of Gram-negative species (Kakirde et al. 2010). The pRS44 plasmid system (Aakvik et al. 2009) has been constructed both as fosmid and BAC system, which enables induction based on control on the vector copy number in *E. coli* and conjugative transfer into other hosts. In addition to the transferable BAC systems, several broad host-range cosmid vectors have also been reported (Craig et al. 2010; Cheng et al. 2014; Wexler et al. 2005).

In order to exploit the benefits of metagenomic library screening in several hosts with complementary features (Martinez et al. 2004; Leis et al. 2015a, b), efficient library transfer between host strains is of high importance. Though library vector isolation by simple plasmid DNA extraction followed by re-transformation into the alternative host is possible, conjugation is in most cases the transformation method of choice. This is generally regardless of whether the library originally was constructed as a cosmid (Wexler et al. 2005; Craig et al. 2010; Cheng et al. 2014), fosmid (Aakvik et al. 2009), or BAC (Kakirde et al. 2010). The conjugative transfer of abovementioned vectors requires the full set of *tra* genes to be present in the donor (F positive) strain. Libraries to be transferred are often large, and therefore library transfer is preferably done in a high throughput fashion, similar to the high throughput conjugation procedure described by Martinez and co-workers (Martinez et al. 2004).

2.3 Expression Host Organisms

As previously mentioned, *E. coli* is presently the most commonly used expression host in metagenomic functional screening efforts (Ekkers et al. 2012; Kennedy et al. 2008; Aakvik et al. 2009; Rondon et al. 2000; Parachin and Gorwa-Grauslund 2011). Several dedicated tools for metagenomic library screening have also been developed for *E. coli*, such as engineered strains suitable for stable replication and copy control of large vectors. These remain at one single copy prior to screening to minimize the potential toxic effects of insert-encoded proteins or other produced metabolites (Taupp et al. 2011). *E. coli* strains have been modified for optimized heterologous expression, e.g., by expression of heterologous sigma factors that allow recognition of a wider range of promoter structures than *E. coli* wild-type strains (Gaida et al. 2015), and for heterologous expression of polyketide synthase (PKS) encoding secondary metabolite gene clusters and production of derivable natural products (Zhang et al. 2015). However, even engineered *E. coli* strains are not in all cases the best-suited hosts with respect to expressing metagenome-encoded functions. This accounts particularly for screening of metagenomic libraries harboring eDNA from extreme environments at conditions not compatible with *E. coli*'s natural lifestyle as a mesophilic human commensal, like very high or low temperatures. In addition, functional expression of genes from species that are phylogenetically distant from *E. coli* can be challenging (Warren et al. 2008). This can be due to, e.g., the differences in codon usage, improper promoter recognition, lack of transcription and/or translation factors, hampered protein folding, absence of cofactors, gene product toxicity, and absence of precursor metabolites. It has been shown that only approximately 40% of all genes can be heterologously expressed in *E. coli* (Gabor et al. 2004). Therefore, the use of multiple, complementary screening hosts has been proposed to express more of the diversity within a metagenomic library (Liebl et al. 2014). Table 2.1 summarizes the most commonly used as well as high potential future host systems for functional metagenome screening.

Table 2.1 Key features of frequently used as well as high potential future hosts for functional expression and screening of metagenomic libraries

Host organism	Key features	Key references
Escherichia coli	+ Well established as a heterologous expression and screening host	Ekkers et al. (2012), Kennedy et al. (2008), Aakvik et al. (2009), and Rondon et al. (2000), Parachin and Gorwa-Grauslund (2011)
	+ Fully developed toolbox for genetic manipulation, easy manageable	Taupp et al. (2011) and Gaida et al. (2015)
	+ Genetic transfer systems well established	
	+ Designed/optimized strains for different screening purposes available	Gabor et al. (2004) and Zhang et al. (2015)
	– Only distantly related to many environmental microbes, codon usage	Warren et al. (2008)
	– Restricted to mesophilic cultivation an screening	Liebl et al. (2014)
	– Limited precursor availability for secondary metabolite formation	
	– Restrictions with respect to cofactor availability	
	– Potential toxicity effects due to limits in protein secretion, Gram-negative	
Thermus thermophilus	+ Allows cultivation and in vivo screening high temperatures	Tabata et al. (1993) and Cava et al. (2009)
	+ Natural competence for DNA uptake	Hidaka et al. (1994)
	+ Efficient transformation protocols available	Schwarzenlander and Averhoff (2006)
	+ Thermostable resistance markers and other genetic tools (promoters, origin of replication, etc.) available	Matsumura and Aiba (1985), Liao et al. (1986), Nakamura et al. (2005), and Tamakoshi et al. (1997)
	+ *T. thermophilus–E. coli* shuttle vectors available	Lasa et al. (1992) and Wayne and Xu (1997)
	+ Chromosomal integration well established	de Grado et al. (1999)
	– Only a limited set of selection markers available	
	– Only a few promoter alternatives developed	
Streptomyces spp.	+ Full set of genetic tools available	Gust et al. (2004), Kieser et al. (2000), and Jones et al. (2013)

(continued)

Table 2.1 (continued)

Host organism	Key features	Key references
Streptomyces coelicolor	+ Optimized strains for heterologous gene cluster expression available	Gomez-Escribano and Bibb (2011, 2012, 2014)
	+ High G + C content, thus complementary to other expression hosts	Gomez-Escribano and Bibb (2014)
	+ Assembly platform for secondary metabolite production machinery	Shima et al. (1996), Okamoto-Hosoya et al. (2000), and Hu et al. (2002)
	+ Natural provision of precursors for sec. metabolite formation	
	+ Gram-positive, efficient protein/ enzyme secretion	
	– *Mycelial growth phenotype, advanced cultivation systems necessary*	Wentzel et al. (2012a)
Rhodobacter capsulatus	+ Suitable for expression of membrane proteins	Liebl et al. (2014)
Gluconobacter oxydans	+ Acid tolerant	Liebl et al. (2014)
Burkholderia graminis	+ *E. coli* alternative	Craig et al. (2010)
Caulobacter vibrioides	+ *E. coli* alternative	Craig et al. (2010)
Pseudomonas putida	+ Stress tolerant	Craig et al. (2010) and Troeschel et al. (2010)
	+ Capable of producing secondary metabolites	Loeschcke and Thies (2015)
Ralstonia metallidurans	+ Robustness at extreme conditions, broad screening spectrum	Craig et al. (2010) and Mergeay et al. (2003)
Pseudomonas fluorescens	+ *E. coli* alternative	Aakvik et al. (2009)
	+ Secretion pathway	Retallack et al. (2006)
Xanthomonas campestris	+ *E. coli* alternative	Aakvik et al. (2009)
	+ Increased stabilities of proteins produced	Leza et al. (1996)
Sinorhizobium meliloti	+ *E. coli* alternative	Cheng et al. (2017)
	+ Chromosomal integration established	Heil et al. (2012)
Agrobacterium tumefaciens	+ *E. coli* alternative, plant symbiont	Craig et al. (2010), Troeschel et al. (2010), and Murai (2013)
	+ Chromosomal integration established	Heil et al. (2012)

Table 2.1 (continued)

Host organism	Key features	Key references
Bacillus subtilis	+ Gram-positive model organism	Biver et al. (2013)
	+ Fully developed toolbox for genetic manipulation, easy manageable	
	+ Low GC, thus complementary to other expression hosts	
	+ Secretion pathway (enzyme production)	Wong (1995) and Zobel et al. (2015)
R. eutropha H16	+ *E. coli* alternative	Gruber et al. (2015)
Saccharomyces cerevisiae	+ Eukaryotic (yeast)	Damon et al. (2011)
	+ Fully developed toolbox for genetic manipulation	
	+ Secretion pathway (enzyme production)	Strausberg and Strausberg (2001)
	+ Protein posttranslational modifications, glycosylation	Holz et al. (2003)

2.3.1 Extremophiles as Expression Hosts for Metagenome Screening

Similar to sampling and cloning of metagenomic DNA from an environment that matches the desired properties of an enzyme, it appears reasonable to use a heterologous expression host for metagenomic library screening that functions optimally under respective conditions, such as in vivo screening for thermostable enzymes at elevated temperature in a thermophilic host. In terms of thermophilic hosts for heterologous gene expression, the hyperthermophilic, Gram-negative bacterium *T. thermophilus* (*Deinococcus-Thermus* phylum), growing optimally at temperatures as high as 85 °C, is the most well-studied species (Tabata et al. 1993; Cava et al. 2009). Several *T. thermophilus* strains have been genome sequenced, and their natural competence (Hidaka et al. 1994; Koyama et al. 1986) renders them very efficient in taking up external DNA without source discrimination (Schwarzenlander and Averhoff 2006). In addition, *T. thermophilus* has been shown to acquire DNA by conjugation, however, not as effectively as by utilizing its natural competence (Ramirez-Arcos et al. 1998; Cava et al. 2007).

A large number of genetic tools have been developed to genetically amend *T. thermophilus* (Tamakoshi et al. 1997; de Grado et al. 1999). In the 1980s a selection marker in the form of a thermostable version of a kanamycin resistance was developed using mutagenesis (Matsumura and Aiba 1985; Liao et al. 1986), allowing antibiotic-based selection for transformed *T. thermophilus* cells. Since then other selectable markers stable at high temperature have been used such as the bleomycin-binding protein conferring bleomycin resistance (Brouns et al. 2005), and a hygromycin B phosphotransferase evolved to thermostability (Nakamura et al. 2005). Several plasmids and vectors are available, like the cryptic pTT8 plasmid (Koyama et al. 1990) used to transfer genes into *T. thermophilus*. pTT8 has also been supplemented with the gene providing thermostable kanamycin resistance described above, resulting in the selectable cloning vector pMKM001 (Mather and Fee 1992) and variants thereof. The *Thermus*-compatible plasmids have also been engineered further into *E. coli–Thermus* shuttle vectors by integration of the cryptic *Thermus* vectors with commonly used *E. coli* plasmids from the pUC series, resulting in several variants, e.g., pMY1-3 and pLU1-4 (Lasa et al. 1992; Wayne and Xu 1997). In addition, there are other plasmids available for *Thermus*, like plasmid pTA103 (Chu et al. 2006), the pS4C, and pL4C plasmids harboring both integrase and transposase (Ruan and Xu 2007) as well as the widely used pMK18 vector carrying the multiple cloning site from pUC18 (de Grado et al. 1999).

As a consequence of the available genetic tools for *T. thermophilus*, these strains have been used as thermophilic cell factories to complement production of certain proteins in, e.g., *E. coli*. *T. thermophilus* has been used to homologously produce *Tth* DNA polymerase more efficiently than in *E. coli* (Moreno et al. 2005), as well as for production of an active thermostable Mn-dependent catalase which failed to express in *E. coli* (Hidalgo et al. 2004). *T. thermophilus* has also been successfully used in metagenomic approaches, e.g., by Angelov and co-workers (2009). In their work, large-insert fosmid libraries were constructed in *E. coli* and transferred to a *T. thermophilus* host. Screening was performed in both species, resulting in different hit spectra. This clearly illustrates the benefits of high-temperature screening for thermostable enzymes. The same authors also constructed a pCC1fos derivative (denoted pCT3FK) which carries *T. thermophilus* HB27 chromosomal DNA sequences which allow integration in the host chromosome by homologous recombination (Angelov et al. 2009). This vector has been used in the screening of a metagenomic library for thermostable esterases in both *E. coli* and *T. thermophilus* hosts, resulting in a higher number of thermostable enzyme candidates in the *T. thermophilus* than in the *E. coli* screening (Leis et al. 2015a, b).

On the opposite end of the temperature range, cold environments provide a large understudied biodiversity. Particularly psychrophilic enzymes from such environments are sought due to their unique characteristics, i.e., high activity at low and moderate temperatures, necessitating lower enzyme concentrations to achieve a similar performance compared to higher temperature homologues. Psychrophilic enzymes are considered to be less stable compared to their mesophilic homologues, as their structural flexibility enables them to function at low temperatures and imparts a decreased thermal stability (Feller 2013). However, the biodiscovery of relevant gene functions from these environments is limited to their expression and function in mesophilic hosts. For instance, the utilization of *E. coli* as a host for the expression of psychrophilic enzymes limits the growth temperature to around 15 °C, which presents a significant barrier to their exploitation in biotechnology (Struvay and Feller 2012).

There are several examples where *E. coli* has been successfully used in the production of cold-adapted enzymes (Cavicchioli et al. 2011; Wang et al. 2010; Zhang and Zeng 2008). However, the total number of such reports is comparably low, reflecting significant challenges. Two strategies to overcome these challenges are (1) low-temperature adaptation of existing mesophilic expression systems and (2) the development of new psychrophilic expression hosts. The former approach includes engineering the mesophilic expression host for sufficient growth at low temperatures to promote correct folding of recombinant proteins. The co-expression of Cpn60 and Cpn10 from *Oleispira antarctica*, cold-adapted homologues of the *E. coli* GroELS chaperonins, provided *E. coli* with an operational folding system at 4–12 °C (Ferrer et al. 2003). This led to improved growth at low temperatures and enhanced solubility of the recombinant proteins produced. Another example is the utilization of cold-shock promoter systems together with solubility partners for psychrophilic genes in *E. coli*. Bjerga and Williamson showed that cspA-driven expression of maltose-binding protein (MBP), thioredoxin (TRX), small ubiquitin-like modifier (SUMO), and trigger factor (TF) encoding gene fusion enabled high level production of soluble protein (Bjerga and Williamson 2015).

Dedicated host-expression systems for the production of cold-adapted products have been developed, such as the pTAUp and pTADw vectors for *Psychrobacter*, found to replicate by rolling circle mechanisms (Tutino et al. 2000). Also, the cryptic replicon plasmid pMtBL from *Pseudoalteromonas* sp. has been used as a psychrophilic expression vector, shown to have a broad host-range profile compatible to not only psychrophiles but also mesophilic species after fusion with a pGEM derivate (Tutino et al. 2001). Other broad host-range vectors for cold-adapted expression include a variant of pJRD215 carrying a regulatory promoter from *Shewanella* and a β-lactamase reporter from *Desulfotalea* (Miyake et al. 2007) and a shuttle vector based on the p54 plasmid originating from a psychrophilic *Arthrobacter* sp. isolated from a Greenland glacier and pUC18. The latter example resulted in a low-temperature expression system transferrable to not only *E. coli* but also some high G + C Gram-positive bacteria (Miteva et al. 2008).

2.3.2 *Actinobacteria as Hosts for Heterologous Natural Product Formation*

The phylum *Actinobacteria* comprises a comprehensive and diverse group of Gram-positive bacteria predominantly with a mycelial lifestyle. They are potent producers of a plethora of natural products with a wide spectrum of medical applications, including antibacterial, antifungal, anthelmintic, and immunosuppressant compounds (Barka et al. 2016). Among them, members of the *Streptomyces* taxon are particularly prolific in this respect, accounting for the majority of antibiotics in medical use today (Hopwood 2007). Actinomycete genomes contain a multitude of secondary metabolite gene clusters (Bentley et al. 2002; Ohnishi et al. 2008; Oliynyk et al. 2007; Udwary et al. 2007) of which, however, only a subset is expressed and the respective compounds produced under laboratory conditions. Hence, the majority of gene clusters remains silent, rendering them cryptic, with functions yet to be discovered. Also among *Actinobacteria*, cultivable strains represent only a minute fraction of the entire diversity (Maldonado et al. 2005), leaving a vast resource of new potential drug candidates untapped, unless new methods to enable cultivation (Zengler et al. 2002), or efficiently allow the heterologous realization of their genetic potential, become available. In that respect, well-described members of the *Actinobacteria* themselves, like the model species *Streptomyces coelicolor*, have been proposed as hosts for the heterologous expression of natural product gene clusters (Gomez-Escribano and Bibb 2011, 2012). Their versatility with respect to expressing complex biosynthetic gene clusters, their high G + C codon usage, and the provision of important precursors necessary to simultaneously form natural products of different compound classes (like polyketides, non-ribosomal peptides, lantibiotics, etc.) are excellent rationales to select such strains for metagenome screening for new bioactive compounds. In addition, these *Streptomyces* spp. strains might prove useful in accessing the potential of cryptic gene clusters of cultivable strains by heterologous expression. In-depth understanding of gene regulation and precursor supply will be instrumental in optimizing model *Actinobacteria* as functional metagenome screening hosts.

S. coelicolor has been extensively studied with respect to the regulation of secondary metabolite production, and all necessary genetic tools for genetic manipulation, like plasmids and inducible promoters, and large-insert library tools for chromosomal integration (Gust et al. 2004; Kieser et al. 2000; Jones et al. 2013), are fully developed. Also, new tools for fast and efficient genome editing, like the CRISPR/Cas system (Garneau et al. 2010), have been optimized and applied to *Actinobacteria* to make deletions and directed genomic mutations (Tong et al. 2015; Huang et al. 2015; Cobb et al. 2015). Though its applicability to introduce larger gene clusters into the *Streptomyces* genome is currently limited, it can be expected that this technology will develop into a powerful tool for reprogramming *Actinobacteria* for the production of new bioactive compounds. Wild-type *S. coelicolor* produces several antibiotic compounds of different classes, including the polyketides actinorhodin (Act, Rudd and Hopwood 1979) and coelimycin (Cpk, Gomez-Escribano et al. 2012), the prodiginine undecylprodigiosin (Red, Feitelson et al. 1985), the lipopeptide calcium-dependent antibiotic (CDA, Hopwood and Wright 1983), and the plasmid-encoded cyclopentanoid methylenomycin (Mmy, Wright and Hopwood 1976). However, its genome sequence revealed a much larger potential of bioactive compounds, represented by more than 20 different, mostly non-expressed gene clusters for secondary metabolites (Bentley et al. 2002). Extensive research has been performed to detect and study cryptic gene clusters (Medema et al. 2011; Nett et al. 2009; Zerikly and Challis 2009; Baltz 2008) and ultimately activate them for product formation (Ochi et al. 2014; Rutledge and Challis 2015; Yoon and Nodwell 2014; Zhu et al. 2014). However, regulation of antibiotic production by *S. coelicolor* is complex and needs to be understood in depth when considering it as a generic cell factory for heterologous natural product formation.

Several factors are involved in triggering antibiotic production in *Streptomyces* in correlation with the species' life cycle (Bibb 2005; van Wezel and McDowall 2011).

Nutrient depletion and cessation of growth induce morphological differentiation and antibiotic production via the stringent response and guanosine tetra- and pentaphosphate (p)ppGpp (Potrykus and Cashel 2008). Programmed cell death and the release of N-acetyl glucosamine (GlcNAc) trigger the onset of development and antibiotic production via the global regulator DasR (Rigali et al. 2006, 2008). Also, induced mycelial fragmentation by overexpression of cell division activator protein SsgA affects antibiotic production in *S. coelicolor* (van Wezel et al. 2009). From responses of the global regulatory network, information is passed on to pathway-specific activators encoded within biosynthetic gene clusters, usually controlled in a growth phase-dependent manner (Wietzorrek and Bibb 1997). Once produced in sufficient amount, these are solely responsible for all further downstream regulation of the biosynthetic gene cluster expression. Removal of pathway-specific regulators (Smanski et al. 2012) or exchange of native promotors (Du et al. 2013) as well as overexpression of export proteins (Huo et al. 2012) have led to improved production yields of platencin, gougerotin, and bottromycin, respectively.

Taking all the different layers of regulation into account will be the key for developing *Streptomyces* into potent heterologous production platforms for natural product discovery, from both silent gene cluster in cultivable microorganisms and realizing the biosynthetic potential in environmental metagenomes. *S. coelicolor* has been extensively used as heterologous expression platform for antibiotic gene clusters as recently reviewed by Gomez-Escribano and Bibb (2014). By successively deleting the biosynthetic gene clusters for Act, Red, CDA, and Cpk in the plasmid-free (thus Mmy negative) wild-type M145 of *S. coelicolor*, a strain (M1146) was obtained with largely reduced background of bioactive compounds produced and secreted to the medium (Gomez-Escribano and Bibb 2011). In the same work, additional introduction of point mutations in the genes rpoB and rpsL, encoding the RNA polymerase β-subunit and the ribosomal protein S12, respectively, (strain M1154) led to a pleiotropic increase in the level of secondary metabolite production. Each of these mutations had previously been shown to enhance antibiotic production levels in *Streptomyces* without negative effects on growth (Shima et al. 1996; Okamoto-Hosoya et al. 2000; Hu et al. 2002) and has been proposed as a new strategy to activate silent gene clusters for new drug discovery (Ochi and Hosaka 2013). M1146 and M1154 have been successfully applied for the heterologous production of numerous antibiotics of diverse classes (Gomez-Escribano and Bibb 2014).

A further optimization of the existing heterologous host strains of *S. coelicolor* as an optimized Superhost for new antibiotics discovery from environmental metagenomes may be guided by the comprehensive knowledge of physiology and gene regulation of antibiotic production, as well as systems biology understanding of this species. A dedicated fermentation strategy for system scale studies of metabolic switching in *S. coelicolor* has been established (Wentzel et al. 2012a), allowing reproducible cultivations of *S. coelicolor* and high-resolution time-scale sampling for full 'omics analysis (Battke et al. 2011). The dynamic architecture of the metabolic switch in *S. coelicolor* was studied at the gene expression (Nieselt et al. 2010), the proteome (Thomas et al. 2012) and the metabolome level (Wentzel et al. 2012b). By studying the effect of different mutations, the complex regulatory interplay of nitrogen and phosphate metabolism was elucidated (Martin et al. 2012; Waldvogel et al. 2011). A genome scale model for *S. coelicolor* is available (Alam et al. 2010), and detailed insight in the structure of the transcription factor mediated regulatory network has been gained (Iqbal et al. 2012).

In addition to *S. coelicolor*, other *Actinobacteria* species have been considered as heterologous expression hosts. *S. avermitilis*, for example, has been engineered as an expression host for heterologous gene clusters (Komatsu et al. 2013), and also *S. lividans* and *S. albus* as well as *Saccharopolyspora* (Baltz 2010) have been used for that purpose. *S. lividans* was used as host organism in successful screening for anti-mycobacterial compounds (Wang et al. 2000), and both *S. lividans* and *S. albus* have been shown to be able to produce products from an introduced Type II PKS pathway (King et al. 2009). *Nonomuraea* sp. ATCC 39727 heterologously produced microbisporicin and planosporicin (Marcone et al. 2010) more efficiently than as *Streptomyces* hosts (Foulston and Bibb 2010; Sherwood and Bibb 2013), indicating potential benefits of using several actinobacterial expression hosts for bioactive compound screening of metagenome libraries. *Streptomyces* spp. have proven to be useful in heterologous gene cluster expression and functional screening for associated bioactivity (Kakirde et al. 2010; Martinez et al. 2005). Screening of a BAC library from soil DNA produced in *E. coli* and transferred to *Pseudomonas putida* (low G + C) and *Streptomyces lividans* (high G + C) resulted in different expression patterns (Martinez et al. 2004), indicating usefulness of the high G + C *Streptomyces* hosts as complement to other metagenome screening platforms for bioactivity, like polyketide production-optimized *E. coli* BTRA (Zhang et al. 2015).

Recently, the "*Tectomicrobia*" candidate phylum including the "*Entotheonella*" candidate genus has been discovered by a combined single cell- and metagenomics-based approach to describe microbial consortia producing bioactive polyketides and peptides in association with the marine sponge species *Theonella swinhoei* (Wilson et al. 2014). This study exemplifies the huge potential of marine environments to identify new compounds produced by non-cultivable microbial strains. The genetic optimization of different actinobacterial model strains for natural product formation will help in establishing a platform of different optimized host strains that in combination can potentially be useful in functional screening also for new natural products from such biodiversity with an increased success rate.

2.3.3 Other Expression Hosts for Metagenome Screening

There are several other species apart from *E. coli* and those discussed above (Sects. 2.3.1 and 2.3.2) that have been considered as hosts for metagenome expression and screening, all with their respective benefits and drawbacks. These species can contribute to building a flexible platform for multi-host expression and screening of microbial metagenomes as suggested before (Liebl et al. 2014).

Mesophilic hosts applied for metagenomic screening, apart from *E. coli* and the *Actinobacteria* covered in detail above (Sect. 2.3.2), include species like *Agrobacterium tumefaciens* (alphaproteobacteria), *Burkholderia graminis* (betaproteobacteria), *Caulobacter vibrioides* (alphaproteobacteria), *Pseudomonas putida* (gammaproteobacteria), and *Ralstonia metallidurans* (betaproteobacteria) that have been used to screen a soil metagenome (Craig et al. 2010). Also, the alphaproteobacterium *Rhizobium leguminosarum* has been used in metagenome screening for alcohol/aldehyde dehydrogenases (Wexler et al. 2005). Other mesophilic host bacteria utilized in metagenomic screening include *Rhodobacter capsulatus* and *Gluconobacter oxydans* (Liebl et al. 2014), where *R. capsulatus* has been shown to be suitable for expression of membrane proteins, and *G. oxydans* to be tolerant to screening at acidic conditions. Also, the low G + C, Gram-positive bacterium *Bacillus subtilis*, widely used for recombinant enzyme production due to its capability to secrete protein in the medium, has been used in metagenome screening (Biver et al. 2013). Similarly, species of *Burkholderia*, *Sphingomonas*, and *Pseudomonas* (Ekkers et al. 2012; Martinez et al. 2004) have been used, and, by using the bacterial symbiont *Sinorhizobium meliloti* as expression host, a greater diversity of clones was found compared to screening in *E. coli* (Lam et al. 2015). In addition, the gammaproteobacteria *Pseudomonas fluorescens* and *Xanthomonas campestris* (Aakvik et al. 2009) as well as integrase-mediated recombination of libraries in hosts *S. meliloti* and *Agrobacterium tumefaciens* (Heil et al. 2012) have been shown to be applicable for functional metagenome screening.

Even though prokaryotic hosts have been applied successfully in screening of metagenomic DNA libraries with content including eukaryotic DNA (Geng et al. 2012), eukaryotic host systems may be an important area for further development of metagenomic tools and expression hosts. Even though much more prokaryotic vector-host systems have been developed and used through history, there are genetic tools available for yeasts such as *Saccharomyces cerevisiae* (e.g., Drew and Kim 2012) and *Pichia pastoris* (e.g., Daly and Hearn 2005), as well as filamentous fungi, for example, *Aspergillus* (Nevalainen et al. 2005). A mutant strain of *S. cerevisiae*, defective in di-/tripeptide uptake, has been used in a functional screening of a soil metagenome library for the identification of novel oligopeptide transporters (Damon et al. 2011), demonstrating the potential of eukaryotic hosts in functional screening of environmental metagenomes.

2.4 In Vitro Expression Systems for Functional Metagenomics

Cell-free protein synthesis (CFPS) covers the in vitro transcription of coding DNA to mRNA and its subsequent translation into polypeptide and functional protein by using cell extracts. CFPS is a field in rapid development with the potential to make large impact in both protein production and screening for new enzyme functions in the future. The first CFPS system of *E. coli* was already introduced in 1961, with the main purpose of studying the process of translation (Matthaei and Nirenberg 1961). Since then, a multitude of advanced CFPS systems using extracts of organisms from all three domains of life, including from Bacteria, Archaea, fungi, plants, insects, and mammals (Zemella et al. 2015), has been developed. With their open nature, CFPS systems bypass a number of limitations existing in cellular, in vivo expression systems, as they are highly flexible with respect to the physicochemical environment, the reaction conditions, and the reaction format for gene expression to take place. In addition, they allow incorporation of nonnatural amino acids/cofactors, avoid biological background, and are not constraint by cell viability in response to toxic proteins being produced. In the absence of membranes to be bypassed, almost unlimited use of substrates for screening of gene libraries is enabled, and library sizes that are not restricted by transformation efficiency of expression host cells. This renders CFPS an increasingly recognized alternative option to cell-based expression systems for both protein screening and production (Catherine et al. 2013).

Several key challenges associated with CFPS have recently been successfully addressed and mitigated, such as low productivities, quality and quantity constraints of DNA templates, posttranslational modifications, and clonal separation for genotype-phenotype coupling. Low productivity has been a major issue due to the rapid depletion of the chemical energy carrier ATP and stoichiometric accumulation of phosphate, binding vital magnesium ions. The development of ATP regeneration methods, in particular utilization of the intact glycolytic pathway to produce ATP from glucose by oxidative phosphorylation (Jewett and Swartz 2004; Calhoun and Swartz 2007; Kim and Kim 2009), represented a major breakthrough in achieving larger protein amounts. Moreover, in situ supply of glucose by hydrolysis of polymeric carbohydrates like maltodextrin or starch could be implemented to control the ATP delivery rate (Wang and Zhang 2009). Other metabolic functions in crude cell extracts for CFPS were used to be beneficial, for example, for the provision of cofactors for produced target enzymes (Kwon et al. 2013).

Several studies have suggested solutions to the challenges connected to high template amounts required, as well as high exonucleolytic degradation of linear DNA template in crude cell extracts. In addition to sufficient template preparation by PCR-based methods (Sawasaki et al. 2002; Endo and Sawasaki 2004), the use of isothermal DNA amplification in connection with CFPS (Kumar and Chernaya 2009) was shown to enable high throughput protein synthesis based on very small amounts of template DNA. mRNA stabilization by inclusion of the terminal stem-loop structures and depletion of extracts from RNase E led to greatly improved protein production (Ahn et al. 2005). More relevant for expression library screening, the protection of linear DNA templates and improved protein production was shown by inhibiting the RecBCD nuclease in *E. coli* extracts by addition of bacteriophage Lambda Gam (Sitaraman et al. 2004). This was also shown to be achieved by using extracts of *E. coli* in which the endonuclease I gene *endA* was removed and the *recBCD* operon was replaced by the Lambda recombination system (Michel-Reydellet et al. 2005). Also the tethering of linear DNA ends to microbeads in an agarose matrix led to improved DNA template stability (Lee et al. 2012).

For posttranslational modifications during in vitro synthesis of eukaryotic proteins, for example, for pharmaceutical applications, several eukaryotic CFPS systems have been developed as recently reviewed by Zemella and co-workers ((Zemella et al. 2015) and references therein). This includes systems based on *S. cerevisiae*, the fall armyworm *Spodoptera frugiperda*, rabbit reticulocytes, CHO cells, and different human cell lines. The set of well-documented eukaryotic CFPS systems also includes plant systems from tobacco BY-2 and the widely used cell-free expression system based on wheat germ embryos which represents a high yield system with correct folding of many protein types, including disulfide-rich proteins (Takai et al. 2010).

In vitro compartmentalization (IVC) represents one possible solution to the demand for clonal separation and genotype–phenotype coupling in cell-free screening systems. Being early addressed by the SIMPLEX approach (Rungpragayphan et al. 2003) using diluted single-molecule templates for PCR and subsequent CFPS in a microtiter format, emulsion-based approaches bear the possibility of substantial library sizes. Small aqueous droplets are prepared in a continuous oil phase to isolate templates in individual micro-reactors for isothermal or PCR-based amplification (Courtois et al. 2008) and CFPS. This represents a promising platform for enzyme activity screening against a wide array of substrates using either FACS- or microfluidics-based screening and sorting methods (Kintses et al. 2010).

The insight in biodiversity and the huge metabolic potential in nature provided by the recent revolutions in next-generation sequencing have renewed attention in the potential of CFPS. Consequently, key improvements have been triggered, greatly expanding the applicability of cell-free systems to HT gene expression and even large-scale protein production (Zemella et al. 2015). CFPS and suitable screening systems may form an ideal platform for the functional screening of enzymes using genomic and metagenomic DNA, independent of the limitations of cell-based systems. In a recent example, a cow rumen metagenomic library was screened for glycoside hydrolases using cell-free expression and utilizing the energy-providing effect of glucose in CFPS extracts (Kim et al. 2011). Energy generation in this case started with the polysaccharides cellulose, xylan, amylose, as well as a small amount of glucose. Enzymatic substrate degradation in a feedback loop then led to increased glucose amounts, ultimately leading to an indicator-detectable pH drop due to acid by-products (Kim et al. 2011).

This example shows that optimized CFPS systems in combination with smart assay design represent a powerful option for expression screening for microbial enzymes with high versatility, in particular when combined with platforms for ultra-high throughput analysis and sorting. Further developments in this field will likely include expansion of CFPS systems to additional microbial species, including from extreme environments, as eDNA from extreme environments may fail to be transcribed or translated by *E. coli* extracts (Angelov et al. 2009). Hence, "unconventional" microbial systems for functional expression are demanded (Liebl et al. 2014). Pure component systems and extracts have already been described for extremophiles from both Bacteria and Archaea (Endoh et al. 2007; Zhou et al. 2012), including *Thermus*, *Pyrococcus*, *Sulfolobus*, and *Thermococcus* (Hethke et al. 1996; Tachibana et al. 1996; Ruggero et al. 2006), which might be a valuable resource for future systems.

2.5 Outlook

Metagenomics has proven to be a powerful tool to describe environmental microbial biodiversity and exploit it for metabolic functions of relevance for commercial applications. With the ever-advancing throughput of next-generation sequencing technologies, (meta)genomic DNA sequence databases are filling rapidly, and based on that, our insight into the huge and diverse metabolic potential existing in nature has never been deeper. However, identification of useful functions is ultimately still dependent on experimental proof. Though in silico predictions are constantly improving, the field of Functional Metagenomics will continue to develop as it directly and efficiently links desired function to its determining source code, the eDNA.

E. coli and genetic tools developed for this species have been the first choice in Functional Metagenomics research, both with respect to library construction, recombinant expression, and functional screening. However, it is presently obvious that *E. coli* has some shortcomings, especially in the light of the growing spectrum of ecological niches and greater microbial diversity being accessed and a broader spectrum of metabolic functions and properties aimed to be exploited. Therefore, along with *E. coli*, which itself is still being improved further as a screening host for specific target classes, other microbial model systems, potentially more suitable for screenings for particular functions of interest, have emerged in recent years. These include, for example, thermophilic and psychrophilic systems for respective enzyme discovery and actinobacterial systems for secondary metabolite gene cluster expression and bioactive compound formation.

New and better tools are demanded and continuously developed to increase efficiency at the different steps of the Functional Metagenomics biodiscovery pipeline (Fig. 2.1). In addition to dedicated sampling and efficient DNA extraction procedures from diverse natural environments and developments within metagenomic (small- and large-insert) library cloning technology, several other aspects are in focus. Vector development for heterologous expression in and transfer between multiple host species (broad host-range) as well as optimization of different host species to heterologously express genes for bioactive functions will likely continue to converge. In particular, a higher efficiency in large-insert cloning of eDNA and its shuffling between different expression hosts allowing screening in in different organisms with complementary features and capabilities has proven to generate complementary hits (Liebl et al. 2014). It is therefore still highly desired to improve the functional metagenomic pipeline for metagenome-based bioactive compound discovery by means of new expression and screening platforms. Several different host organisms may be included, and shuffling of metagenomic libraries between these, connected to multiple host screening, is of potentially high value. It can be expected that newly developed expression systems aim to be optimal within screening for specific targeted applications (specific enzyme functions or bioactive compounds) or product properties. The concept of specifically accessing environments providing desired properties (e.g., of an enzyme of choice) and subsequently using screening hosts that perform optimally at similar conditions can be expected to produce further valuable output in the future. In addition, within this concept, the metabolic optimization of the host species from different phyla or even domains (including Archaea and Eukaryotes) may be pursued. The integration of new host species of phyla other than the *Actinobacteria* may expand the options to access biodiversity for medical compound discovery and thus mitigate the threat of antibiotic resistance, as well as help fighting deadly diseases, including cancer.

System biology understanding, the application of new genome editing tools, and synthetic biology principles will guide new approaches to optimize host strains for heterologous expression of metagenomic genes and formation of new natural products. Optimized Superhosts for bioactivity screening based on different model *Actinobacteria* will enable heterologous expression of biosynthetic gene clusters and compound formation from uncultured bacteria. Well-established thermophilic and psychrophilic host species will be good candidates for further optimization with respect to high- and low-temperature screening. Thereby, optimal hosts should attribute, among others, stable cloning vector maintenance, sensitivity toward relevant antibiotics for selection purposes, and suitable transcription and translation machinery. In addition, they should ensure correct folding, cofactor provision and insertion, relevant precursor supply, as well as counteract toxic effects from product formation (e.g., by product export mechanisms).

In vivo systems for Functional Metagenomics come with their inherent set of challenges, like limitations in achievable library sizes and the spectrum of usable substrates for screening. Consequently, cell-free (in vitro) expression systems have lately emerged as a potential alternative in functional metagenome screening for enzymatic functions (Sect. 2.4). In vitro expression systems still have their own limitations, in particular regarding large-scale production, which, however, is not very relevant for screening and biodiscovery, requiring only small amounts of product. Key challenges are constantly being addressed with new research, and solutions to key bottlenecks have already been found. An expanded spectrum of CFPS species, similar to the diversification of in vivo expression systems, as well as hybrid systems combining beneficial components of different species, can be expected to become available soon. Thus, in combination with ongoing developments of compartmentalization and miniaturization of screening technology, as achievable by, e.g., using advanced microfluidics devices, in vitro systems may become a potential future alternative to in vivo systems in Functional Metagenomics.

References

Aakvik T, Degnes KF, Dahlsrud R, Schmidt F, Dam R, Yu L, Volker U, Ellingsen TE, Valla S (2009) A plasmid RK2-based broad-host-range cloning vector useful for transfer of metagenomic libraries to a variety of bacterial species. FEMS Microbiol Lett 296(2):149–158. doi:10.1111/j.1574-6968.2009.01639.x

Ahn JH, Chu HS, Kim TW, Oh IS, Choi CY, Hahn GH, Park CG, Kim DM (2005) Cell-free synthesis of recombinant proteins from PCR-amplified genes at a comparable productivity to that

of plasmid-based reactions. Biochem Biophys Res Commun 338(3):1346–1352. doi:10.1016/j.bbrc.2005.10.094
Alam MT, Merlo ME, Consortium S, Hodgson DA, Wellington EM, Takano E, Breitling R (2010) Metabolic modeling and analysis of the metabolic switch in Streptomyces coelicolor. BMC Genomics 11:202. doi:10.1186/1471-2164-11-202
Angelov A, Mientus M, Liebl S, Liebl W (2009) A two-host fosmid system for functional screening of (meta)genomic libraries from extreme thermophiles. Syst Appl Microbiol 32(3):177–185. doi:10.1016/j.syapm.2008.01.003
Baltz RH (2008) Renaissance in antibacterial discovery from actinomycetes. Curr Opin Pharmacol 8(5):557–563. doi:10.1016/j.coph.2008.04.008
Baltz RH (2010) Streptomyces and Saccharopolyspora hosts for heterologous expression of secondary metabolite gene clusters. J Ind Microbiol Biotechnol 37(8):759–772. doi:10.1007/s10295-010-0730-9
Barka EA, Vatsa P, Sanchez L, Gaveau-Vaillant N, Jacquard C, Klenk HP, Clement C, Ouhdouch Y, van Wezel GP (2016) Taxonomy, physiology, and natural products of actinobacteria. Microbiol Mol Biol Rev 80(1):1–43. doi:10.1128/MMBR.00019-15
Battke F, Herbig A, Wentzel A, Jakobsen ØM, Bonin M, Hodgson DA, Wohlleben W, Ellingsen TE, Consortium S, Nieselt K (2011) A technical platform for generating reproducible expression data from Streptomyces coelicolor batch cultivations. In: Arabnia HR, Tran QN (eds) Software tools and algorithms for biological systems. Springer, New York, pp 3–15. doi:10.1007/978-1-4419-7046-6_1
Bentley SD, Chater KF, Cerdeno-Tarraga AM, Challis GL, Thomson NR, James KD, Harris DE, Quail MA, Kieser H, Harper D, Bateman A, Brown S, Chandra G, Chen CW, Collins M, Cronin A, Fraser A, Goble A, Hidalgo J, Hornsby T, Howarth S, Huang CH, Kieser T, Larke L, Murphy L, Oliver K, O'Neil S, Rabbinowitsch E, Rajandream MA, Rutherford K, Rutter S, Seeger K, Saunders D, Sharp S, Squares R, Squares S, Taylor K, Warren T, Wietzorrek A, Woodward J, Barrell BG, Parkhill J, Hopwood DA (2002) Complete genome sequence of the model actinomycete Streptomyces coelicolor A3(2). Nature 417(6885):141–147. doi:10.1038/417141a
Bibb MJ (2005) Regulation of secondary metabolism in streptomycetes. Curr Opin Microbiol 8(2):208–215. doi:10.1016/j.mib.2005.02.016
Biver S, Steels S, Portetelle D, Vandenbol M (2013) Bacillus subtilis as a tool for screening soil metagenomic libraries for antimicrobial activities. J Microbiol Biotechnol 23(6):850–855. doi:10.4014/jmb.1212.12008
Bjerga GEK, Williamson AK (2015) Cold shock induction of recombinant Arctic environmental genes. BMC Biotechnol 15(1):1–12. doi:10.1186/s12896-015-0185-1
Brady SF (2007) Construction of soil environmental DNA cosmid libraries and screening for clones that produce biologically active small molecules. Nat Protoc 2(5):1297–1305. doi:10.1038/nprot.2007.195
Brouns SJ, Wu H, Akerboom J, Turnbull AP, de Vos WM, van der Oost J (2005) Engineering a selectable marker for hyperthermophiles. J Biol Chem 280(12):11422–11431. doi:10.1074/jbc.M413623200
Calhoun KA, Swartz JR (2007) Energy systems for ATP regeneration in cell-free protein synthesis reactions. Methods Mol Biol 375:3–17. doi:10.1007/978-1-59745-388-2_1
Catherine C, Lee KH, Oh SJ, Kim DM (2013) Cell-free platforms for flexible expression and screening of enzymes. Biotechnol Adv 31(6):797–803. doi:10.1016/j.biotechadv.2013.04.009
Cava F, Laptenko O, Borukhov S, Chahlafi Z, Blas-Galindo E, Gomez-Puertas P, Berenguer J (2007) Control of the respiratory metabolism of Thermus thermophilus by the nitrate respiration conjugative element NCE. Mol Microbiol 64(3):630–646. doi:10.1111/j.1365-2958.2007.05687.x
Cava F, Hidalgo A, Berenguer J (2009) Thermus thermophilus as biological model. Extremophiles 13(2):213–231. doi:10.1007/s00792-009-0226-6
Cavicchioli R, Charlton T, Ertan H, Mohd Omar S, Siddiqui KS, Williams TJ (2011) Biotechnological uses of enzymes from psychrophiles. Microb Biotechnol 4(4):449–460. doi:10.1111/j.1751-7915.2011.00258.x

Cheng J, Pinnell L, Engel K, Neufeld JD, Charles TC (2014) Versatile broad-host-range cosmids for construction of high quality metagenomic libraries. J Microbiol Methods 99:27–34. doi:10.1016/j.mimet.2014.01.015

Cheng J, Romantsov T, Engel K, Doxey AC, Rose DR, Neufeld JD, Charles TC (2017) Functional metagenomics reveals novel beta-galactosidases not predictable from gene sequences. PLOS ONE 12(3):e0172545. doi:10.1371/journal.pone.0172545

Chu SF, Shu HY, Lin LC, Chen MY, Tsay SS, Lin GH (2006) Characterization of a rolling-circle replication plasmid from Thermus aquaticus NTU103. Plasmid 56(1):46–52. doi:10.1016/j.plasmid.2006.01.005

Cobb RE, Wang YJ, Zhao HM (2015) High-efficiency multiplex genome editing of streptomyces species using an engineered crispr/cas system. ACS Synth Biol 4(6):723–728. doi:10.1021/sb500351f

Courtois F, Olguin LF, Whyte G, Bratton D, Huck WT, Abell C, Hollfelder F (2008) An integrated device for monitoring time-dependent in vitro expression from single genes in picolitre droplets. Chembiochem 9(3):439–446. doi:10.1002/cbic.200700536

Craig JW, Chang FY, Kim JH, Obiajulu SC, Brady SF (2010) Expanding small-molecule functional metagenomics through parallel screening of broad-host-range cosmid environmental dna libraries in diverse proteobacteria. Appl Environ Microbiol 76(5):1633–1641. doi:10.1128/Aem.02169-09

Daly R, Hearn MT (2005) Expression of heterologous proteins in Pichia pastoris: a useful experimental tool in protein engineering and production. J Mol Recognit 18(2):119–138. doi:10.1002/jmr.687

Damon C, Vallon L, Zimmermann S, Haider MZ, Galeote V, Dequin S, Luis P, Fraissinet-Tachet L, Marmeisse R (2011) A novel fungal family of oligopeptide transporters identified by functional metatranscriptomics of soil eukaryotes. ISME J 5(12):1871–1880. doi:10.1038/ismej.2011.67

Danhorn T, Young CR, DeLong EF (2012) Comparison of large-insert, small-insert and pyrosequencing libraries for metagenomic analysis. ISME J 6:2056–2066. doi:10.1038/ismej.2012.35

de Grado M, Castan P, Berenguer J (1999) A high-transformation-efficiency cloning vector for Thermus thermophilus. Plasmid 42(3):241–245. doi:10.1006/plas.1999.1427

Drew D, Kim H (2012) Preparation of saccharomyces cerevisiae expression plasmids. Methods Mol Biol 866:41–46. doi:10.1007/978-1-61779-770-5_4

Du D, Zhu Y, Wei JH, Tian YQ, Niu G, Tan HR (2013) Improvement of gougerotin and nikkomycin production by engineering their biosynthetic gene clusters. Appl Microbiol Biotechnol 97(14):6383–6396. doi:10.1007/s00253-013-4836-7

Ekkers DM, Cretoiu MS, Kielak AM, van Elsas JD (2012) The great screen anomaly-a new frontier in product discovery through functional metagenomics. Appl Microbiol Biotechnol 93(3):1005–1020. doi:10.1007/s00253-011-3804-3

Endo Y, Sawasaki T (2004) High-throughput, genome-scale protein production method based on the wheat germ cell-free expression system. J Struct Funct Genom 5(1–2):45–57. doi:10.1023/B:JSFG.0000029208.83739.49

Endoh T, Kanai T, Imanaka T (2007) A highly productive system for cell-free protein synthesis using a lysate of the hyperthermophilic archaeon, Thermococcus kodakaraensis. Appl Microbiol Biotechnol 74(5):1153–1161. doi:10.1007/s00253-006-0753-3

Feitelson JS, Malpartida F, Hopwood DA (1985) Genetic and biochemical characterization of the red gene cluster of Streptomyces coelicolor A3(2). J Gen Microbiol 131(9):2431–2441. doi:10.1099/00221287-131-9-2431

Feller G (2013) Psychrophilic enzymes: from folding to function and biotechnology. Scientifica (Cairo) 2013:28. doi:10.1155/2013/512840

Fernandez-Arrojo L, Guazzaroni ME, Lopez-Cortes N, Beloqui A, Ferrer M (2010) Metagenomic era for biocatalyst identification. Curr Opin Biotechnol 21(6):725–733. doi:10.1016/j.copbio.2010.09.006

Ferrer M, Chernikova TN, Yakimov MM, Golyshin PN, Timmis KN (2003) Chaperonins govern growth of Escherichia coli at low temperatures. Nat Biotechnol 21(11):1266–1267. doi:10.1038/nbt1103-1266b
Ferrer M, Beloqui A, Golyshina OV, Plou FJ, Neef A, Chernikova TN, Fernandez-Arrojo L, Ghazi I, Ballesteros A, Elborough K, Timmis KN, Golyshin PN (2007) Biochemical and structural features of a novel cyclodextrinase from cow rumen metagenome. Biotechnol J 2(2):207–213. doi:10.1002/biot.200600183
Ferrer M, Martinez-Martinez M, Bargiela R, Streit WR, Golyshina OV, Golyshin PN (2016) Estimating the success of enzyme bioprospecting through metagenomics: current status and future trends. Microb Biotechnol 9(1):22–34. doi:10.1111/1751-7915.12309
Foulston LC, Bibb MJ (2010) Microbisporicin gene cluster reveals unusual features of lantibiotic biosynthesis in actinomycetes. Proc Natl Acad Sci U S A 107(30):13461–13466. doi:10.1073/pnas.1008285107
Gabor EM, Alkema WB, Janssen DB (2004) Quantifying the accessibility of the metagenome by random expression cloning techniques. Environ Microbiol 6(9):879–886. doi:10.1111/j.1462-2920.2004.00640.x
Gaida SM, Sandoval NR, Nicolaou SA, Chen Y, Venkataramanan KP, Papoutsakis ET (2015) Expression of heterologous sigma factors enables functional screening of metagenomic and heterologous genomic libraries. Nat Commun 6:7045. doi:10.1038/ncomms8045
Garneau JE, Dupuis ME, Villion M, Romero DA, Barrangou R, Boyaval P, Fremaux C, Horvath P, Magadan AH, Moineau S (2010) The CRISPR/Cas bacterial immune system cleaves bacteriophage and plasmid DNA. Nature 468(7320):67–71. doi:10.1038/nature09523
Geng A, Zou G, Yan X, Wang Q, Zhang J, Liu F, Zhu B, Zhou Z (2012) Expression and characterization of a novel metagenome-derived cellulase Exo2b and its application to improve cellulase activity in Trichoderma reesei. Appl Microbiol Biotechnol 96(4):951–962. doi:10.1007/s00253-012-3873-y
Godiska R, Mead D, Dhodda V, Wu C, Hochstein R, Karsi A, Usdin K, Entezam A, Ravin N (2010) Linear plasmid vector for cloning of repetitive or unstable sequences in Escherichia coli. Nucleic Acids Res 38(6):e88. doi:10.1093/nar/gkp1181
Gomez-Escribano JP, Bibb MJ (2011) Engineering Streptomyces coelicolor for heterologous expression of secondary metabolite gene clusters. Microb Biotechnol 4(2):207–215. doi:10.1111/j.1751-7915.2010.00219.x
Gomez-Escribano JP, Bibb MJ (2012) Streptomyces coelicolor as an expression host for heterologous gene clusters. Methods Enzymol 517:279–300. doi:10.1016/B978-0-12-404634-4.00014-0
Gomez-Escribano JP, Bibb MJ (2014) Heterologous expression of natural product biosynthetic gene clusters in Streptomyces coelicolor: from genome mining to manipulation of biosynthetic pathways. J Ind Microbiol Biotechnol 41(2):425–431. doi:10.1007/s10295-013-1348-5
Gomez-Escribano JP, Song LJ, Fox DJ, Yeo V, Bibb MJ, Challis GL (2012) Structure and biosynthesis of the unusual polyketide alkaloid coelimycin P1, a metabolic product of the cpk gene cluster of Streptomyces coelicolor M145. Chem Sci 3(9):2716–2720. doi:10.1039/c2sc20410j
Gruber S, Schwab H, Koefinger P (2015) Versatile plasmid-based expression systems for gram-negative bacteria—general essentials exemplified with the bacterium ralstonia eutropha H16. New Biotechnol 32(6):552–558. doi:10.1016/j.nbt.2015.03.015
Guazzaroni M-E, Silva-Rocha R, Ward RJ (2015) Synthetic biology approaches to improve biocatalyst identification in metagenomic library screening. Microb Biotechnol 8(1):52–64. doi:10.1111/1751-7915.12146
Gust B, Chandra G, Jakimowicz D, Tian YQ, Bruton CJ, Chater KF (2004) Lambda red-mediated genetic manipulation of antibiotic-producing Streptomyces. Adv Appl Microbiol 54:107–128. doi:10.1016/S0065-2164(04)54004-2
Hain T, Otten S, von Both U, Chatterjee SS, Technow U, Billion A, Ghai R, Mohamed W, Domann E, Chakraborty T (2008) Novel bacterial artificial chromosome vector pUvBBAC for use in studies of the functional genomics of Listeria spp. Appl Environ Microbiol 74(6):1892–1901. doi:10.1128/AEM.00415-07

Handelsman J, Rondon MR, Brady SF, Clardy J, Goodman RM (1998) Molecular biological access to the chemistry of unknown soil microbes: a new frontier for natural products. Chem Biol 5(10):R245–R249. doi:10.1016/S1074-5521(98)90108-9

Heil JR, Cheng J, Charles TC (2012) Site-specific bacterial chromosome engineering: PhiC31 integrase mediated cassette exchange (IMCE). J Vis Exp 61. doi:10.3791/3698

Hethke C, Geerling AC, Hausner W, de Vos WM, Thomm M (1996) A cell-free transcription system for the hyperthermophilic archaeon Pyrococcus furiosus. Nucleic Acids Res 24(12):2369–2376. doi:10.1093/nar/24.12.2369

Hidaka Y, Hasegawa M, Nakahara T, Hoshino T (1994) The entire population of Thermus thermophilus cells is always competent at any growth phase. Biosci Biotechnol Biochem 58(7):1338–1339. doi:10.1271/bbb.58.1338

Hidalgo A, Betancor L, Moreno R, Zafra O, Cava F, Fernandez-Lafuente R, Guisan JM, Berenguer J (2004) Thermus thermophilus as a cell factory for the production of a thermophilic Mn-dependent catalase which fails to be synthesized in an active form in Escherichia coli. Appl Environ Microbiol 70(7):3839–3844. doi:10.1128/AEM.70.7.3839-3844.2004

Holz C, Prinz B, Bolotina N, Sievert V, Büssow K, Simon B, Stahl U, Lang C (2003) Establishing the yeast Saccharomyces cerevisiae as a system for expression of human proteins on a proteome-scale. J Struct Funct Genom 4(2–3):97–108. doi:10.1023/A:1026226429429

Hopwood DA (2007) Streptomyces in nature and medicine: the antibiotic makers. Oxford University Press, New York

Hopwood DA, Wright HM (1983) CDA is a new chromosomally determined antibiotic from Streptomyces coelicolor A3(2). J Gen Microbiol 129:3575–3579. doi:10.1099/00221287-129-12-3575

Hu H, Zhang Q, Ochi K (2002) Activation of antibiotic biosynthesis by specified mutations in the rpoB gene (encoding the RNA polymerase beta subunit) of Streptomyces lividans. J Bacteriol 184(14):3984–3991. doi:10.1128/JB.184.14.3984-3991.2002

Hu XP, Heath C, Taylor MP, Tuffin M, Cowan DA (2012) A novel extremely alkaliphilic and cold-active esterase from Antarctic desert soil. Extremophiles 16:79–86. doi:10.1007/s00792-011-0407-y

Huang H, Zheng GS, Jiang WH, Hu H, Lu YH (2015) One-step high-efficiency CRISPR/Cas9-mediated genome editing in Streptomyces. Acta Biochim Biophys Sin 47(4):231–243. doi:10.1093/abbs/gmv007

Huo L, Rachid S, Stadler M, Wenzel SC, Muller R (2012) Synthetic biotechnology to study and engineer ribosomal bottromycin biosynthesis. Chem Biol 19(10):1278–1287. doi:10.1016/j.chembiol.2012.08.013

Iqbal M, Mast Y, Amin R, Hodgson DA, Consortium S, Wohlleben W, Burroughs NJ (2012) Extracting regulator activity profiles by integration of de novo motifs and expression data: characterizing key regulators of nutrient depletion responses in Streptomyces coelicolor. Nucleic Acids Res 40(12):5227–5239. doi:10.1093/nar/gks205

Jewett MC, Swartz JR (2004) Mimicking the Escherichia coli cytoplasmic environment activates long-lived and efficient cell-free protein synthesis. Biotechnol Bioeng 86(1):19–26. doi:10.1002/bit.20026

Jones AC, Gust B, Kulik A, Heide L, Buttner MJ, Bibb MJ (2013) Phage p1-derived artificial chromosomes facilitate heterologous expression of the FK506 gene cluster. PLoS One 8(7):e69319. doi:10.1371/journal.pone.0069319

Kakirde KS, Parsley LC, Liles MR (2010) Size does matter: application-driven approaches for soil metagenomics. Soil Biol Biochem 42(11):1911–1923. doi:10.1016/j.soilbio.2010.07.021

Kennedy J, Marchesi JR, Dobson AD (2008) Marine metagenomics: strategies for the discovery of novel enzymes with biotechnological applications from marine environments. Microb Cell Factories 7:27. doi:10.1186/1475-2859-7-27

Kieser T, Bibb MJ, Buttner MJ, Chater KF, Hopwood DA (2000) Practical Streptomyces genetics. John Innes Foundation, Norwich

Kim HC, Kim DM (2009) Methods for energizing cell-free protein synthesis. J Biosci Bioeng 108(1):1–4. doi:10.1016/j.jbiosc.2009.02.007

Kim UJ, Shizuya H, de Jong PJ, Birren B, Simon MI (1992) Stable propagation of cosmid sized human DNA inserts in an F factor based vector. Nucleic Acids Res 20(5):1083–1085. doi:10.1093/nar/20.5.1083

Kim HC, Kim TW, Kim DM (2011) Prolonged production of proteins in a cell-free protein synthesis system using polymeric carbohydrates as an energy source. Process Biochem 2011(46):1366–1369. doi:10.1016/j.procbio.2011.03.008

King RW, Bauer JD, Brady SF (2009) An environmental DNA-derived type II polyketide biosynthetic pathway encodes the biosynthesis of the pentacyclic polyketide erdacin. Angew Chem 48:6257–6261. doi:10.1002/anie.200901209

Kintses B, van Vliet LD, Devenish SR, Hollfelder F (2010) Microfluidic droplets: new integrated workflows for biological experiments. Curr Opin Chem Biol 14(5):548–555. doi:10.1016/j.cbpa.2010.08.013

Komatsu M, Komatsu K, Koiwai H, Yamada Y, Kozone I, Izumikawa M, Hashimoto J, Takagi M, Omura S, Shin-ya K, Cane DE, Ikeda H (2013) Engineered Streptomyces avermitilis host for heterologous expression of biosynthetic gene cluster for secondary metabolites. ACS Synth Biol 2(7):384–396. doi:10.1021/sb3001003

Kotlar HK, Lewin A, Johansen J, Throne-Holst M, Haverkamp T, Markussen S, Winnberg A, Ringrose P, Aakvik T, Ryeng E, Jakobsen K, Drabløs F, Valla S (2011) High coverage sequencing of DNA from microorganisms living in an oil reservoir 2.5 kilometres subsurface. Environ Microbiol Rep 3(6):674–681. doi:10.1111/j.1758-2229.2011.00279.x

Koyama Y, Hoshino T, Tomizuka N, Furukawa K (1986) Genetic transformation of the extreme thermophile Thermus thermophilus and of other Thermus spp. J Bacteriol 166(1):338–340. doi: 0021-9193/86/040338-03$02.00/0

Koyama Y, Okamoto S, Furukawa K (1990) Cloning of alpha- and beta-galactosidase genes from an extreme thermophile, Thermus strain T2, and their expression in Thermus thermophilus HB27. Appl Environ Microbiol 56(7):2251–2254. doi: 0099-2240/90/072251-04$02.00/0

Kumar G, Chernaya G (2009) Cell-free protein synthesis using multiply-primed rolling circle amplification products. BioTechniques 47:637–639. doi:10.2144/000113171

Kwon YC, Oh IS, Lee N, Lee KH, Yoon YJ, Lee EY, Kim BG, Kim DM (2013) Integrating cell-free biosyntheses of heme prosthetic group and apoenzyme for the synthesis of functional P450 monooxygenase. Biotechnol Bioeng 110(4):1193–1200. doi:10.1002/bit.24785

Lam KN, Charles TC (2015) Strong spurious transcription likely a cause of DNA insert bias in typical metagenomic clone libraries. Microbiome 3:22. doi:10.1101/013763

Lam KN, Cheng J, Engel K, Neufeld JD, Charles TC (2015) Current and future resources for functional metagenomics. Front Microbiol 6:1196. doi:10.3389/fmicb.2015.01196

Lammle K, Zipper H, Breuer M, Hauer B, Buta C, Brunner H, Rupp S (2007) Identification of novel enzymes with different hydrolytic activities by metagenome expression cloning. J Biotechnol 127(4):575–592. doi:10.1016/j.jbiotec.2006.07.036

Lasa I, de Grado M, de Pedro MA, Berenguer J (1992) Development of Thermus-Escherichia shuttle vectors and their use for expression of the Clostridium thermocellum celA gene in Thermus thermophilus. J Bacteriol 174(20):6424–6431. doi: 0021-9193/92/206424-08$02.00/0

Lee KH, Lee KY, Byun JY, Kim BG, Kim DM (2012) On-bead expression of recombinant proteins in an agarose gel matrix coated on a glass slide. Lab Chip 12(9):1605–1610. doi:10.1039/c2lc21239k

Leis B, Angelov A, Mientus M, Li HJ, Pham VTT, Lauinger B, Bongen P, Pietruszka J, Goncalves LG, Santos H, Liebl W (2015a) Identification of novel esterase-active enzymes from hot environments by use of the host bacterium Thermus thermophilus. Front Microbiol 6:275. doi:10.3389/frricb.2015.00275

Leis B, Heinze S, Angelov A, Pham VT, Thürmer A, Jebbar M, Golyshin PN, Streit WR, Daniel R, Liebl W (2015b) Functional screening of hydrolytic activities reveals an extremely thermostable cellulase from a deep-sea archaeon. Front Bioeng Biotechnol 3:95. doi:10.3389/fbioe.2015.00095

Lewin A, Wentzel A, Valla S (2013) Metagenomics of microbial life in extreme temperature environments. Curr Opin Biotechnol 24(3):516–525. doi:10.1016/j.copbio.2012.10.012

Leza A, Palmeros B, García JO, Galindo E, Soberón-Chávez G (1996) Xanthomonas campestris as a host for the production of recombinant Pseudomonas aeruginosa lipase. J Ind Microbiol 16(1):22–28. doi:10.1007/BF01569917

Li C, Zhang F, Kelly WL (2011) Heterologous production of thiostrepton A and biosynthetic engineering of thiostrepton analogs. Mol Biosyst 7(1):82–90. doi:10.1039/c0mb00129e

Liao H, McKenzie T, Hageman R (1986) Isolation of a thermostable enzyme variant by cloning and selection in a thermophile. Proc Natl Acad Sci U S A 83(3):576–580. doi: pnas00307-0057

Liebl W, Angelov A, Juergensen J, Chow J, Loeschcke A, Drepper T, Classen T, Pietruszka J, Ehrenreich A, Streit WR, Jaeger KE (2014) Alternative hosts for functional (meta) genome analysis. Appl Microbiol Biotechnol 98(19):8099–8109. doi:10.1007/s00253-014-5961-7

Liles MR, Williamson LL, Rodbumrer J, Torsvik V, Goodman RM, Handelsman J (2008) Recovery, purification, and cloning of high-molecular-weight DNA from soil microorganisms. Appl Environ Microbiol 74(10):3302–3305. doi:10.1128/AEM.02630-07

Loeschcke A, Thies S (2015) Pseudomonas putida—a versatile host for the production of natural products. Appl Microbiol Biotechnol 99(15):6197–6214. doi:10.1007/s00253-015-6745-4

Maldonado LA, Stach JE, Pathom-aree W, Ward AC, Bull AT, Goodfellow M (2005) Diversity of cultivable actinobacteria in geographically widespread marine sediments. Antonie Van Leeuwenhoek 87(1):11–18. doi:10.1007/s10482-004-6525-0

Marcone GL, Foulston L, Binda E, Marinelli F, Bibb M, Beltrametti F (2010) Methods for the genetic manipulation of Nonomuraea sp. ATCC 39727. J Ind Microbiol Biotechnol 37(10):1097–1103. doi:10.1007/s10295-010-0807-5

Martin JF, Santos-Beneit F, Rodriguez-Garcia A, Sola-Landa A, Smith MC, Ellingsen TE, Nieselt K, Burroughs NJ, Wellington EM (2012) Transcriptomic studies of phosphate control of primary and secondary metabolism in Streptomyces coelicolor. Appl Microbiol Biotechnol 95(1):61–75. doi:10.1007/s00253-012-4129-6

Martinez A, Kolvek SJ, Yip CL, Hopke J, Brown KA, MacNeil IA, Osburne MS (2004) Genetically modified bacterial strains and novel bacterial artificial chromosome shuttle vectors for constructing environmental libraries and detecting heterologous natural products in multiple expression hosts. Appl Environ Microbiol 70(4):2452–2463. doi:10.1128/AEM.70.4.2452-2463.2004

Martinez A, Kolvek SJ, Hopke J, Yip CL, Osburne MS (2005) Environmental DNA fragment conferring early and increased sporulation and antibiotic production in Streptomyces species. Appl Environ Microbiol 71(3):1638–1641. doi:10.1128/AEM.71.3.1638-1641.2005

Mather MW, Fee JA (1992) Development of plasmid cloning vectors for Thermus thermophilus HB8: expression of a heterologous, plasmid-borne kanamycin nucleotidyltransferase gene. Appl Environ Microbiol 58(1):421–425. doi: 0099-2240/92/010421-05$02.0O/O

Matsumura M, Aiba S (1985) Screening for thermostable mutant of kanamycin nucleotidyltransferase by the use of a transformation system for a thermophile, Bacillus stearothermophilus. J Biol Chem 260(28):15298–15303. doi: 260/28/15298

Matthaei JH, Nirenberg MW (1961) The dependence of cell-free protein synthesis in *E. coli* upon RNA prepared from ribosomes. Biochem Biophys Res Commun 28(4):404–408. doi: pnas00214-0066

McMahon MD, Guan C, Handelsman J, Thomas MG (2012) Metagenomic analysis of Streptomyces lividans reveals host-dependent functional expression. Appl Environ Microbiol 78(10):3622–3629. doi:10.1128/AEM.00044-12

Medema MH, Breitling R, Bovenberg R, Takano E (2011) Exploiting plug-and-play synthetic biology for drug discovery and production in microorganisms. Nat Rev Microbiol 9(2):131–137. doi:10.1038/nrmicro2478

Mergeay M, Monchy S, Vallaeys T, Auquier V, Benotmane A, Bertin P, Taghavi S, Dunn J, van der Lelie D, Wattiez R (2003) Ralstonia metallidurans, a bacterium specifically adapted to toxic metals: towards a catalogue of metal-responsive genes. FEMS Microbiol Rev 27(2–3):385–410. doi:10.1016/s0168-6445(03)00045-7

Michel-Reydellet N, Woodrow K, Swartz J (2005) Increasing PCR fragment stability and protein yields in a cell-free system with genetically modified Escherichia coli extracts. J Mol Microbiol Biotechnol 9(1):26–34. doi:10.1159/000088143

Miteva V, Lantz S, Brenchley J (2008) Characterization of a cryptic plasmid from a Greenland ice core Arthrobacter isolate and construction of a shuttle vector that replicates in psychrophilic high G+C Gram-positive recipients. Extremophiles 12(3):441–449. doi:10.1007/s00792-008-0149-7

Miyake R, Kawamoto J, Wei YL, Kitagawa M, Kato I, Kurihara T, Esaki N (2007) Construction of a low-temperature protein expression system using a cold-adapted bacterium, Shewanella sp. strain Ac10, as the host. Appl Environ Microbiol 73(15):4849–4856. doi:10.1128/AEM.00824-07

Moreno R, Haro A, Castellanos A, Berenguer J (2005) High-level overproduction of His-tagged Tth DNA polymerase in Thermus thermophilus. Appl Environ Microbiol 71(1):591–593. doi:10.1128/AEM.71.1.591-593.2005

Mullany P (2014) Functional metagenomics for the investigation of antibiotic resistance. Virulence 5(3):443–447. doi:10.4161/viru.28196

Murai N (2013) Review: plant binary vectors of Ti plasmid in agrobacterium tumefaciens with a broad host-range replicon of pRK2, pRi, pSa or pVS1. AJPS 4:932–939. doi:10.4236/ajps.2013.44115

Nakamura A, Takakura Y, Kobayashi H, Hoshino T (2005) In vivo directed evolution for thermostabilization of Escherichia coli hygromycin B phosphotransferase and the use of the gene as a selection marker in the host-vector system of Thermus thermophilus. J Biosci Bioeng 100(2):158–163. doi:10.1263/jbb.100.158

Nett M, Ikeda H, Moore BS (2009) Genomic basis for natural product biosynthetic diversity in the actinomycetes. Nat Prod Rep 26(11):1362–1384. doi:10.1039/b817069j

Nevalainen KM, Te'o VS, Bergquist PL (2005) Heterologous protein expression in filamentous fungi. Trends Biotechnol 23(9):468–474. doi:10.1016/j.tibtech.2005.06.002

Nieselt K, Battke F, Herbig A, Bruheim P, Wentzel A, Jakobsen OM, Sletta H, Alam MT, Merlo ME, Moore J, Omara WA, Morrissey ER, Juarez-Hermosillo MA, Rodriguez-Garcia A, Nentwich M, Thomas L, Iqbal M, Legaie R, Gaze WH, Challis GL, Jansen RC, Dijkhuizen L, Rand DA, Wild DL, Bonin M, Reuther J, Wohlleben W, Smith MC, Burroughs NJ, Martin JF, Hodgson DA, Takano E, Breitling R, Ellingsen TE, Wellington EM (2010) The dynamic architecture of the metabolic switch in Streptomyces coelicolor. BMC Genomics 11:10. doi:10.1186/1471-2164-11-10

Novakova J, Farkasovsky M (2013) Bioprospecting microbial metagenome for natural products. Biologia 68(6):1079–1086. doi:10.2478/s11756-013-0246-7

Ochi K, Hosaka T (2013) New strategies for drug discovery: activation of silent or weakly expressed microbial gene clusters. Appl Microbiol Biotechnol 97(1):87–98. doi:10.1007/s00253-012-4551-9

Ochi K, Tanaka Y, Tojo S (2014) Activating the expression of bacterial cryptic genes by rpoB mutations in RNA polymerase or by rare earth elements. J Ind Microbiol Biotechnol 41(2):403–414. doi:10.1007/s10295-013-1349-4

Ohnishi Y, Ishikawa J, Hara H, Suzuki H, Ikenoya M, Ikeda H, Yamashita A, Hattori M, Horinouchi S (2008) Genome sequence of the streptomycin-producing microorganism Streptomyces griseus IFO 13350. J Bacteriol 190(11):4050–4060. doi:10.1128/JB.00204-08

Okamoto-Hosoya Y, Sato TA, Ochi K (2000) Resistance to paromomycin is conferred by rpsL mutations, accompanied by an enhanced antibiotic production in Streptomyces coelicolor A3(2). J Antibiot 53(12):1424–1427. doi:10.7164/antibiotics.53.1424

Oliynyk M, Samborskyy M, Lester JB, Mironenko T, Scott N, Dickens S, Haydock SF, Leadlay PF (2007) Complete genome sequence of the erythromycin-producing bacterium Saccharopolyspora erythraea NRRL23338. Nat Biotechnol 25(4):447–453. doi:10.1038/nbt1297

Parachin NS, Gorwa-Grauslund MF (2011) Isolation of xylose isomerases by sequence- and function-based screening from a soil metagenomic library. Biotechnol Biofuels 4:9. doi:10.1186/1754-6834-4-9

Parks RJ, Graham FL (1997) A helper-dependent system for adenovirus vector production helps define a lower limit for efficient dna packaging. J Virol 71(4):3293–3298. doi: 0022-538X/97/$04.0010

Pel J, Broemeling D, Mai L, Poon HL, Tropini G, Warren RL, Holt RA, Marziali A (2009) Nonlinear electrophoretic response yields a unique parameter for separation of biomolecules. Proc Natl Acad Sci U S A 106(35):14796–14801. doi:10.1073/pnas.0907402106

Potrykus K, Cashel M (2008) (p)ppGpp: still magical? Annu Rev Microbiol 62:35–51. doi:10.1146/annurev.micro.62.081307.162903

Ramirez-Arcos S, Fernandez-Herrero LA, Marin I, Berenguer J (1998) Anaerobic growth, a property horizontally transferred by an Hfr-like mechanism among extreme thermophiles. J Bacteriol 180(12):3137–3143. doi: 0021-9193/98/$04.0010

Retallack D, Schneider JC, Chew L, Ramseier T, Allen J, Patkar A, Squires C, Talbot H, Mitchell J (2006) Pseudomonas fluorescens—a robust expression platform for pharmaceutical protein production. Microb Cell Factories 5(1):1–1. doi:10.1186/1475-2859-5-s1-s28

Riesenfeld CS, Goodman RM, Handelsman J (2004) Uncultured soil bacteria are a reservoir of new antibiotic resistance genes. Environ Microbiol 6(9):981–989. doi:10.1111/j.1462-2920.2004.00664.x

Rigali S, Nothaft H, Noens EE, Schlicht M, Colson S, Muller M, Joris B, Koerten HK, Hopwood DA, Titgemeyer F, van Wezel GP (2006) The sugar phosphotransferase system of Streptomyces coelicolor is regulated by the GntR-family regulator DasR and links N-acetylglucosamine metabolism to the control of development. Mol Microbiol 61(5):1237–1251. doi:10.1111/j.1365-2958.2006.05319.x

Rigali S, Titgemeyer F, Barends S, Mulder S, Thomae AW, Hopwood DA, van Wezel GP (2008) Feast or famine: the global regulator DasR links nutrient stress to antibiotic production by Streptomyces. EMBO Rep 9(7):670–675. doi:10.1038/embor.2008.83

Rodrigue S, Malmstrom RR, Berlin AM, Birren BW, Henn MR, Chisholm SW (2009) Whole genome amplification and de novo assembly of single bacterial cells. PLoS One 4(9):e6864. doi:10.1371/journal.pone.0006864

Rondon MR, August PR, Bettermann AD, Brady SF, Grossman TH, Liles MR, Loiacono KA, Lynch BA, MacNeil IA, Minor C, Tiong CL, Gilman M, Osburne MS, Clardy J, Handelsman J, Goodman RM (2000) Cloning the soil metagenome: a strategy for accessing the genetic and functional diversity of uncultured microorganisms. Appl Environ Microbiol 66(6):2541–2547. doi:10.1128/AEM.66.6.2541-2547.2000

Ruan L, Xu X (2007) Sequence analysis and characterizations of two novel plasmids isolated from Thermus sp. 4C. Plasmid 58(1):84–87. doi:10.1016/j.plasmid.2007.04.001

Rudd BA, Hopwood DA (1979) Genetics of actinorhodin biosynthesis by Streptomyces coelicolor A3(2). J Gen Microbiol 114(1):35–43. doi:10.1099/00221287-114-1-35

Ruggero D, Creti R, Londei P (2006) In vitro translation of archaeal natural mRNAs at high temperature. FEMS Micobiol Lett 107(1):89–94. doi:10.1111/j.1574-6968.1993.tb06009.x

Rungpragayphan S, Nakano H, Yamane T (2003) PCR-linked in vitro expression: a novel system for high-throughput construction and screening of protein libraries. FEBS Lett 540(1–3):147–150. doi:10.1016/S0014-5793(03)00251-5

Rutledge PJ, Challis GL (2015) Discovery of microbial natural products by activation of silent biosynthetic gene clusters. Nat Rev Microbiol 13(8):509–523. doi:10.1038/nrmicro3496

Sabree ZL, Rondon MR, Handelsman J (2009) Metagenomics. In: Schaechter M (ed) Encyclopedia of microbiology. Elsevier, Amsterdam, pp. 622–632. doi:10.1016/B978-012373944-5.00034-1

Sawasaki T, Hasegawa Y, Tsuchimochi M, Kamura N, Ogasawara T, Kuroita T, Endo Y (2002) A bilayer cell-free protein synthesis system for high-throughput screening of gene products. FEBS Lett 514(1):102–105. doi:10.1016/S0014-5793(02)02329-3

Schwarzenlander C, Averhoff B (2006) Characterization of DNA transport in the thermophilic bacterium Thermus thermophilus HB27. FEBS J 273(18):4210–4218. doi:10.1111/j.1742-4658.2006.05416.x
Sherwood EJ, Bibb MJ (2013) The antibiotic planosporicin coordinates its own production in the actinomycete Planomonospora alba. Proc Natl Acad Sci U S A 110(27):E2500–E2509. doi:10.1073/pnas.1305392110
Shima J, Hesketh A, Okamoto S, Kawamoto S, Ochi K (1996) Induction of actinorhodin production by rpsL (encoding ribosomal protein S12) mutations that confer streptomycin resistance in Streptomyces lividans and Streptomyces coelicolor A3(2). J Bacteriol 178(24):7276–7284. doi: 0021-9193/96/$04.0010
Shizuya H, Birren B, Kim UJ, Mancino V, Slepak T, Tachiiri Y, Simon M (1992) Cloning and stable maintenance of 300-kilobase-pair fragments of human DNA in Escherichia coli using an F-factor-based vector. Proc Natl Acad Sci U S A 89(18):8794–8797. doi: pnas01092-0395
Sitaraman K, Esposito D, Klarmann G, Le Grice SF, Hartley JL, Chatterjee DK (2004) A novel cell-free protein synthesis system. J Biotechnol 110(3):257–263. doi:10.1016/j.jbiotec.2004.02.014
Smanski MJ, Casper J, Peterson RM, Yu Z, Rajski SR, Shen B (2012) Expression of the platencin biosynthetic gene cluster in heterologous hosts yielding new platencin congeners. J Nat Prod 75(12):2158–2167. doi:10.1021/np3005985
Sosio M, Giusino F, Cappellano C, Bossi E, Puglia AM, Donadio S (2000) Artificial chromosomes for antibiotic-producing actinomycetes. Nat Biotechnol 18(3):343–345. doi:10.1038/73810
Strausberg RL, Strausberg SL (2001) Overview of protein expression in saccharomyces cerevisiae. In: Current protocols in protein science. Wiley, Hoboken. doi:10.1002/0471140864.ps0506s02
Struvay C, Feller G (2012) Optimization to low temperature activity in psychrophilic enzymes. Int J Mol Sci 13(9):11643–11665. doi:10.3390/ijms130911643
Tabata K, Kosuge T, Nakahara T, Hoshino T (1993) Physical map of the extremely thermophilic bacterium Thermus thermophilus HB27 chromosome. FEBS Lett 331(1–2):81–85. doi:10.1016/0014-5793(93)80301-A
Tachibana A, Tanaka T, Taniguchi M, Oi S (1996) Evidence for farnesol-mediated isoprenoid synthesis regulation in a halophilic archaeon, Haloferax volcanii. FEBS Lett 379(1):43–46. doi:10.1016/0014-5793(95)01479-9
Takai K, Sawasaki T, Endo Y (2010) The wheat-germ cell-free expression system. Curr Pharm Biotechnol 11(3):272–278. doi:10.1016/j.febslet.2014.05.061
Tamakoshi M, Uchida M, Tanabe K, Fukuyama S, Yamagishi A, Oshima T (1997) A new Thermus-Escherichia coli shuttle integration vector system. J Bacteriol 179(15):4811–4814. doi: 0021-9193/97/$04.0010
Taupp M, Mewis K, Hallam SJ (2011) The art and design of functional metagenomic screens. Curr Opin Biotechnol 22(3):465–472. doi:10.1016/j.copbio.2011.02.010
Tebbe CC, Vahjen W (1993) Interference of humic acids and DNA extracted directly from soil in detection and transformation of recombinant-DNA from bacteria and a yeast. Appl Environ Microbiol 59(8):2657–2665. doi: aem00037-0325
Terron-Gonzalez L, Medina C, Limon-Mortes MC, Santero E (2013) Heterologous viral expression systems in fosmid vectors increase the functional analysis potential of metagenomic libraries. Sci Rep 3:1107. doi:10.1038/srep01107
Thomas L, Hodgson DA, Wentzel A, Nieselt K, Ellingsen TE, Moore J, Morrissey ER, Legaie R, Consortium S, Wohlleben W, Rodriguez-Garcia A, Martin JF, Burroughs NJ, Wellington EM, Smith MC (2012) Metabolic switches and adaptations deduced from the proteomes of Streptomyces coelicolor wild type and phoP mutant grown in batch culture. Mol Cell Proteomics 11(2):M111.013797. doi:10.1074/mcp.M111.013797
Tong YJ, Charusanti P, Zhang LX, Weber T, Lee SY (2015) CRISPR-Cas9 based engineering of actinomycetal genomes. ACS Synth Biol 4(9):1020–1029. doi:10.1021/acssynbio.5b00038
Troeschel S, Drepper T, Leggewie C, Streit W, Jaeger K-E (2010) Novel tools for the functional expression of metagenomic DNA. In: Streit WR, Daniel R (eds) Metagenomics, Methods in molecular biology, vol 668. Humana Press, New York, pp 117–139. doi:10.1007/978-1-60761-823-2_8

Tutino ML, Duilio A, Moretti MA, Sannia G, Marino G (2000) A rolling-circle plasmid from Psychrobacter sp. TA144: evidence for a novel rep subfamily. Biochem Biophys Res Commun 274(2):488–495. doi:10.1006/bbrc.2000.3148

Tutino ML, Duilio A, Parrilli E, Remaut E, Sannia G, Marino G (2001) A novel replication element from an Antarctic plasmid as a tool for the expression of proteins at low temperature. Extremophiles 5(4):257–264. doi:10.1007/s007920100203

Uchiyama T, Miyazaki K (2009) Functional metagenomics for enzyme discovery: challenges to efficient screening. Curr Opin Biotechnol 20(6):616–622. doi:10.1016/j.copbio.2009.09.010

Udwary DW, Zeigler L, Asolkar RN, Singan V, Lapidus A, Fenical W, Jensen PR, Moore BS (2007) Genome sequencing reveals complex secondary metabolome in the marine actinomycete Salinispora tropica. Proc Natl Acad Sci U S A 104(25):10376–10381. doi:10.1073/pnas.0700962104

van Wezel GP, McDowall KJ (2011) The regulation of the secondary metabolism of Streptomyces: new links and experimental advances. Nat Prod Rep 28(7):1311–1333. doi:10.1039/c1np00003a

van Wezel GP, McKenzie NL, Nodwell JR (2009) Chapter 5. Applying the genetics of secondary metabolism in model actinomycetes to the discovery of new antibiotics. Methods Enzymol 458:117–141. doi:10.1016/S0076-6879(09)04805-8

Vester JK, Glaring MA, Stougaard P (2015) Improved cultivation and metagenomics as new tools for bioprospecting in cold environments. Extremophiles 19(1):17–29. doi:10.1007/s00792-014-0704-3

Waldvogel E, Herbig A, Battke F, Amin R, Nentwich M, Nieselt K, Ellingsen TE, Wentzel A, Hodgson DA, Wohlleben W, Mast Y (2011) The PII protein GlnK is a pleiotropic regulator for morphological differentiation and secondary metabolism in Streptomyces coelicolor. Appl Microbiol Biotechnol 92(6):1219–1236. doi:10.1007/s00253

Wang F, Hao J, Yang C, Sun M (2010) Cloning, expression, and identification of a novel extracellular cold-adapted alkaline protease gene of the marine bacterium strain YS-80-122. Appl Biochem Biotechnol 162(5):1497–1505. doi:10.1007/s12010-010-8927-y

Wang GYS, Graziani E, Waters B, Pan W, Li X, McDermott J, Meurer G, Saxena G, Andersen RJ, Davies J (2000) Novel natural products from soil DNA libraries in a Streptomycete host. Organic Letters 2(16):2401–2404. doi: 10.1021/ol005860z

Wang Y, Zhang YH (2009) Cell-free protein synthesis energized by slowly-metabolized maltodextrin. BMC Biotechnol 9:58. doi:10.1186/1472-6750-9-58

Warburton P, Roberts AP, Allan E, Seville L, Lancaster H, Mullany P (2009) Characterization of tet(32) genes from the oral metagenome. Antimicrob Agents Chemother 53(1):273–276. doi:10.1128/AAC.00788-08

Warren RL, Freeman JD, Levesque RC, Smailus DE, Flibotte S, Holt RA (2008) Transcription of foreign DNA in Escherichia coli. Genome Res 18(11):1798–1805. doi:10.1101/gr.080358.108

Wayne J, Xu SY (1997) Identification of a thermophilic plasmid origin and its cloning within a new Thermus-E. coli shuttle vector. Gene 195(2):321–328. doi:10.1016/S0378-1119(97)00191-1

Wentzel A, Bruheim P, Overby A, Jakobsen OM, Sletta H, Omara WA, Hodgson DA, Ellingsen TE (2012a) Optimized submerged batch fermentation strategy for systems scale studies of metabolic switching in Streptomyces coelicolor A3(2). BMC Syst Biol 6:59. doi:10.1186/1752-0509-6-59

Wentzel A, Sletta H, Consortium S, Ellingsen TE, Bruheim P (2012b) Intracellular metabolite pool changes in response to nutrient depletion induced metabolic switching in streptomyces coelicolor. Metabolites 2(1):178–194. doi:10.3390/metabo2010178

Wexler M, Bond PL, Richardson DJ, Johnston AW (2005) A wide host-range metagenomic library from a waste water treatment plant yields a novel alcohol/aldehyde dehydrogenase. Environ Microbiol 7(12):1917–1926. doi:10.1111/j.1462-2920.2005.00854.x

Wietzorrek A, Bibb M (1997) A novel family of proteins that regulates antibiotic production in streptomycetes appears to contain an OmpR-like DNA-binding fold. Mol Microbiol 25(6):1181–1184. doi:10.1046/j.1365-2958.1997.5421903.x

Wild J, Hradecna Z, Szybalski W (2002) Conditionally amplifiable BACs: switching from single-copy to high-copy vectors and genomic clones. Genome Res 12:1434–1444. doi:10.1101/gr.130502

Wilson MC, Mori T, Ruckert C, Uria AR, Helf MJ, Takada K, Gernert C, Steffens UA, Heycke N, Schmitt S, Rinke C, Helfrich EJ, Brachmann AO, Gurgui C, Wakimoto T, Kracht M, Crusemann M, Hentschel U, Abe I, Matsunaga S, Kalinowski J, Takeyama H, Piel J (2014) An environmental bacterial taxon with a large and distinct metabolic repertoire. Nature 506(7486):58–62. doi:10.1038/nature12959

Wong S-L (1995) Advances in the use of Bacillus subtilis for the expression and secretion of heterologous proteins. Curr Opin Biotechnol 6(5):517–522. doi:10.1016/0958-1669(95)80085-9

Wright LF, Hopwood DA (1976) Identification of the antibiotic determined by the SCP1 plasmid of Streptomyces coelicolor A3(2). J Gen Microbiol 95(1):96–106. doi:10.1099/00221287-95-1-96

Yoon V, Nodwell JR (2014) Activating secondary metabolism with stress and chemicals. J Ind Microbiol Biotechnol 41(2):415–424. doi:10.1007/s10295-013-1387-y

Yoon MY, Lee KM, Yoon Y, Go J, Park Y, Cho YJ, Tannock GW, Yoon SS (2013) Functional screening of a metagenomic library reveals operons responsible for enhanced intestinal colonization by gut commensal microbes. Appl Environ Microbiol 79(12):3829–3838. doi:10.1128/AEM.00581-13

Zemella A, Thoring L, Hoffmeister C, Kubick S (2015) Cell-free protein synthesis: pros and cons of prokaryotic and eukaryotic systems. Chembiochem 16(17):2420–2431. doi:10.1002/cbic.201500340

Zengler K, Toledo G, Rappe M, Elkins J, Mathur EJ, Short JM, Keller M (2002) Cultivating the uncultured. Proc Natl Acad Sci U S A 99(24):15681–15686. doi:10.1073/pnas.252630999

Zerikly M, Challis GL (2009) Strategies for the discovery of new natural products by genome mining. Chembiochem 10(4):625–633. doi:10.1002/cbic.200800389

Zhang JW, Zeng RY (2008) Molecular cloning and expression of a cold-adapted lipase gene from an Antarctic deep sea psychrotrophic bacterium Pseudomonas sp 7323. Mar Biotechnol 10(5):612–621. doi:10.1007/s10126-008-9099-4

Zhang K, Martiny AC, Reppas NB, Barry KW, Malek J, Chisholm SW, Church GM (2006) Sequencing genomes from single cells by polymerase cloning. Nat Biotechnol 24(6):680–686. doi:10.1038/nbt1214

Zhang G, Li Y, Fang L, Pfeifer BA (2015) Tailoring pathway modularity in the biosynthesis of erythromycin analogs heterologously engineered in *E. coli*. Sci Adv 1(4):e1500077. doi:10.1126/sciadv.1500077

Zhou JZ, Bruns MA, Tiedje JM (1996) DNA recovery from soils of diverse composition. Appl Environ Microbiol 62(2):316–322. doi: 0099-2240/96/$04.0010

Zhou Y, Asahara H, Gaucher EA, Chong S (2012) Reconstitution of translation from Thermus thermophilus reveals a minimal set of components sufficient for protein synthesis at high temperatures and functional conservation of modern and ancient translation components. Nucleic Acids Res 40(16):7932–7945. doi:10.1093/nar/gks568

Zhu H, Sandiford SK, van Wezel GP (2014) Triggers and cues that activate antibiotic production by actinomycetes. J Ind Microbiol Biotechnol 41(2):371–386. doi:10.1007/s10295-013-1309-z

Zobel S, Kumpfmüller J, Süssmuth R, Schweder T (2015) Bacillus subtilis as heterologous host for the secretory production of the non-ribosomal cyclodepsipeptide enniatin. Appl Microbiol Biotechnol 99(2):681–691. doi:10.1007/s00253-014-6199-0

Chapter 3
Engineering of *E. coli* for Heterologous Expression of Secondary Metabolite Biosynthesis Pathways Recovered from Metagenomics Libraries

Lei Fang, Guojian Zhang, and Blaine A. Pfeifer

Abstract A key component of the functional metagenomics approach for complex natural product discovery is the host system chosen to screen environmental DNA. The host must provide technical simplicity to enable high throughput assessment of the target compounds of interest. Furthermore, intracellular support is crucial to allowing biosynthesis of those compounds with the most chemical and bioactivity diversity. This chapter examines the characteristics of functional metagenomics screening hosts, including those historically used in discovery applications. An emphasis is placed on identifying desirable features of selected hosts and how engineering strategies may be applied to further enable the goals of compound discovery. A special emphasis is placed on the use of *Streptomyces* spp. and *Escherichia coli* as screening hosts and how the parallel field of heterologous biosynthesis engineering with these two hosts has interfaced with past and present objectives in functional metagenomics. Other screening hosts and future prospects for this component of metagenomics-based discovery are also discussed.

3.1 Introduction

Functional metagenomics offers the potential to tap the vast chemical diversity available from nature for societally beneficial outcomes. The process requires environmental DNA (eDNA), the collection of which is a focal point of other chapters within this book. In this chapter, we will focus on the host systems used to functionally express environmentally derived pathways. In particular, the chapter will be dedicated to the host cell requirements needed for complex natural product formation.

L. Fang • G. Zhang • B.A. Pfeifer (✉)
Department of Chemical and Biological Engineering, University at Buffalo, The State University of New York, Buffalo, NY 14260-4200, USA
e-mail: blainepf@buffalo.edu

T.C. Charles et al. (eds.), *Functional Metagenomics: Tools and Applications*,
DOI 10.1007/978-3-319-61510-3_3

The historical, medicinal, and economic impact of natural products have shifted world events (penicillin's influence on World War II (Demain and Sanchez 2009; Kardos and Demain 2011)), led to blockbuster anticancer treatments (Taxol (Frense 2007; Kingston 1994; Jennewein and Croteau 2001)), and drastically altered quality of life (infectious disease mortality dropped from ~33% at the end of the nineteenth century to 4% by the end of the twentieth century, leading to a drastic increase in life expectancy (Kardos and Demain 2011; Lederberg 2000; Nicolaou and Montagnon 2008; Verdine 1996; Demain 2009)). Natural products have produced or inspired numerous clinical therapies (Cragg and Newman 2013; Newman and Cragg 2012; Newman et al. 2009; Blunt et al. 2013; Rahman et al. 2010). Perhaps the most impressive part of these outcomes is the fact that the products and responsible organisms represent only a small fraction of the possible number of environmental compounds available (Torsvik et al. 1990; Amann et al. 1995; Embley 1996; Rappe and Giovannoni 2003; Pace et al. 1986), indicating that a wealth of molecular diversity still exists in nature (Davies 2011; Davies and Ryan 2012). Accessing the virtually untapped resource of the uncultivated "natural product parvome" (Davies 2011; Davies and Ryan 2012) is, in our opinion, the ultimate objective in natural product discovery research. The significance of this proposition is reflected in the recognized impact natural products have had and the promise of continued discovery from a vast environmental repository.

However, the challenges in natural product discovery and application have provided both financial and technical disincentives to continued research (Li and Vederas 2009; Pelaez 2006; Fox 2006; Katz et al. 2006; Berdy 2012). The result has been a drastic reduction in commercial industrial investment into new compounds like antibiotics at a time when multidrug-resistant pathogens are becoming more and more common and dangerous (Baker et al. 2007; Koehn and Carter 2005; Li and Vederas 2009; Molinari 2009; Roemer et al. 2011; Andersson and Hughes 2010; Baltz 2006; Barnes and Jinks 2008; Cars et al. 2011; Fischbach and Walsh 2009; Boucher et al. 2009; Spizek et al. 2010). Still, the discovery of new natural products and the inevitable development of resistance can be viewed as a challenge to which solutions can be posed. Essentially, the discovery process must be made more efficient, effective, and economical. By doing so, there will be alternative answers for a natural products discovery field sorely in need of new compounds. Advances in functional metagenomics have this potential. Furthermore, this particular approach can be redirected at numerous other pressing therapeutic challenges (e.g., cancer) by adjusting the focus of the discovery effort.

Closely connected to the potential of functional metagenomics is the screening host used in the discovery process (Fig. 3.1). The use of surrogate hosts for screening and discovery purposes is an example of heterologous biosynthesis. Heterologous biosynthesis was developed to circumvent the challenges of working with native natural product producers (Pfeifer and Khosla 2001; Zhang et al. 2008, 2011) and is the key step in translating sequencing and metagenomic information during natural product discovery. Those organisms demonstrating innate biological advantages in growth kinetics and cellular morphology were considered ideal heterologous hosts.

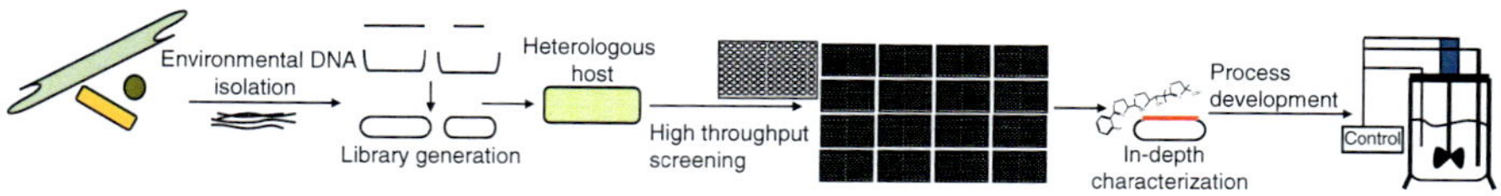

Fig. 3.1 Functional metagenomics natural product discovery scheme featuring a heterologous host needed to convert template information to complex natural products

It was no surprise that these same hosts had the most advanced molecular biology tools and were continuously featured in genetic, metabolic, and process engineering applications. These features are considered the first required characteristic of the heterologous host component of the functional metagenomics process. In other words, the screening host must provide experimental flexibility in terms of transformation competency and culture capability.

In the goal to discover the most chemically diverse new compounds, the second requirement of a heterologous host is the ability to support the intracellular process of complex natural product biosynthesis. The primary challenge in completing biosynthesis through otherwise ideal hosts is the lack of innate cellular support in the form of required substrates, posttranslational modification, gene expression and protein folding machinery, or some combination of deficiencies. Without this capability, the chemical space afforded by natural product pathways will be significantly reduced, leading to a much diminished discovery potential.

A final trait of a heterologous host, which is not absolutely required but highly desirable, is the potential for eventual scaled production. This feature would allow the host to be used from discovery to a final production process, as indicated as the final step in Fig. 3.1. The end result would be streamlined access to newly discovered environmental chemicals.

Another key element of the functional metagenomics approach is the screen applied which will dictate the outcome of the discovery process. This is dependent on the availability of high-throughput technology to be coupled to a screen capable of specific compound selection. Common functional metagenomics targets have included both small molecules and enzymes spanning a range of bioactivities (Table 3.1). Additional target compounds and bioactivities are limited only by the effectiveness and throughput capability of the screens.

3.2 Screening Hosts Used Previously

Table 3.1 summarizes those screening hosts used previously segmented by application, timeframe, and type of compound discovered. In particular, Table 3.1 features primarily bacterial host options, as eukaryotic systems have not been commonly used previously (to be discussed further below). The table also includes applications in addition to natural product discovery, such as those efforts dedicated to enzyme discovery.

Table 3.1 Summary of previous functional metagenomics efforts

Host	Type of compound	Time frame	Reference
E. coli	Antibiotic	*2005*	Brady and Clardy (2005a)
	Antibiotic	*2000*	August et al. (2000)
	Antibiotic	*2002*	Brady et al. (2002, 2004)
	Antibiotic	*2004*	Brady and Clardy (2004)
	Antibiotic	*2002*	Gillespie et al. (2002)
	Lipases	*2000*	Henne et al. (2000)
	Oxidoreductase	*2003*	Knietsch et al. (2003)
	Amylases	*2004*	Yun et al. (2004)
	Membrane protein	*2001*	Majernik et al. (2001)
	Biocatalysts	*1999*	Henne et al. (1999)
	Amidase	*2004*	Gabor et al. (2004)
Ralstonia metallidurans	Antibiotic	*2009*	Craig et al. (2009)
Streptomyces lividans	Unknown (terragine)	*2000*	Wang et al. (2000)
Agrobacterium tumefaciens	Biofilm inhibitor	*2009*	Schipper et al. (2009)
Saccharomyces cerevisiae	Acid phosphatases	*2010*	Kellner et al. (2011)
Pseudomonas putida	Antibiotics	*2004*	Martinez et al. (2004)

When analyzing Table 3.1, several trends are observed. Generally, the heterologous host is a key component to a successful metagenomic drug discovery effort and necessary to convert the encoded eDNA information into chemical readout. To do so, the following criteria must be met: (1) successful gene expression, (2) sufficient enzymatic levels, and (3) active final enzymes. The foreign genetic material must be transcribed and translated to provide enzymatic products. This first step is complicated by the range of eDNA to be introduced to the heterologous host and the differences in gene expression machinery of the heterologous host when compared to that of the original host. Separately, there is growing evidence that even with successful expression within a heterologous host, insufficient levels of biosynthetic enzymes will result in reduced or undetectable activity (Wang and Pfeifer 2008; Zhang et al. 2009, 2010a), which has prompted the use of copy-up shuttle plasmids now being utilized in functional metagenomics applications (Wild et al. 1996, 2001, 2002; Wild and Szybalski 2004; Aakvik et al. 2009; Kakirde et al. 2011). Finally, the enzymes resulting from gene expression must be in an active position to support biosynthesis. Factors affecting final enzymatic activity include protein folding mechanisms, biosynthetic metabolic substrate support, and posttranslational modification (Zhang et al. 2011).

Early in metagenomic drug discovery attempts, *E. coli* was a commonly used heterologous host (Rondon et al. 2000; Brady et al. 2001; Gillespie et al. 2002; MacNeil et al. 2001). Initially, this stemmed from the experimental simplicity of this organism. Well-documented molecular biology techniques (Swartz 1996) allowed the rapid and efficient transformation of eDNA for the purpose of screening. Simple microbiological properties (particularly a rapid doubling time) allowed timely and cost-effective screens. Thinking further into the complete drug production process, *E. coli* also offers unparalleled metabolic and process engineering tools to optimize production once discovery has been confirmed (Lee 1996; Shiloach and Fass 2005). As a result, *E. coli* is a compelling choice as a heterologous host during metagenomic drug discovery.

Unfortunately, *E. coli* suffers from a significant weakness: the cell does not natively support substantial production of complex natural products. Thus, there are insufficient intracellular support mechanisms to foster successful discovery of such compounds. Alternatively, the lack of a significant natural product biosynthetic background offers advantages in reducing metabolic overlap or "cross talk" between native and introduced pathways and contamination issues due to native complex natural product formation. It came as no surprise then that the only compounds discovered through *E. coli* were simple and derived from standard cellular building blocks (fatty acids, amino acids, etc.) (Brady and Clardy 2000; Brady et al. 2001, 2002; 2004; Brady and Clardy 2004, 2005a, b; August et al. 2000; Gillespie et al. 2002; Guan et al. 2007; Lim et al. 2005; MacNeil et al. 2001).

Efforts then shifted to alternative heterologous screening hosts that could better support complex natural products and hence extend the chemical range of products to be discovered (Martinez et al. 2004; Chang and Brady 2013; Courtois et al. 2003; Craig et al. 2009, 2010; Wang et al. 2000). Among these options were *Streptomyces* spp. due to a renowned reputation for producing complex natural products. As such, the intrinsic support for biosynthesis would be provided. However, such options provide more technical challenges (longer doubling times, more complicated transformation protocols and media formulations) than *E. coli*, potentially limiting throughput when using this host. Interestingly, despite better metabolic support, functional metagenomic attempts using *Streptomyces* (in addition to other host options) have also resulted in only simple chemical hits, most likely linked to the lack of large insert eDNA clones capable of encoding complete natural product pathways (as detailed in other chapters of this book).

Functional metagenomics studies have also tested *Ralstonia metallidurans* and *Pseudomonas putida* as additional bacterial screening hosts. Both organisms offer genetic tractability and simple growth properties. These hosts also have background metabolism and support capabilities (such as aromatic compound support and cytochrome P450 activity in *Ralstonia* (Craig et al. 2009) and acetyl-CoA precursor distribution and phosphopantetheine transferase activity in *Pseudomonas* (Wenzel et al. 2005)) that may serve the objectives of foreign natural product biosynthesis during screening efforts. As a β-Proteobacterium, *R. metallidurans* also provides variety in the division of bacteria used during screening. As described below, variation in screening hosts, coupled with the diversity of eDNA to be screened, may enhance the output of discovered compounds. Other hosts, such as *Agrobacterium tumefaciens* were useful as system that could be directly coupled to a particular assay, such as quorum sensing interference. Similarly, deletion mutants of *Saccharomyces cerevisiae* allowed an inbuilt screen for complementation enzymes identified from a soil library. This last example provides a rare use of a eukaryotic host during functional genomics; however, the procedure required the additional step of cDNA library construction.

3.3 Engineering *E. coli* as a Heterologous Host System in the Context of Functional Metagenomics

In this section, the development of *E. coli* as a heterologous host will be detailed. In particular, the progress with *E. coli* will be placed in context to similar efforts with *Streptomyces* spp. As mentioned earlier, the use of certain screening hosts within functional metagenomics is a specialized application of heterologous biosynthesis. As such, the following text will outline the development of first *Streptomyces* spp. and then *E. coli* as heterologous hosts for complex natural product biosynthesis. Insight from these efforts sets the stage for implementation into next-generation functional metagenomics efforts.

Though heavily featured across the range of functional metagenomics attempts to date (Table 3.1), *E. coli* emerged later in natural product heterologous biosynthesis. The initiation of heterologous biosynthesis began with basic and applied efforts to understand organisms (notably, actinomycetes) responsible for the majority of isolated therapeutic natural compounds. Driving early efforts in this regard was Sir David Hopwood in characterizing and utilizing *Streptomyces coelicolor* A3(2) (Glauert and Hopwood 1959, 1960, 1961; Hopwood 1960; Hopwood and Glauert 1960). Concisely summarizing all of Professor Hopwood's research is not possible within the limits of this chapter (Hopwood 1997, 1999, 2003). Instead, an emphasis will be placed on those accomplishments that fueled the heterologous biosynthetic field. This summary is also not meant to exclude others who had meaningful contributions to *Streptomyces* genetics and molecular tool development (Hopwood 1999, 2003, 2006). These research efforts culminated with the complete sequencing of the *S. coelicolor* A3(2) host (Bentley et al. 2002).

The general microbiology principles established with *Streptomyces* genetics (Kieser et al. 2000) enabled a series of recombinant DNA technology tools that were crucial for attempts at biosynthetic engineering. Among these were transformation protocols, expression plasmids, and gene expression control elements. Identification of the fertility SCP1, SCP2, and other plasmids (and their ability to harbor antibiotic biosynthetic genes in the case of SCP1 (Kirby et al. 1975)) enabled plasmid-based recombinant DNA transfer within *Streptomyces* hosts for engineering applications (Hopwood and Wright 1973; Bibb et al. 1977; Schrempf et al. 1975; Kieser et al. 1982; Lydiate et al. 1985). Closely tied to the development of plasmids for recombinant purposes was a general transformation protocol for *Streptomyces* spp. based upon polyethylene glycol (PEG) treatment of protoplasts (Bibb et al. 1978). The establishment of such tools initiated earnest attempts at directed genetic manipulations of natural product biosynthesis in *Streptomyces* hosts. These tools were also the precursors needed for metagenomics efforts using similar hosts.

Tools for genetic manipulation prompted applied efforts at the heterologous production of natural products through *Streptomyces* hosts (Kirby et al. 1975; Feitelson and Hopwood 1983; Malpartida and Hopwood 1984; Zhang et al. 2011). The experimental insight, tools, and approaches that had been developed for *S. coelicolor*

A3(2) were then applied to other *Streptomyces* spp., such as *S. lividans*, *S. venezuelae*, and *S. albus*, and are still commonly and successfully utilized (Zhang et al. 2011). As such, researchers now had a growing variety of *Streptomyces* spp. with which they could test in the context of heterologous biosynthesis (as well as functional metagenomics). A more challenging task was to develop a similar recombinant system for complex natural product biosynthesis using a phylogenetically distinct heterologous host like *E. coli*.

Why attempt heterologous production of complex natural products in *E. coli* if so much effort had been successfully spent in developing a *Streptomyces coelicolor* A3(2) expression system? Over time, the answer to this question has evolved from a comparison of the limitations or advantages associated with each host to one of potential. Given the developments in engineering *E. coli*, to be further elaborated upon below, this alternative host offers unique options when trying to establish and direct heterologous biosynthesis. The example below outlines the design of heterologous biosynthesis by focusing on the production of the antibiotic erythromycin and its polyketide core termed 6-deoxyerythronolide B (6dEB). This system was representative of complex natural products in general with specific requirements that included (1) the expression of three large genes (*eryAI*, *eryAII*, *eryAIII*; ~10 kb each), the protein products of which (termed the deoxyerythronolide B polyketide synthase [DEBS]) were responsible for generating 6dEB, (2) posttranslational modification of the DEBS enzymes through the addition of a 4′-phosphopantetheine group; (3) cellular precursors (propionyl-CoA, (2*S*)-methylmalonyl-CoA, NADPH, and other metabolites) needed for biosynthesis; (4) the expression of 17 additional genes needed to convert 6dEB to erythromycin; and (5) coordinated expression machinery to ensure proper enzyme activity (Table 3.2).

However, at the time research commenced, there was a more basic reasoning for attempting to adopt *E. coli*: simplicity. For all the advances made with *S. coelicolor*, there were seemingly night-and-day differences in the ease of experimental protocols when comparing the two hosts. Furthermore and more supportive from an engi-

Table 3.2 Summary of engineering to *E. coli* to enable complex natural product formation

Category	Rationale	Specific example
Metabolic engineering	Precursors needed for biosynthetic support	The engineering of a "propionate" pathway to provide propionyl- and (2*S*)-methylmalonyl-CoA
Gene expression	Coordinated expression of multiple and large pathway genes	Systematic individual and operon expression of the 20 erythromycin biosynthetic pathway genes, including three ~10 kb polyketide synthase genes
Protein folding	Enzymatic activity of the biosynthetic pathway	GroES/GroEL protein folding chaperones to enable full biosynthesis
Posttranslational modification	Enzymatic activity of the biosynthetic pathway	Introduction of the Sfp phosphopantetheine transferase to activate the polyketide synthase enzymes required for 6dEB formation

neering perspective, success was achieved in producing recombinant polyketide synthase enzymes from *E. coli* as a preface to in vitro biochemical studies. Initially, these efforts were confined to single type I and II polyketide synthase domains through protocols that were standard in *E. coli* recombinant protein production (Carreras et al. 1997; Carreras and Khosla 1998; Gokhale et al. 1999a; Dreier et al. 1999). However, additional efforts began to test the potential of expressing individual modular sequences or entire multimodular genes from type I systems, such as the DEBS system associated with erythromycin formation (Roberts et al. 1993; Gokhale et al. 1999b). With success came the notion that full complex natural product biosynthetic activity could be reconstituted using *E. coli*.

However, the experimental strategy needed for success would require a number of the same considerations that allowed the development of the *S. coelicolor* A3(2) heterologous system. The primary difference was as a result of the innate properties of the two hosts and previously established protocols for each. As mentioned, *Streptomyces* spp. were logically well-suited to provide the metabolic support required for complex polyketide natural product formation; however, lacking were the basic molecular biology and microbiology tools needed for recombinant production efforts. The reverse situation held for *E. coli*. By this time, strong molecular biology tools existed (though not necessarily designed for the unique features of complex natural product gene clusters), but the cell lacked the evolved metabolism to support robust polyketide or other natural product biosyntheses.

Fortunately, efforts using *E. coli* were in a position to leverage the knowledge gained from working with polyketide systems through *Streptomyces* hosts (Kao et al. 1994). In other words, research was at a crossroads of knowledge (accumulated with *Streptomyces* spp.) and technology (associated with *E. coli*). The first modification was to the *E. coli* cell. Within *Streptomyces* hosts, there is often the concern of background natural product biosynthesis that could compete with or contaminate heterologous production efforts, and this innate metabolism would be targeted for genetic removal (Kao et al. 1994). However, in the case of *E. coli*, efforts were not attempting to remove metabolism that would potentially compete with or contaminate the output of a pathway to be introduced. Instead, engineering was dedicated to introducing metabolism needed to allow production of a target compound. To do so, a hybrid metabolic pathway was designed composed of native *E. coli* genes and genes transplanted from *S. coelicolor* A3(2). This particular pathway was chosen as a result of serendipitous research completed by Professor Hugo Gramajo, who had recently biochemically characterized the propionyl-CoA carboxylase (PCC) genes from *S. coelicolor* (Rodriguez and Gramajo 1999). The result was a "propionate pathway" that required exogenous propionate to be converted intracellularly to propionyl-CoA and (2*S*)-methylmalonyl-CoA, the two precursors required for 6dEB formation. The manipulation of the native propionyl-CoA synthetase (PrpE) gene was completed through the introduction of a promiscuous 4′-phosphopantetheine transferase gene from *Bacillus subtilis* (Lambalot et al. 1996; Quadri et al. 1998) which simultaneously eliminated the remainder of the *prp* operon to minimize potential catabolism of exogenously fed propionate. The *sfp* gene was introduced to *E. coli* strain BL21(DE3) through homologous recombination, and the process resulted in both *sfp* and the *prpE* genes under inducible T7 promoter control (a point to be further elaborated upon below). Early confirmation of this strain, termed BAP1, included diagnostic PCR verification and Sfp activity by in vivo phosphopantetheinylation of individual polyketide synthase modules that previously required posttranslational modification through the use of a helper plasmid containing *sfp*. At this point, a logical question would be why not use the native *S. coelicolor* phosphopantetheine transferase instead of a counterpart from *B. subtilis*? This occurred for two reasons: (1) without the complete sequence of *S. coelicolor* A3(2) having yet been completed, there was not simple access to the native phosphopantetheine transferase genes and (2) another fortunate collaboration with and timely series of research accomplishments by Christopher Walsh's group at Harvard University provided the *sfp* gene as an alternative. After completing these steps, the BAP1 strain could successfully posttranslationally modify polyketide synthase enzymes (via Sfp) and convert exogenous propionate to propionyl-CoA (via PrpE) and (2*S*)-methylmalonyl-CoA (via PCC).

The challenge then turned to a recombinant expression and transfer system that could rival the functionality of the plasmids developed and utilized for *Streptomyces* spp. Here, the *E. coli* expression systems that were becoming commonplace for biochemical studies, and used successfully in the work cited above for certain type I and II polyketide systems, were still only tenuously suitable for the full type I DEBS system responsible for 6dEB formation. First, there were very few if any reports of individual genes of this size successfully expressed through *E. coli*, and there were no reports of attempts to express three such genes simultaneously. Furthermore, concerns of foreign codon usage, appropriate promoter choice, and proper protein folding were just beginning to be addressed through developing knowledge and options.

Initial plasmid expression systems were designed based upon those that had been used successfully for individual polyketide synthase enzymes (Gokhale et al. 1999b), and early designs were crude but functional. The pET expression plasmids that accompanied BL21(DE3) (Studier and Moffatt 1986) were altered to accept more than one gene per plasmid. Expression plasmids (such as the Novagen Duet vectors) are commercially available today in formats to allow multiple plasmids per cell, a feature that much more readily allows for the cloning of a biosynthetic pathway. These currently available multi-cystronic plasmids are also compatible with one another, a feature that did not accompany the original design for 6dEB formation. Instead, two incompatible, but separately selectable, plasmids (based upon pET21c and pET28a) were used to introduce *eryAI*, *eryAII*, *eryAIII*, and *pcc*. As a result, two selection markers would allow the transformation of both plasmids to the BAP1 strain, providing the basis for the first attempts at the heterologous production of a complex type I polyketide in *E. coli*.

Plasmid expression relied on the T7 promoter under control of a *lac* operator. As such, the "natural" regulatory features that accompanied *Streptomyces* hosts and expression vectors were completely removed in line with the inducible T7 expression system. The single plasmid used to establish 6dEB formation in *S. coelicolor* featured the *eryAI*, *II*, and *III* genes in the same operon configuration found in the native *Saccharopolyspora erythraea* chromosome. For expression in *E. coli*, using a similar arrangement was problematic. The low-copy SCP2 fertility plasmid was natively ~30 kb and could therefore be expected to be altered to carry the ~30 kb DEBS gene sequence. However, there was no indication of the gene carrying capacity for the pET expression vectors. Cloning became much more difficult with even single DEBS genes, and placing three ~10 kb genes into one pET vector seemed highly unlikely. These practical considerations and other cloning limitations led to the "crude" expression design between two pET plasmids. In one of the plasmids, the *pcc* genes and *eryAI* each had their own dedicated promoter. In the other plasmid, *eryAII* and *III* were introduced as an operon. The operon format was also risky because it was unknown whether a transcript of this length would be possible in *E. coli*. From this vantage point, it may have been fortunate that the system relied on a strong and processive T7 expression platform (Golomb and Chamberlin 1974; McAllister et al. 1981).

However imperfect or unknown, the new system allowed for the production of 6dEB from *E. coli* (Pfeifer et al. 2001). Success also opened the possibility for the production of similarly complex natural compounds which would pose similar challenges in heterologous pathway transfer and reconstitution. For the current system, the next substantial challenge was introducing the remaining biosynthetic genes required for complete erythromycin biosynthesis. In previous efforts with *Streptomyces* heterologous hosts, gene clusters were often transferred with little engineering of the expression control elements associated with the biosynthetic pathway. This was as a result of the close phylogenetic relationship between the native and heterologous hosts being utilized. In the case of *E. coli*, care had to be taken to ensure promoters, repressors, and terminator sequences had been introduced to facilitate expression within the new cellular background and that expression had been confirmed. This step was taken in establishing 6dEB production and was also used in the attempt to reconstitute full erythromycin A biosynthesis. The process complicates biosynthetic reconstitution efforts within *E. coli*, but in our experience, confirmation of gene expression has been a positive harbinger for successful biosynthetic reconstitution.

With >20 genes needed to allow for *E. coli* erythromycin production, additional considerations were required to allow successful pathway transfer and coordinated expression. Namely, a variation on the plasmid design utilized for 6dEB formation was extended to erythromycin production. To introduce the 17 deoxysugar biosynthetic, self-resistance, and tailoring genes needed to convert 6dEB to erythromycin A, a sequential operon construction method was employed similar to attempts by the Gramajo and Khosla groups to build upon the 6dEB framework (Peiru et al. 2005; Lee and Khosla 2007). In our case, we confirmed individual gene expression from both pET21 and pET28 plasmids, noting that in certain cases, we observed expression from the pET28 plasmid background but not from pET21. We suspect that the leader sequence in pET28 was beneficial in foreign gene expression and used a combination of the successful expression cassettes when building operons. The result was four multi-cystronic plasmids (two responsible for 6dEB biosynthesis and two required for conversion to erythromycin), with the addition of two helper plasmids (one contained the *E. coli* GroES/GroEL chaperonin genes and the other contained an additional copy of *eryK* which encoded for a late-stage tailoring hydroxylation). Without the addition of these helper plasmids, we did not observe erythromycin production (in the case of the GroEL/GroES plasmid) or complete erythromycin formation (in the case of the *eryK* plasmid) (Zhang et al. 2010b). Despite the complexity in final design, the system managed to accomplish a significant extension of the original effort to produce 6dEB by accounting for both individual megasynthase gene expression and the coordinated expression of the entire erythromycin A pathway.

The success described above for complex natural product biosynthesis in heterologous hosts has been extended significantly over the last 15–20 years (Zhang et al. 2011; Ongley et al. 2013). Further engineering has enabled the production of complex natural products such as nonribosomal peptides and isoprenoids using ideal hosts such as *E. coli* (Zhang et al. 2010b; Pfeifer et al. 2003; Ajikumar et al. 2010; Watanabe et al. 2006; Mutka et al. 2006). These advanced hosts are now available to complement more traditional screening hosts in new metagenomic natural product discovery efforts with the promise of discovering similarly complex new natural products.

3.4 Future Directions

The first future direction to be outlined is the implementation of the newest screening hosts developed for the production of complex natural products into current metagenomics strategies. This has the potential to capture significantly more chemical diversity from products encoded in eDNA. Complementing this updated approach is emerging technology (outlined in remaining chapters) associated with eDNA capture. Using the erythromycin example above, complex natural products will typically require contiguous DNA stretches of 20–100 kb. Prior to emerging DNA capture technology, clones of this size were not possible, such that even with the implementation of advanced screening hosts, the discovery of increasingly complex natural products could not be expected. With these concurrently emerging technologies, there are revised prospects for maximizing the chemical diversity emerging from metagenomic natural product discovery platforms.

Returning to the question of screening host choice, the implementation of advanced *E. coli* strains that have added metabolic support for complex natural products (as introduced above) allows a potentially significant upgrade to functional metagenomics discovery objectives. However, there are still remaining questions associated with gene expression and enzyme activity for foreign clusters introduced to the host. On one hand, the bacterial nature of introduced clusters may allow for the current machinery of *E. coli* to function properly. However, if problems are encountered with the gene expression process (including codon usage, expression elements [promoters, ribosomal binding sites, terminator sequences, etc.], transcript stability) or protein activity (including protein folding and posttranslational modifications), strategies similar to equipping the screening host for metabolic support can be utilized. Namely, cellular engineering to implement rare tRNA molecules, expanded transcription factors, posttranslational enzymes, or chaperonins can hypothetically aid the gene expression process prior to encountering the metabolic support needed for biosynthesis. Many of these alterations exist within commercial *E. coli* strains designed to aid expression of foreign genes (typically used in the context of biochemical protein characterization). As such, it is reasonable to utilize this same knowledge, as well as insight into the gene expression challenges associated with complex natural product pathways, to further design an ideal *E. coli* host for continually advanced metagenomic discovery formats.

Another future direction for the functional metagenomics community is the strategic implementation of screening hosts in line with the phylogenetic signatures of the environmental samples to be screened. Characterization of the environmental samples in relation to their microbial content, in particular, will help to profile potential screening hosts most suitable for metagenomic discovery efforts. For example, a previous assessment of soil microbial content found the distribution between five major divisions of bacteria: *Actinobacteria* (such as *Streptomyces* spp.), *Acidobacteria*, and α-, β (*R. metallidurans*)-, and γ (*E. coli* and *Pseudomonas putida*)-*Proteobacteria* (Fierer et al. 2007). A more recent phylogenetic evaluation of soil eDNA, together with new heterologous hosts (in particular, *Acidobacteria* (George et al. 2011) and *Sinorhizobium meliloti* (Aneja et al. 2004)), is also being developed and combined for screening purposes. Thus, a series of different screening hosts could be used for the same eDNA sample to maximize discovery. Of course, this strategy would require the availability of multiple screening hosts which would necessitate many of the same technical developments outlined above for current options such as *Streptomyces* spp. and *E. coli*.

Another frontier for functional metagenomics is the application of the approach to a wider environmental cellular content. Namely, current technology is directed primarily at bacterial DNA from the environment, which, as outlined above, is substantial and laden with biological and chemical diversity. However, potential would be further heightened if the approach could be extended to more complex organisms, such as eukaryotic microbial and plant sources, which also possess extensive natural product production capabilities (Vederas 2014; Song et al. 2014). The fundamental difference in cell biology between prokaryotic and eukaryotic organisms suggests that a screening host would have to adopt congruent cellular features with eDNA content resulting from more complex organisms. Options would include various fungal hosts (such as *S. cerevisae* or *Aspergillus* spp.) which meet this criterion but also retain prerequisite growth and genetic tractability requirements to allow implementation into functional metagenomic screening.

Finally, from a long-term research perspective, a future goal would be to build a completely synthetic cell designed only for natural product biosynthesis. By doing so, biology is only selectively retained for the purpose of enabling the directed production of encoded natural products from eDNA. The synthetic cell could be a mas-

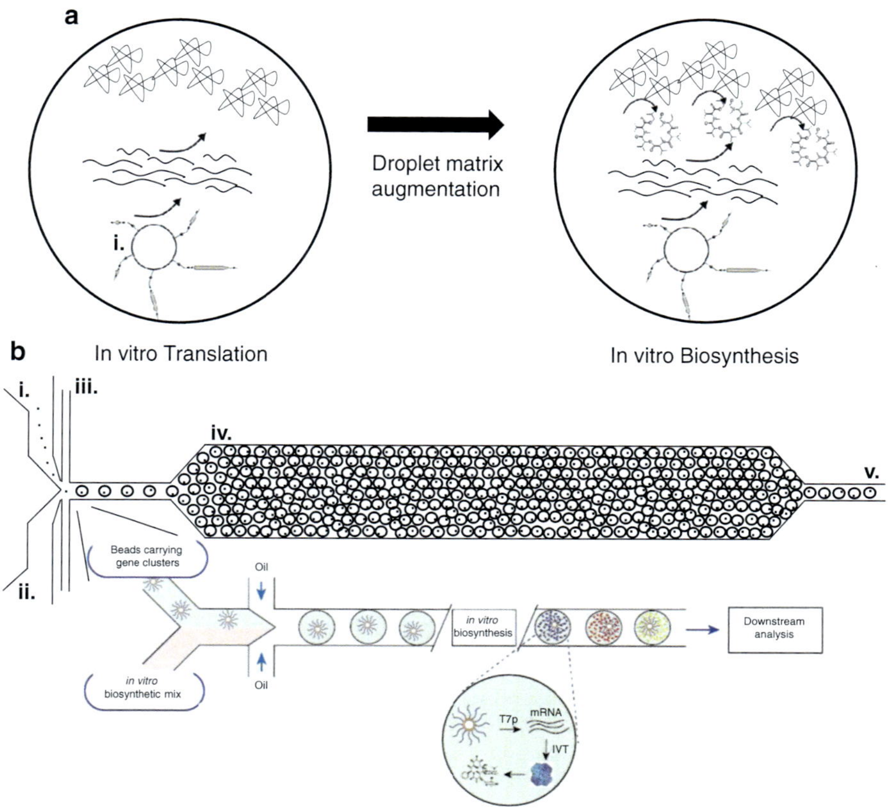

Fig. 3.2 Synthetic cell concept for functional metagenomics. (**a**) Droplet in vitro transcription and translation technology to allow complex natural product biosynthesis. Beads containing affixed eDNA fragments (*i*) would be sequestered to microfluidic drops to initiate the discovery process. (**b**) A microfluidic format featuring (*i*) stream for bead-gene cluster input, (*ii*) stream for biosynthetic matrix, (*iii*) oil inlet to generate droplets, (*iv*) biosynthetic incubation chamber, (*v*) outlet to analysis

sively redesigned biological host (such as *E. coli* as a starting template) or, alternatively, could be built based upon in vitro transcription and translation technology with an additional emphasis on natural product biosynthesis. In tandem, the evolving prospects of extensive designer gene synthesis and pathway assembly allow the potential to build each target natural produce cluster from scratch, thus allowing a template-driven means of addressing concerns of subsequent gene expression by standardizing or addressing potential transcription challenges. Such an approach could then also leverage high throughput microfluidic technology in coordination with metagenomics screening (Fig. 3.2). Though much effort and research still remains to test the feasibility of this approach, success would potentially enable a truly universal approach and "host" for complex natural product metagenomics discovery.

References

Aakvik T, Degnes KF, Dahlsrud R, Schmidt F, Dam R, Yu L, Volker U, Ellingsen TE, Valla S (2009) A plasmid RK2-based broad-host-range cloning vector useful for transfer of metagenomic libraries to a variety of bacterial species. FEMS Microbiol Lett 296(2):149–158. doi:10.1111/j.1574-6968.2009.01639.x

Ajikumar PK, Xiao WH, Tyo KE, Wang Y, Simeon F, Leonard E, Mucha O, Phon TH, Pfeifer B, Stephanopoulos G (2010) Isoprenoid pathway optimization for Taxol precursor overproduction in Escherichia coli. Science 330(6000):70–74. doi:10.1126/science.1191652. 330/6000/70 [pii]

Amann RI, Ludwig W, Schleifer KH (1995) Phylogenetic identification and in situ detection of individual microbial cells without cultivation. Microbiol Rev 59(1):143–169

Andersson DI, Hughes D (2010) Antibiotic resistance and its cost: is it possible to reverse resistance? Nat Rev Microbiol 8(4):260–271. doi:10.1038/nrmicro2319. nrmicro2319 [pii]

Aneja P, Dai M, Lacorre DA, Pillon B, Charles TC (2004) Heterologous complementation of the exopolysaccharide synthesis and carbon utilization phenotypes of Sinorhizobium meliloti Rm1021 polyhydroxyalkanoate synthesis mutants. FEMS Microbiol Lett 239(2):277–283. doi:10.1016/j.femsle.2004.08.045

August PR, Grossman TH, Minor C, Draper MP, MacNeil IA, Pemberton JM, Call KM, Holt D, Osburne MS (2000) Sequence analysis and functional characterization of the violacein biosynthetic pathway from Chromobacterium violaceum. J Mol Microbiol Biotechnol 2(4):513–519

Baker DD, Chu M, Oza U, Rajgarhia V (2007) The value of natural products to future pharmaceutical discovery. Nat Prod Rep 24(6):1225–1244. doi:10.1039/b602241n

Baltz RH (2006) Marcel Faber Roundtable: is our antibiotic pipeline unproductive because of starvation, constipation or lack of inspiration? J Ind Microbiol Biotechnol 33(7):507–513. doi:10.1007/s10295-005-0077-9

Barnes TA, Jinks A (2008) Methicillin-resistant Staphylococcus aureus: the modern-day challenge. Br J Nurs 17(16):1012, 1014, 1016–1018

Bentley SD, Chater KF, Cerdeno-Tarraga AM, Challis GL, Thomson NR, James KD, Harris DE, Quail MA, Kieser H, Harper D, Bateman A, Brown S, Chandra G, Chen CW, Collins M, Cronin A, Fraser A, Goble A, Hidalgo J, Hornsby T, Howarth S, Huang CH, Kieser T, Larke L, Murphy L, Oliver K, O'Neil S, Rabbinowitsch E, Rajandream MA, Rutherford K, Rutter S, Seeger K, Saunders D, Sharp S, Squares R, Squares S, Taylor K, Warren T, Wietzorrek A, Woodward J, Barrell BG, Parkhill J, Hopwood DA (2002) Complete genome sequence of the model actinomycete Streptomyces coelicolor A3(2). Nature 417(6885):141–147. doi:10.1038/417141a. 417141a [pii]

Berdy J (2012) Thoughts and facts about antibiotics: where we are now and where we are heading. J Antibiot (Tokyo) 65(8):441. doi:10.1038/ja.2012.54

Bibb MJ, Freeman RF, Hopwood DA (1977) Physical and genetic characterization of a 2nd sex factor, Scp2, for streptomyces-coelicolor A3(2). Mol Gen Genet 154(2):155–166. doi:10.1007/Bf00330831

Bibb MJ, Ward JM, Hopwood DA (1978) Transformation of plasmid DNA into Streptomyces at high frequency. Nature 274(5669):398–400

Blunt JW, Copp BR, Keyzers RA, Munro MH, Prinsep MR (2013) Marine natural products. Nat Prod Rep 30(2):237–323. doi:10.1039/c2np20112g

Boucher HW, Talbot GH, Bradley JS, Edwards JE, Gilbert D, Rice LB, Scheld M, Spellberg B, Bartlett J (2009) Bad bugs, no drugs: no ESKAPE! An update from the Infectious Diseases Society of America. Clin Infect Dis 48(1):1–12. doi:10.1086/595011

Brady SF, Clardy J (2000) Long-chain N-acyl amino acid antibiotics isolated from heterologously expressed environmental DNA. JACS 122:12903

Brady SF, Clardy J (2004) Palmitoylputrescine, an antibiotic isolated from the heterologous expression of DNA extracted from bromeliad tank water. J Nat Prod 67(8):1283–1286

Brady SF, Clardy J (2005a) Cloning and heterologous expression of isocyanide biosynthetic genes from environmental DNA. Angew Chem Int Ed Engl 44(43):7063–7065

Brady SF, Clardy J (2005b) N-acyl derivatives of arginine and tryptophan isolated from environmental DNA expressed in Escherichia coli. Org Lett 7(17):3613–3616
Brady SF, Chao CJ, Handelsman J, Clardy J (2001) Cloning and heterologous expression of a natural product biosynthetic gene cluster from eDNA. Org Lett 3(13):1981–1984
Brady SF, Chao CJ, Clardy J (2002) New natural product families from an environmental DNA (eDNA) gene cluster. J Am Chem Soc 124(34):9968–9969
Brady SF, Chao CJ, Clardy J (2004) Long-chain N-acyltyrosine synthases from environmental DNA. Appl Environ Microbiol 70(11):6865–6870. doi:10.1128/AEM.70.11.6865-6870.2004
Carreras CW, Khosla C (1998) Purification and in vitro reconstitution of the essential protein components of an aromatic polyketide synthase. Biochemistry 37(8):2084–2088. doi:10.1021/bi972919+
Carreras CW, Gehring AM, Walsh CT, Khosla C (1997) Utilization of enzymatically phosphopantetheinylated acyl carrier proteins and acetyl-acyl carrier proteins by the actinorhodin polyketide synthase. Biochemistry 36(39):11757–11761
Cars O, Hedin A, Heddini A (2011) The global need for effective antibiotics-moving towards concerted action. Drug Resist Updat 14(2):68–69. doi:10.1016/j.drup.2011.02.006
Chang FY, Brady SF (2013) Discovery of indolotryptoline antiproliferative agents by homology-guided metagenomic screening. Proc Natl Acad Sci U S A 110(7):2478–2483. doi:10.1073/pnas.1218073110
Courtois S, Cappellano CM, Ball M, Francou FX, Normand P, Helynck G, Martinez A, Kolvek SJ, Hopke J, Osburne MS, August PR, Nalin R, Guerineau M, Jeannin P, Simonet P, Pernodet JL (2003) Recombinant environmental libraries provide access to microbial diversity for drug discovery from natural products. Appl Environ Microbiol 69(1):49–55
Cragg GM, Newman DJ (2013) Natural products: a continuing source of novel drug leads. Biochim Biophys Acta 1830(6):3670–3695. doi:10.1016/j.bbagen.2013.02.008
Craig JW, Chang FY, Brady SF (2009) Natural products from environmental DNA hosted in Ralstonia metallidurans. ACS Chem Biol 4(1):23–28. doi:10.1021/cb8002754
Craig JW, Chang FY, Kim JH, Obiajulu SC, Brady SF (2010) Expanding small-molecule functional metagenomics through parallel screening of broad-host-range cosmid environmental DNA libraries in diverse proteobacteria. Appl Environ Microbiol 76(5):1633–1641. doi:10.1128/AEM.02169-09
Davies J (2011) How to discover new antibiotics: harvesting the parvome. Curr Opin Chem Biol 15(1):5–10. doi:10.1016/j.cbpa.2010.11.001
Davies J, Ryan KS (2012) Introducing the parvome: bioactive compounds in the microbial world. ACS Chem Biol 7(2):252–259. doi:10.1021/cb200337h
Demain AL (2009) Antibiotics: natural products essential to human health. Med Res Rev 29(6):821–842. doi:10.1002/med.20154
Demain AL, Sanchez S (2009) Microbial drug discovery: 80 years of progress. J Antibiot (Tokyo) 62(1):5–16. doi:10.1038/ja.2008.16
Dreier J, Shah AN, Khosla C (1999) Kinetic analysis of the actinorhodin aromatic polyketide synthase. J Biol Chem 274(35):25108–25112
Embley TM (1996) Molecular Approaches to Environmental Microbiology. Prentice Hall, Essex
Feitelson JS, Hopwood DA (1983) Cloning of a Streptomyces gene for an O-methyltransferase involved in antibiotic biosynthesis. Mol Gen Genet 190(3):394–398
Fierer N, Bradford MA, Jackson RB (2007) Toward an ecological classification of soil bacteria. Ecology 88(6):1354–1364
Fischbach MA, Walsh CT (2009) Antibiotics for emerging pathogens. Science 325(5944):1089–1093. doi:10.1126/science.1176667. 325/5944/1089 [pii]
Fox JL (2006) The business of developing antibacterials. Nat Biotechnol 24(12):1521–1528. doi:10.1038/nbt1206-1521
Frense D (2007) Taxanes: perspectives for biotechnological production. Appl Microbiol Biotechnol 73(6):1233–1240
Gabor EM, de Vries EJ, Janssen DB (2004) Construction, characterization, and use of small-insert gene banks of DNA isolated from soil and enrichment cultures for the recovery of novel amidases. Environ Microbiol 6(9):948–958. doi:10.1111/j.1462-2920.2004.00643.x

George IF, Hartmann M, Liles MR, Agathos SN (2011) Recovery of as-yet-uncultured soil acidobacteria on dilute solid media. Appl Environ Microbiol 77(22):8184–8188. doi:10.1128/AEM.05956-11

Gillespie DE, Brady SF, Bettermann AD, Cianciotto NP, Liles MR, Rondon MR, Clardy J, Goodman RM, Handelsman J (2002) Isolation of antibiotics turbomycin a and B from a metagenomic library of soil microbial DNA. Appl Environ Microbiol 68(9):4301–4306

Glauert AM, Hopwood DA (1959) A membranous component of the cytoplasm in Streptomyces coelicolor. J Biophys Biochem Cytol 6:515–516

Glauert AM, Hopwood DA (1960) The fine structure of Streptomyces coelicolor. I. The cytoplasmic membrane system. J Biophys Biochem Cytol 7:479–488

Glauert AM, Hopwood DA (1961) The fine structure of Streptomyces violaceoruber (S. coelicolor). III. The walls of the mycelium and spores. J Biophys Biochem Cytol 10:505–516

Gokhale RS, Hunziker D, Cane DE, Khosla C (1999a) Mechanism and specificity of the terminal thioesterase domain from the erythromycin polyketide synthase. Chem Biol 6(2):117–125

Gokhale RS, Tsuji SY, Cane DE, Khosla C (1999b) Dissecting and exploiting intermodular communication in polyketide synthases. Science 284(5413):482–485

Golomb M, Chamberlin M (1974) Characterization of T7-specific ribonucleic acid polymerase. IV. Resolution of the major in vitro transcripts by gel electrophoresis. J Biol Chem 249(9):2858–2863

Guan C, Ju J, Borlee BR, Williamson LL, Shen B, Raffa KF, Handelsman J (2007) Signal mimics derived from a metagenomic analysis of the gypsy moth gut microbiota. Appl Environ Microbiol 73(11):3669–3676. doi:10.1128/AEM.02617-06

Henne A, Daniel R, Schmitz RA, Gottschalk G (1999) Construction of environmental DNA libraries in Escherichia coli and screening for the presence of genes conferring utilization of 4-hydroxybutyrate. Appl Environ Microbiol 65(9):3901–3907

Henne A, Schmitz RA, Bomeke M, Gottschalk G, Daniel R (2000) Screening of environmental DNA libraries for the presence of genes conferring lipolytic activity on Escherichia coli. Appl Environ Microbiol 66(7):3113–3116

Hopwood DA (1960) Phase-contrast observations on Streptomyces coelicolor. J Gen Microbiol 22:295–302

Hopwood DA (1997) Genetic contributions to understanding polyketide synthases. Chem Rev 97(7):2465–2498

Hopwood DA (1999) Forty years of genetics with Streptomyces: from in vivo through in vitro to in silico. Microbiology 145(Pt 9):2183–2202

Hopwood DA (2003) Streptomyces genes: from Waksman to Sanger. J Ind Microbiol Biotechnol 30(8):468–471. doi:10.1007/s10295-003-0031-7

Hopwood DA (2006) Soil to genomics: the Streptomyces chromosome. Annu Rev Genet 40:1–23. doi:10.1146/annurev.genet.40.110405.090639

Hopwood DA, Glauert AM (1960) The fine structure of Streptomyces coelicolor. II. The nuclear material. J Biophys Biochem Cytol 8:267–278

Hopwood DA, Wright HM (1973) A plasmid of Streptomyces coelicolor carrying a chromosomal locus and its inter-specific transfer. J Gen Microbiol 79(2):331–342

Jennewein S, Croteau R (2001) Taxol: biosynthesis, molecular genetics, and biotechnological applications. Appl Microbiol Biotechnol 57(1–2):13–19

Kakirde KS, Wild J, Godiska R, Mead DA, Wiggins AG, Goodman RM, Szybalski W, Liles MR (2011) Gram negative shuttle BAC vector for heterologous expression of metagenomic libraries. Gene 475(2):57–62. doi:10.1016/j.gene.2010.11.004

Kao CM, Katz L, Khosla C (1994) Engineered biosynthesis of a complete macrolactone in a heterologous host. Science 265(5171):509–512

Kardos N, Demain AL (2011) Penicillin: the medicine with the greatest impact on therapeutic outcomes. Appl Microbiol Biotechnol 92(4):677–687. doi:10.1007/s00253-011-3587-6

Katz ML, Mueller LV, Polyakov M, Weinstock SF (2006) Where have all the antibiotic patents gone? Nat Biotechnol 24(12):1529–1531. doi:10.1038/nbt1206-1529

Kellner H, Luis P, Portetelle D, Vandenbol M (2011) Screening of a soil metatranscriptomic library by functional complementation of Saccharomyces cerevisiae mutants. Microbiol Res 166(5):360–368. doi:10.1016/j.micres.2010.07.006

Kieser T, Hopwood DA, Wright HM, Thompson CJ (1982) pIJ101, a multi-copy broad host-range Streptomyces plasmid: functional analysis and development of DNA cloning vectors. Mol Gen Genet 185(2):223–228

Kieser T, Bibb MJ, Buttner MJ, Chater KF, Hopwood DA (2000) Practical Streptomyces genetics. The John Innes Foundation, Norwich

Kingston DG (1994) Taxol: the chemistry and structure-activity relationships of a novel anticancer agent. Trends Biotechnol 12(6):222–227

Kirby R, Wright LF, Hopwood DA (1975) Plasmid-determined antibiotic synthesis and resistance in Streptomyces coelicolor. Nature 254(5497):265–267

Knietsch A, Waschkowitz T, Bowien S, Henne A, Daniel R (2003) Metagenomes of complex microbial consortia derived from different soils as sources for novel genes conferring formation of carbonyls from short-chain polyols on Escherichia coli. J Mol Microbiol Biotechnol 5 (1):46–56. doi:68724

Koehn FE, Carter GT (2005) The evolving role of natural products in drug discovery. Nat Rev Drug Discov 4(3):206–220

Lambalot RH, Gehring AM, Flugel RS, Zuber P, LaCelle M, Marahiel MA, Reid R, Khosla C, Walsh CT (1996) A new enzyme superfamily—the phosphopantetheinyl transferases. Chem Biol 3(11):923–936

Lederberg J (2000) Infectious history. Science 288(5464):287–293

Lee SY (1996) High cell-density culture of Escherichia coli. Trends Biotechnol 14(3):98–105. doi:10.1016/0167-7799(96)80930-9. 0167-7799(96)80930-9 [pii]

Lee HY, Khosla C (2007) Bioassay-guided evolution of glycosylated macrolide antibiotics in Escherichia coli. PLoS Biol 5(2):e45

Li JW, Vederas JC (2009) Drug discovery and natural products: end of an era or an endless frontier? Science 325(5937):161–165. doi:10.1126/science.1168243

Lim HK, Chung EJ, Kim JC, Choi GJ, Jang KS, Chung YR, Cho KY, Lee SW (2005) Characterization of a forest soil metagenome clone that confers indirubin and indigo production on Escherichia coli. Appl Environ Microbiol 71(12):7768–7777. doi:10.1128/AEM.71.12.7768-7777.2005

Lydiate DJ, Malpartida F, Hopwood DA (1985) The Streptomyces plasmid SCP2*: its functional analysis and development into useful cloning vectors. Gene 35(3):223–235

MacNeil IA, Tiong CL, Minor C, August PR, Grossman TH, Loiacono KA, Lynch BA, Phillips T, Narula S, Sundaramoorthi R, Tyler A, Aldredge T, Long H, Gilman M, Holt D, Osburne MS (2001) Expression and isolation of antimicrobial small molecules from soil DNA libraries. J Mol Microbiol Biotechnol 3(2):301–308

Majernik A, Gottschalk G, Daniel R (2001) Screening of environmental DNA libraries for the presence of genes conferring Na(+)(Li(+))/H(+) antiporter activity on Escherichia coli: characterization of the recovered genes and the corresponding gene products. J Bacteriol 183(22):6645–6653. doi:10.1128/JB.183.22.6645-6653.2001

Malpartida F, Hopwood DA (1984) Molecular cloning of the whole biosynthetic pathway of a Streptomyces antibiotic and its expression in a heterologous host. Nature 309(5967):462–464

Martinez A, Kolvek SJ, Yip CL, Hopke J, Brown KA, MacNeil IA, Osburne MS (2004) Genetically modified bacterial strains and novel bacterial artificial chromosome shuttle vectors for constructing environmental libraries and detecting heterologous natural products in multiple expression hosts. Appl Environ Microbiol 70(4):2452–2463

McAllister WT, Morris C, Rosenberg AH, Studier FW (1981) Utilization of bacteriophage T7 late promoters in recombinant plasmids during infection. J Mol Biol 153(3):527–544

Molinari G (2009) Natural products in drug discovery: present status and perspectives. Adv Exp Med Biol 655:13–27. doi:10.1007/978-1-4419-1132-2_2

Mutka SC, Carney JR, Liu Y, Kennedy J (2006) Heterologous production of epothilone C and D in Escherichia coli. Biochemistry 45(4):1321–1330

Newman DJ, Cragg GM (2012) Natural products as sources of new drugs over the 30 years from 1981 to 2010. J Nat Prod 75(3):311–335. doi:10.1021/np200906s
Newman DJ, Cragg GM, Battershill CN (2009) Therapeutic agents from the sea: biodiversity, chemo-evolutionary insight and advances to the end of Darwin's 200th year. Diving Hyperb Med 39(4):216–225
Nicolaou KC, Montagnon T (2008) Molecules that changed the world. Wiley, Weinheim
Ongley SE, Bian X, Neilan BA, Muller R (2013) Recent advances in the heterologous expression of microbial natural product biosynthetic pathways. Nat Prod Rep 30(8):1121–1138. doi:10.1039/c3np70034h
Pace NR, Stahl DA, Lane DJ, Olsen GJ (1986) The analysis of natural microbial-populations by ribosomal-RNA sequences. Adv Microb Ecol 9:1–55
Peiru S, Menzella HG, Rodriguez E, Carney J, Gramajo H (2005) Production of the potent antibacterial polyketide erythromycin C in Escherichia coli. Appl Environ Microbiol 71(5):2539–2547
Pelaez F (2006) The historical delivery of antibiotics from microbial natural products—can history repeat? Biochem Pharmacol 71(7):981–990. doi:10.1016/j.bcp.2005.10.010
Pfeifer BA, Khosla C (2001) Biosynthesis of polyketides in heterologous hosts. Microbiol Mol Biol Rev 65(1):106–118
Pfeifer BA, Admiraal SJ, Gramajo H, Cane DE, Khosla C (2001) Biosynthesis of complex polyketides in a metabolically engineered strain of E. coli. Science 291(5509):1790–1792
Pfeifer BA, Wang CC, Walsh CT, Khosla C (2003) Biosynthesis of Yersiniabactin, a complex polyketide-nonribosomal peptide, using Escherichia coli as a heterologous host. Appl Environ Microbiol 69(11):6698–6702
Quadri LE, Weinreb PH, Lei M, Nakano MM, Zuber P, Walsh CT (1998) Characterization of Sfp, a Bacillus subtilis phosphopantetheinyl transferase for peptidyl carrier protein domains in peptide synthetases. Biochemistry 37(6):1585–1595
Rahman H, Austin B, Mitchell WJ, Morris PC, Jamieson DJ, Adams DR, Spragg AM, Schweizer M (2010) Novel anti-infective compounds from marine bacteria. Mar Drugs 8(3):498–518. doi:10.3390/md8030498
Rappe MS, Giovannoni SJ (2003) The uncultured microbial majority. Annu Rev Microbiol 57:369–394. doi:10.1146/annurev.micro.57.030502.090759
Roberts GA, Staunton J, Leadlay PF (1993) Heterologous expression in Escherichia coli of an intact multienzyme component of the erythromycin-producing polyketide synthase. Eur J Biochem 214(1):305–311
Rodriguez E, Gramajo H (1999) Genetic and biochemical characterization of the alpha and beta components of a propionyl-CoA carboxylase complex of Streptomyces coelicolor A3(2). Microbiology 145(Pt 11):3109–3119
Roemer T, Xu D, Singh SB, Parish CA, Harris G, Wang H, Davies JE, Bills GF (2011) Confronting the challenges of natural product-based antifungal discovery. Chem Biol 18(2):148–164. doi:10.1016/j.chembiol.2011.01.009
Rondon MR, August PR, Bettermann AD, Brady SF, Grossman TH, Liles MR, Loiacono KA, Lynch BA, MacNeil IA, Minor C, Tiong CL, Gilman M, Osburne MS, Clardy J, Handelsman J, Goodman RM (2000) Cloning the soil metagenome: a strategy for accessing the genetic and functional diversity of uncultured microorganisms. Appl Environ Microbiol 66(6):2541–2547
Schipper C, Hornung C, Bijtenhoorn P, Quitschau M, Grond S, Streit WR (2009) Metagenome-derived clones encoding two novel lactonase family proteins involved in biofilm inhibition in Pseudomonas aeruginosa. Appl Environ Microbiol 75(1):224–233. doi:10.1128/AEM.01389-08
Schrempf H, Bujard H, Hopwood DA, Goebel W (1975) Isolation of covalently closed circular deoxyribonucleic acid from Streptomyces coelicolor A3(2). J Bacteriol 121(2):416–421
Shiloach J, Fass R (2005) Growing E. coli to high cell density—a historical perspective on method development. Biotechnol Adv 23(5):345–357. doi:10.1016/j.biotechadv.2005.04.004. S0734-9750(05)00046-7 [pii]

Song MC, Kim EJ, Kim E, Rathwell K, Nam SJ, Yoon YJ (2014) Microbial biosynthesis of medicinally important plant secondary metabolites. Nat Prod Rep 31(11):1497–1509. doi:10.1039/c4np00057a

Spizek J, Novotna J, Rezanka T, Demain AL (2010) Do we need new antibiotics? The search for new targets and new compounds. J Ind Microbiol Biotechnol 37(12):1241–1248. doi:10.1007/s10295-010-0849-8

Studier FW, Moffatt BA (1986) Use of bacteriophage T7 RNA polymerase to direct selective high-level expression of cloned genes. J Mol Biol 189(1):113–130

Swartz JR (1996) Escherichia coli recombinant DNA technology. In: Neidhardt FC (ed) Escherichia coli and Salmonella typhimurium, vol 2, 2nd edn. American Society for Microbiology, Washington, DC, pp 1693–1711

Torsvik V, Goksoyr J, Daae FL (1990) High diversity in DNA of soil bacteria. Appl Environ Microbiol 56(3):782–787

Vederas JC (2014) Explorations of fungal biosynthesis of reduced polyketides—a personal viewpoint. Nat Prod Rep 31(10):1253–1259. doi:10.1039/c4np00091a

Verdine GL (1996) The combinatorial chemistry of nature. Nature 384(6604 Suppl):11–13. doi:10.1038/384011a0

Wang Y, Pfeifer BA (2008) 6-Deoxyerythronolide B production through chromosomal localization of the deoxyerythronolide B synthase genes in E. coli. Metab Eng 10(1):33–38

Wang GY, Graziani E, Waters B, Pan W, Li X, McDermott J, Meurer G, Saxena G, Andersen RJ, Davies J (2000) Novel natural products from soil DNA libraries in a streptomycete host. Org Lett 2(16):2401–2404

Watanabe K, Hotta K, Praseuth AP, Koketsu K, Migita A, Boddy CN, Wang CC, Oguri H, Oikawa H (2006) Total biosynthesis of antitumor nonribosomal peptides in Escherichia coli. Nat Chem Biol 2(8):423–428

Wenzel SC, Gross F, Zhang Y, Fu J, Stewart AF, Muller R (2005) Heterologous expression of a myxobacterial natural products assembly line in pseudomonads via red/ET recombineering. Chem Biol 12(3):349–356

Wild J, Szybalski W (2004) Copy-control pBAC/oriV vectors for genomic cloning. Methods Mol Biol 267:145–154. doi:10.1385/1-59259-774-2:145

Wild J, Hradecna Z, Posfai G, Szybalski W (1996) A broad-host-range in vivo pop-out and amplification system for generating large quantities of 50- to 100-kb genomic fragments for direct DNA sequencing. Gene 179(1):181–188

Wild J, Hradecna Z, Szybalski W (2001) Single-copy/high-copy (SC/HC) pBAC/oriV novel vectors for genomics and gene expression. Plasmid 45:142

Wild J, Hradecna Z, Szybalski W (2002) Conditionally amplifiable BACs: switching from single-copy to high-copy vectors and genomic clones. Genome Res 12(9):1434–1444

Yun J, Kang S, Park S, Yoon H, Kim MJ, Heu S, Ryu S (2004) Characterization of a novel amylolytic enzyme encoded by a gene from a soil-derived metagenomic library. Appl Environ Microbiol 70(12):7229–7235. doi:10.1128/AEM.70.12.7229-7235.2004

Zhang H, Wang Y, Pfeifer BA (2008) Bacterial hosts for natural product production. Mol Pharm 5(2):212–225

Zhang H, Wang Y, Boghigian B, Pfeifer BA (2009) Probing the heterologous metabolism supporting 6-deoxyerythronolide B biosynthesis in E. coli. Microb Biotechnol 2(3):390–394

Zhang H, Boghigian BA, Pfeifer BA (2010a) Investigating the role of native propionyl-CoA and methylmalonyl-CoA metabolism on heterologous polyketide production in Escherichia coli. Biotechnol Bioeng 105(3):567–573. doi:10.1002/bit.22560

Zhang H, Wang Y, Wu J, Skalina K, Pfeifer BA (2010b) Complete biosynthesis of erythromycin A and designed analogs using E. coli as a heterologous host. Chem Biol 17(11):1232–1240. doi:10.1016/j.chembiol.2010.09.013

Zhang H, Boghigian BA, Armando J, Pfeifer BA (2011) Methods and options for the heterologous production of complex natural products. Nat Prod Rep 28(1):125–151. doi:10.1039/c0np00037j

Chapter 4
Functional Analysis in Metagenomics Using MEGAN 6

Sina Beier, Rewati Tappu, and Daniel H. Huson

Abstract Early microbiome studies focused on estimating the taxonomic composition of an assemblage of microbes using amplicon sequencing. With improved throughput and decreased cost of sequencing, whole genome shotgun (WGS) sequencing of environmental samples has become a standard procedure in microbial studies. This allows a more detailed analysis of the taxonomic composition and the analysis of the functional potential of a microbiome. Typical metagenomic projects may involve hundreds of samples and billions of reads. Fast sequence alignment tools and powerful analysis methods are an important requirement for any metagenomic study. Here we describe how to efficiently perform functional analysis of large-scale metagenomic datasets using a pipeline consisting of DIAMOND for sequencing alignment, MEGAN 6 for interactive exploration and analysis, and MeganServer for easy access to files.

4.1 Introduction

High-throughput, cost-efficient sequencing methods allow large-scale genomic studies, including detailed studies of environmental samples by metagenomic sequencing. Metagenomic projects are undertaken with the aim of determining the community composition and functional capacity of microbiomes in different settings, including (personalized) medicine, agricultural research, or wastewater treatment (Qin et al. 2010; Fierer et al. 2012), to name a few.

While early work in metagenomics focused on taxonomic assessment, there is increased interest in the functional characterization of unculturable microbes (Eiler et al. 2014) and in the assessment of the metabolic differences between multiple related samples (Greenblum et al. 2012). Many projects aim at identifying new genes with novel characteristics in environmental samples or try to relate changes in the functional content of a microbiome to changes in community composition or in

S. Beier • R. Tappu • D.H. Huson (✉)
Algorithms in Bioinformatics, University of Tübingen, Tübingen, Germany
e-mail: daniel.huson@uni-tuebingen.de

T.C. Charles et al. (eds.), *Functional Metagenomics: Tools and Applications*,
DOI 10.1007/978-3-319-61510-3_4

associated technological, biological, or experimental parameters (so-called metadata). The analysis of metagenomic data in the context of informative metadata can lead to insights into the role of a microbe in its environment. An in-depth functional classification of metagenomic data allows one to study the roles of different metabolic pathways in a community.

Current next-generation sequencing technologies used in metagenomics typically produce short reads (between 100 and 300 bp). Methods for analyzing such reads can be categorized as alignment based, composition based, or phylogenetic-marker based (Cummings and Bazinet 2012). While all three types of approaches can be employed in the taxonomic analysis of metagenomic samples, functional analysis is usually addressed with an alignment-based approach. In the past, the alignment of metagenomic reads to a protein reference base such as NCBI-nr posed a significant computational bottleneck; however, this problem has been alleviated by the introduction of a new high-throughput DNA-to-protein alignment tool called DIAMOND (Buchfink et al. 2015). Taxonomic and functional analysis of a sample can be performed efficiently by post-processing the output of DIAMOND either using MEGAN (Huson et al. 2011) or a new associated command line tool called Meganizer.

MEGAN is an interactive program for the taxonomic and functional analysis of one or multiple metagenomic or metatranscriptomic samples. Written in Java, it runs on all three major operating systems, Linux, Mac OS X, and Windows. It was originally published in 2007 (Huson et al. 2007) and has developed together with the field, recently resulting in a major update to MEGAN 6 (Huson et al. 2016).

In a preprocessing step, all reads are aligned against a reference database such as NCBI-nr or GenBank. To analyze a set of reads, MEGAN imports all reads and all associated alignments. The program performs taxonomic binning of all reads using the lowest common ancestor (LCA) algorithm, which places each read on the lowest taxonomic node that is a common ancestor of all organisms for which the read has a top scoring alignment. MEGAN considers an alignment to be top scoring, if its bit score exceeds a given threshold (minScore parameter, 50 by default) and if its bit score lies within a fixed percentage (topPercent parameter, 10 by default) of the best bit score seen for that read.

The program performs functional binning using the InterPro2GO (Mitchell et al. 2014; Hunter et al. 2014), KEGG (Kanehisa and Goto 2000), SEED (Overbeek et al. 2013), and eggNOG (Powell et al. 2012) classifications. In more detail, for a given functional classification, the program considers each read in turn and determines the highest-scoring alignment for which the functional class of the corresponding reference sequence is known and assigns the read to that functional class. Functional classes for InterPro2GO, eggNOG, SEED, and KEGG are InterPro families, COGs, functional roles, and KEGG orthology groups, respectively.

MEGAN allows one to combine, visualize, and compare multiple samples simultaneously, in a so-called comparison document. The program provides methods for analyzing the alpha or beta diversity of samples and for computing PCoA plots, bi-plots, and tri-plots, as discussed in more detail in Sect. 4.2.2.1. MEGAN 6 facilitates the incorporation of metadata. The program supports various input formats for alignment data and can export analysis results and selected input data in standardized text formats as well as graphic representations.

4.2 Workflow

In the following, we describe a standard workflow for the taxonomic and functional analysis of metagenomic shotgun data, which uses DIAMOND for read alignment and MEGAN 6 for taxonomic and functional analysis and MeganServer to serve the results.

4.2.1 *Using DIAMOND for Sequence Alignment*

As both the number and size of samples considered in metagenomic studies continue to increase, the first step in alignment-based analysis is computationally challenging, especially when using large reference databases such as NCBI-nr or InterPro. We propose to use DIAMOND for alignment of shotgun metagenomic datasets to large reference databases. DIAMOND is an open-source algorithm based on double indexing and spaced seeds that runs up to 20,000 times faster than BLASTX on short reads while achieving an adequate degree of sensitivity (Buchfink et al. 2015).

The tool can be easily coupled with MEGAN 6 in order to speed up the process of metagenome analysis. The first step for alignment with DIAMOND is to build a DIAMOND database index from the FASTA file of the protein database sequences using the makedb command. The resulting database file can then be used for alignment of protein (BLASTP mode) or nucleotide (BLASTX mode) sequences. DIAMOND BLASTX mode is used for metagenomic datasets and offers a fast and sensitive alignment similar to BLASTX. DIAMOND writes output to a file in DAA ("DIAMOND alignment archive") format, which contains all significant alignments of reads to references. This file can either be imported directly into MEGAN or can be processed by a command line tool called Meganizer. The latter program augments or "meganizes" DAA files so they can be directly opened in MEGAN. A meganized DAA file, containing all aligned reads, alignments, classifications, and classification-specific indices, is approximately the same size as the original uncompressed FastQ file that was used as input for DIAMOND.

4.2.2 *Taxonomic and Functional Classification with MEGAN 6*

Alignment is the first step toward computational analysis of a metagenomic sample. The next step is to bin all reads taxonomically and functionally, based on their alignments to reference sequences. MEGAN and Meganizer perform taxonomic analysis using the LCA algorithm to assign each read to the lowest common ancestor node in the NCBI taxonomy that lies above all organisms for which the read has a significant alignment (Huson et al. 2007). For functional classification, reads are placed into the KEGG, SEED, eggNOG, and InterPro2GO hierarchy, using mapping files that link GI numbers (or RefSeq ids) to identifiers in these classifications, in this case using the best alignment for which a functional assignment is available.

Importing alignment files into MEGAN can be time-consuming; therefore, we recommend using MEGAN command line tools such as blast2rma to compute so-called RMA files or Meganizer to compute meganized DAA files, on a server. Those files can then be opened in MEGAN individually or together in a comparative document. Comparative documents can be saved as a summary file (*.megan) that contains the classification binning counts or "abundances," but not reads or alignments, making these files small and easily portable, e.g., via email. After creating one or more RMA files, or meganized DAA files, the user can interactively apply MEGAN 6 to inspect and analyze the samples, as well as to visualize the results. This includes basic metagenomic analysis methods such as rarefaction curves and the Shannon-Weaver and Simpson diversity measures. In addition, PCoA or hierarchical clustering using different ecological indices can be applied both to taxonomic and functional profiles.

4.2.2.1 Visualization of the Functional Information

One way to display functional assignments is to use a tree representation of the functional classification hierarchy, with each node scaled to represent the number of reads assigned to the associated function. Such a tree can be collapsed at a higher level to provide an overview of the functional capacity shown or can be uncollapsed to a lower level to allow a detailed investigation of genes of interest, as indicated in Fig. 4.1.

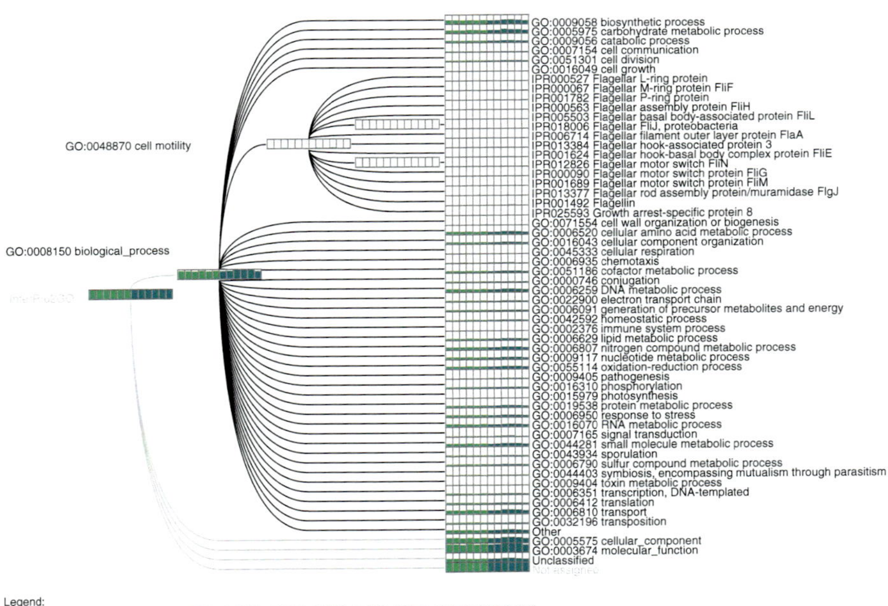

Fig. 4.1 InterPro2GO comparison of 12 samples, as displayed in MEGAN. Each GOslim node is drawn as a *bar chart* indicating the number of reads assigned to the class for each sample; for each of the 12 samples, *colors* stand for the two different individuals. The sample data has been taken from Willmann et al. (2015) and is available on the public MeganServer instance

MEGAN 6 offers a number of different charts and plots for summarizing the content of samples (see Figs. 4.2 and 4.3). The user is required to select the nodes representing the genes or functions of interest, and then a button on the toolbar allows the user to select a specific visualization.

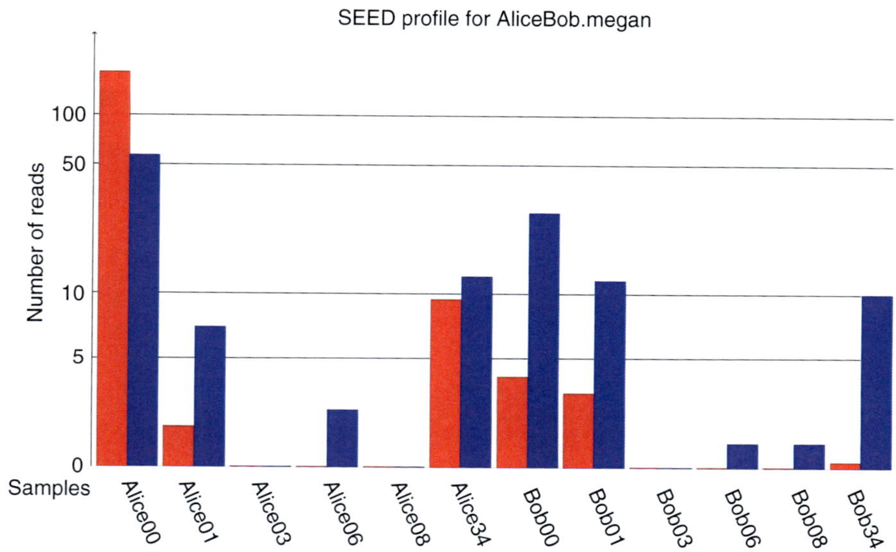

Fig. 4.2 *Bar chart* for type IV conjugative transfer system (*red*) and type III secretion system (*blue*) of bacteria as it is classified by SEED. Both functions are reduced during antibiotic intake (day 3, 6, and 8) in both individuals (Alice and Bob) but increase again at day 34. Bacterial secretion is potentially correlating with pathogenicity

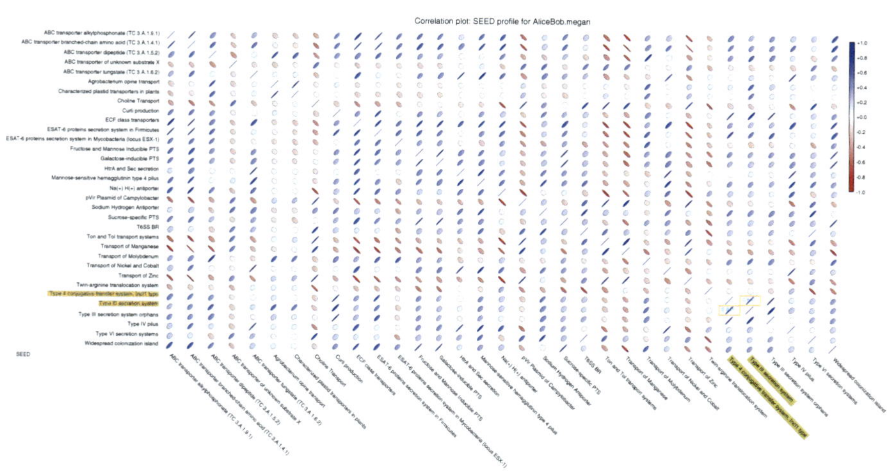

Fig. 4.3 Correlation plot for SEED pathways of bacterial secretion. *Blue* and *red ovals* represent positive and negative correlated pairs of functional roles, respectively. *Skinny ovals* and *higher color saturation* indicate higher correlation values, while *circular ovals* imply values close to zero. Here, type III secretion systems, which are related to pathogenicity, show a significant correlation with type IV conjugative secretion systems, related to genetic plasticity of an organism

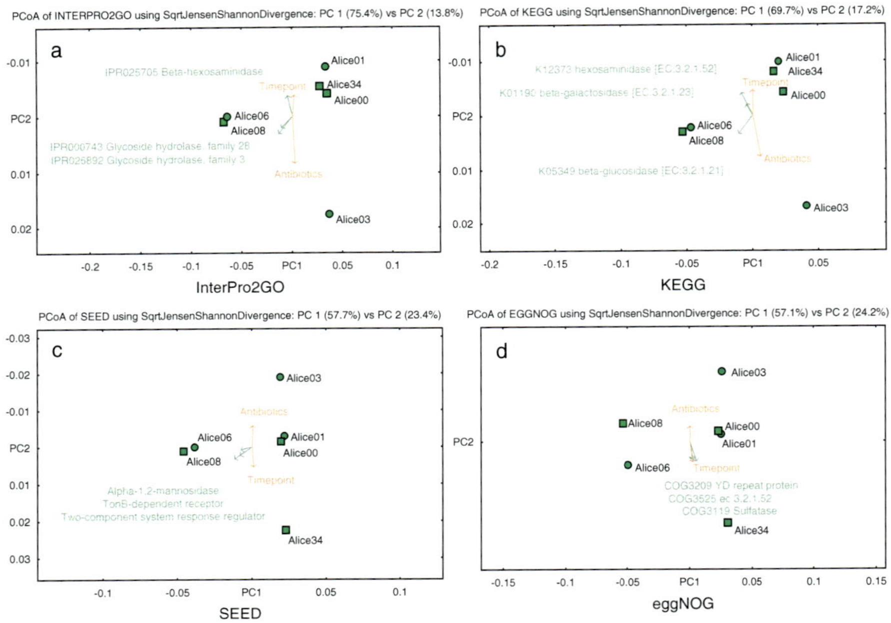

Fig. 4.4 PCoA analysis of six samples associated with "Alice," based on the KEGG, SEED, InterPro2GO, and eggNOG functional classifications, as implemented in MEGAN. Bi-plot vectors (*green*) indicate which three functional classes correlate the most with the differences seen in the PCoA plot, whereas tri-plot vectors (*orange*) indicate correlation with the two main sample attributes, time point and treatment

MEGAN 6 supports the computation of a bi-plot and tri-plot (Fig. 4.4) in which vectors indicate which taxonomic or functional classes and which metadata attributes show the highest correlation to the differences indicated in the PCoA plot.

4.2.2.2 Exporting Read Counts

Sophisticated statistical analysis may be required in order to test the hypotheses made before carrying out the metagenomic analysis. For example, the goal of the experiment may be to assess whether there exists a significant difference in the functional composition between healthy and diseased samples and to determine which features are responsible for this difference. Statistical software packages like R or MATLAB offer a variety of functions and packages for hypothesis testing. MEGAN 6 allows one to export data such that it can be used as input for statistical software packages. Abundances can be exported as assigned (only the reads directly assigned to the specific node) or summarized (reads assigned to the specific node and all nodes below in the hierarchy) counts or percentages of each sample.

4.2.2.3 Studying Sequences from a Specific Functional Category

In MEGAN 6, all reads assigned to a specific taxonomic group or functional category can be extracted by selecting that node and using the Export Reads option. For example, a KEGG Orthology (KO) group of interest can be selected, and then all reads that matched this KO group can be exported into a text file. Hence, the binning performed in MEGAN 6 can be used as a filter to select reads of interest. The user can interactively filter the data based on a selected function of interest and then inspect this subset further. The taxonomic groups that are associated with a given function are also easily visible, and the corresponding genes can be studied in further detail. For example, the reads can be assembled into contigs. Figure 4.5a shows the MEGAN "inspector window" which can be used to drill down to individual reads and matches.

All matches and alignments associated with a given functional group can also be extracted, which enables the user to generate a summary or RMA file from a subset of interesting functions and use it to study the taxa that are responsible for these different functions in the microbiome. This data can only be extracted from RMA files and meganized DAA files, not from summary files and multi-sample comparisons. Thus, from an RMA file, one can export reads, contigs, counts, alignments, and matches.

4.2.2.4 Assembling Binned Reads into Contigs

For detailed analysis of specific genes, MEGAN 6 offers the option to assemble mapped reads into fragments spanning complete genes, thus enabling better annotation and detection of variants. This assembly is guided by a "minimum overlap" parameter, and the resulting contigs can be exported for downstream analysis. Using the option File.. Export.., assembly exports contigs for selected taxonomic of functional nodes. Another option is to assemble only the reads aligned to a reference sequence using the Show Alignment viewer.

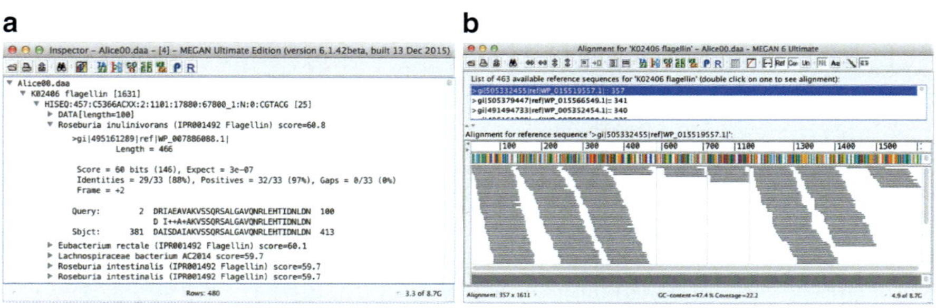

Fig. 4.5 (**a**) The inspector window shows the reads for KO2406 with their matches and alignments. (**b**) The Alignment view shows the reads aligned to one reference sequence assigned to KO2406

This viewer (Fig. 4.5b) can be opened for any node in a taxonomic or functional hierarchy. It shows all reference sequences mapping to this node, and the user can select one reference sequence to show all alignments of reads to this reference. Selecting File.. Export.. Contigs in this dialog will assemble only the reads from the currently shown alignment into contigs. This can be used to retrieve specific gene-centric assemblies for a gene of interest.

4.2.2.5 MEGAN 6 Server

When working with a number of large RMA files or Meganized DAA files, downloading them onto a desktop or a laptop is inconvenient. In order to address this problem, a light-weight web service called MeganServer that runs on a server has been developed. Depending on the setup, this web service can be accessed via a local network or the internet. MEGAN 6 is able to connect to one or more instances of MeganServer, and this can facilitate the analysis and comparison of multiple MEGAN files without the need to have them present on the local computer. This enables the analysis of large metagenomic experiments to be shared with multiple users. As access is read-only, no user will risk to introduce changes in the parameters or results for any other user while still being able to run analyses and generate visualizations for their own area of interest.

4.3 Challenges

New sequencing technologies are moving toward a much increased read length, currently at the cost of decreased accuracy (Greninger et al. 2015). The arising data will allow more specific taxonomic and functional analysis. Like most tools currently available for metagenomic analysis, MEGAN 6 was designed for shotgun sequencing data and thus has to be adapted to improve analysis of long reads. Currently, functional assignments are based on the highest-scoring hit to a protein in the used reference database. While this is sufficient for short reads, with increasing read length, the chance of covering multiple proteins with one read increases. This should be kept in mind when using MEGAN 6 for functional assignment of long sequences, such as produced by Oxford Nanopore Technologies or by metagenomic assembly.

For the comparison of multiple metagenomic datasets, there are a number of caveats. Generally, due to DNA extraction and sequencing constraints, when one attempts to sequence samples to the same coverage, the resulting read counts will be different. Hence, count data cannot be directly compared for a valid statistical analysis of differences in taxonomic and functional diversity. Even in a single dataset, the taxonomic and functional quantification is biased by genome size and gene size, respectively (Beszteri et al. 2010). For generalized function, it can be even biased by the number of genes in a pathway, as pathways including more steps will present themselves as overrepresented when compared to smaller pathways.

4.4 Summary

MEGAN 6 is a powerful, interactive tool for analyzing both taxonomic and functional features of metagenomic data. We described how this tool could be used for mining useful functional information from metagenomes and discussed the types of output it produces that can be used for further analysis in different tools. In conclusion, fast alignment tools like DIAMOND coupled with an easy-to-use GUI-based tool like MEGAN 6 make functional characterization of metagenomes easier while still offering functionality for an in-depth analysis of the data.

References

Beszteri B, Temperton B, Frickenhaus S, Giovannoni SJ (2010) Average genome size: a potential source of bias in comparative metagenomics. ISME J 4:1075–1077

Buchfink B, Xie C, Huson DH (2015) Fast and sensitive protein alignment using diamond. Nat Methods 12:59–60. Published online 17 November 2014

Cummings MP, Bazinet AL (2012) A comparative evaluation of sequence classification programs. BMC Bioinformatics 13:92. PubMed Central PMCID: PMC3428669

Eiler A, Zaremba-Niedzwiedzka K et al (2014) Productivity and salinity structuring of the microplankton revealed by comparative freshwater metagenomics. Environ Microbiol 16(9):2682–2698

Fierer N, Leff J, Adams B et al (2012) Cross-biome metagenomic analysis of soil microbial communities and their functional attributes. PNAS 109(52):21390–21395

Greenblum S, Turnbaugh PJ, Elhanan B (2012) Metagenomic systems biology of the human gut microbiome reveals topological shifts associated with obesity and inflammatory bowel disease. PNAS 109(2):594–599

Greninger AL, Naccache SN, Federman S et al (2015) Rapid metagenomic identification of viral pathogens in clinical samples by real-time nanopore sequencing analysis. Genome Med 7:99

Hunter S, Corbett M, Denise H, Fraser M, Gonzalez-Beltran A, Hunter C, Jones P, Leinonen R, McAnulla C, Maguire E, Maslen J, Mitchell A, Nuka G, Oisel A, Pesseat S, Radhakrishnan R, Rocca-Serra P, Scheremetjew M, Sterk P, Vaughan D, Cochrane G, Field D, Sansone SA (2014) Ebi metagenomics–a new resource for the analysis and archiving of metagenomic data. Nucleic Acids Res 42(Database issue):D600–D606. doi:10.1093/nar/gkt961

Huson DH, Auch AF, Qi J, Schuster SC (2007) Megan analysis of metagenomic data. Genome Res 17:377–386

Huson DH, Beier S, Flade I, Górska A, El-Hadidi M, Mitra S, Ruscheweyh H-J, Tappu R, Poisot T (2016) MEGAN Community Edition - Interactive Exploration and Analysis of Large-Scale Microbiome Sequencing Data. PLOS Comput Biol 12(6):e1004957

Huson DH, Mitra S, Weber N, Ruscheweyh H-J, Schuster SC (2011) Integrative analysis of environmental sequences using megan4. Genome Res 21:1552–1560

Kanehisa M, Goto S (2000) Kegg: kyoto encyclopedia of genes and genomes. Nucleic Acid Res 28(1):27–30

Mitchell A, Chang HY, Daugherty L, Fraser M, Hunter S, Lopez R, McAnulla C, Mc- Menamin C, Nuka G, Pesseat S, Sangrador-Vegas A, Scheremetjew M, Rato C, Yong SY, Bateman A, Punta M, Attwood TK, Sigrist CJ, Redaschi N, Rivoire C, Xenarios I, Kahn D, Guyot D, Bork P, Letunic I, Gough J, Oates M, Haft D, Huang H, Natale DA, Wu CH, Orengo C, Sillitoe I, Mi H, Thomas PD, Finn RD (2015) The InterPro protein families database: the classification resource after 15 years. Nucleic Acids Res 43(Database issue):D213–D221. doi:10.1093/nar/gku1243

Overbeek R, Olson R, Pusch GD, Olsen GJ, Davis JJ, Disz T, Edwards RA, Gerdes S, Parrello B, Shukla M, Vonstein V, Wattam AR, Xia F, Stevens R (2014) The SEED and the Rapid Annotation of microbial genomes using Subsystems Technology (RAST). Nucleic Acids Res 42(Database issue):D206–D214. doi:10.1093/nar/gkt1226

Powell S, Szklarczyk D, Trachana K, Roth A, Kuhn M, Muller J, Arnold R, Rattei T, Letunic I, Doerks T, Jensen LJ, von Mering C, Bork P (2012) eggnog v3.0: orthologous groups covering 1133 organisms at 41 different taxonomic ranges. Nucleic Acids Res 40(D1):D284–D289

Qin J, Li R, Raes J et al (2010) A human gut microbial gene catalogue established by metagenomic sequencing. Nature 464:59–65

Willmann M, El-Hadidi M, Huson DH et al (2015) Antibiotic selection pressure determination through sequence-based metagenomics. Antimicrob Agents Chemother 59(12):7335–7345

Chapter 5
Enhancing Metagenomic Approaches Through Synthetic Biology

Luana de Fátima Alves, Rafael Silva-Rocha, and María-Eugenia Guazzaroni

Abstract Bioactive compounds and enzymes with tolerance to process-specific parameters or improved catalytic performance play a crucial role in the development of applications in the chemical and pharmaceutical industry or energy production. Metagenomics takes advantage of the wealth of biochemical diversity present in the genomes of microorganisms found in environmental samples and provides a set of new technologies directed toward screening for new genes with potential in biotechnological applications. However, despite the vast number of published successful studies using this approach, metagenomic strategies typically have low rates of target discovery, and a number of issues need to be addressed in order to improve the screening efficiency of metagenomic libraries. Current limitations include biases imposed by expression in foreign host organisms, low vector performance in particular hosts, and the absence of suitable screening strategies for many targets. These restrictions cannot be overcome by using a single approach but rather require the synergetic implementation of multiple methodologies. In this chapter, we review some of the principal constraints regarding the discovery of new genes with potential use in biotechnology in metagenomic libraries and discuss how these might be resolved using synthetic biology methods. In addition, we review the state of art of synthetic biology approaches directed to improve the recovery of target genes in metagenomic screenings.

L.de.F. Alves • M.-E. Guazzaroni (✉)
Departamento de Bioquimica, FMRP-University of São Paulo,
Ribeirão Preto, SP, Brazil
e-mail: meguazzaroni@gmail.com

R. Silva-Rocha
Departamento de Biologia Celular, FMRP—University of São Paulo,
Ribeirão Preto, São Paulo, Brazil

T.C. Charles et al. (eds.), *Functional Metagenomics: Tools and Applications*,
DOI 10.1007/978-3-319-61510-3_5

5.1 Introduction

Nowadays there is an imperative necessity to sort out a number of boundaries in diverse fields such as medicine, pharmacy, energy generation, agriculture, and industry. In this regard, biotechnology, defined as the use of living systems and organisms to develop or make products, plays an essential role in proving solutions for those limitations. In the industrial context, many manufacturing processes that previously depended strictly on complex (and frequently harmful) chemical reactions have been superseded by much simpler and safer enzyme-based catalysis (Kirk et al. 2002). The introduction of biotechnology in industrial processes does not only generate a reduction in the final amount and toxicity of effluents but can also considerably reduce costs (Herrera 2004). The number of biotechnology applications has expanded in recent years, and this has created a growing demand for biocatalysts with superior performance or tolerance to extreme application-specific conditions (Lorenz et al. 2002; Schloss and Handelsman 2003). This is particularly true in those industries that produce bulk commodities such as detergents (Maurer 2004). Similarly, fine-chemical industries require multiple biocatalysts in order to perform highly diverse transformations for the production of new compounds (Homann et al. 2004).

The impact of the use of enzymes in industrial processes has stimulated an increased interest both in identifying new variants with enhanced kinetic parameters and in modifying previously characterized enzymes to increase their suitability for industrial applications (Lorenz et al. 2002; Schloss and Handelsman 2003). Parameters such as activity, efficacy, specificity, and stability are used to characterize and select enzymes for different purposes (Lorenz and Eck 2005). Enzymes used in industry have been identified from different sources through a combination of two major strategies: (1) the identification of novel enzymes from cultured microorganisms and (2) molecular evolution by DNA shuffling and rational design (Lynd et al. 2002; Percival Zhang et al. 2006; Krogh et al. 2010; Ward 2011; Chen et al. 2012). However, enzymes suitable for a given biotechnology application need to work efficiently within specified parameters, and since those currently used are frequently not optimized, the industrial processes have to be adjusted to accommodate these suboptimal catalysts (Warnecke and Hess 2009). As a consequence, there is an increasing demand for new biocatalysts with improved properties for industrial applications, such as higher catalytic efficiency on insoluble substrates (as in the case of cellulases used in the production of second generation bioethanol), increased stability at elevated temperature and at defined pH, and higher tolerance to end-product inhibition (Ward 2011; Singhania et al. 2013).

Microorganisms play a central role in biotechnology, not only as tools in molecular biology techniques but also as the major source of biocatalysts for industrial applications (Fernandez-Arrojo et al. 2010). Although prokaryotes represent the largest proportion of individual living organisms with estimated 10^3–10^5 microbial species in 1 g of soil (Schloss and Handelsman 2006), only a small proportion can be cultured using existing methodologies (Sleator et al. 2008). If we assume that a single genome encodes 4000 proteins (as is the case for *Escherichia coli*), then 4×10^8 potential proteins might be expected in just 1 g of soil. Supposing that 40% of these proteins display catalytic activity (Dinsdale et al. 2008), we might expect to find 1.6×10^8 biocatalysts, which highlights the vast inventory of biological functions available in nature. In this sense, metagenomics avoids the necessity of isolation and laboratory cultivation of individual microorganisms and has become a powerful tool for accessing and exploring the biological and molecular diversity present in different natural environments. Over the past decades, many studies using metagenomic approaches have proven to be successful for the recovery of novel enzymes with potential use in industrial applications (Lorenz and Eck 2005; Fernandez-Arrojo et al. 2010).

On the other hand, recovery of new bioactive compounds is of high interest since they are used for several medical, pharmaceutical, and biotechnological applications (Houssen and Jaspars 2012). Biotechnological potential of these compounds is huge since they efficiently interact with proteins, DNA, and other biological molecules to produce a desired outcome, which could be exploited for designing natural product-derived therapeutic agents (Ajikumar et al. 2008). Marine microorganisms continue to be a major focus of many natural product research efforts, with 491 new compounds reported in 2013, an increase of 14% from 2012 (Blunt et al. 2015). It is believed that specific physiochemical properties of the marine environment, such as pressure, temperature, pH, osmolarity, and uncommon functional groups, may result in the production of bioactive substances with different properties from those found in terrestrial habitats (Bharate et al. 2013; Bhatnagar and Kim 2010; Trincone 2011). In fact, it is estimated that the ocean contains the highest percentage of prokaryotic cells on Earth, with a reported number of 10^{30} cells (Whitman et al. 1998). This relatively unexploited reservoir of bioactivity present in the marine environment has allowed the identification of a number of new bioactive compounds, using both culture-dependent and -independent isolation of marine microorganisms (Rocha-Martin et al. 2014). However, isolation and cultivation of a novel marine microorganism is the main bottleneck in the discovery of new marine natural products (Rocha-Martin et al. 2014), as only a minor fraction of present microorganisms can be studied by classical microbiological methods. As metagenomics eliminates the requirement of isolation and cultivation of individual microorganisms, this approach has provided great progress in the discovery of new genes coding for antimicrobials and other bioactive molecules over the last decade (Jackson et al. 2015). As a result, several examples in literature show that metagenomics has been successful for the identification of novel biosynthetic genes and pathways coding for bioactive compounds both in terrestrial (Feng et al. 2011; Brady et al. 2009) and aquatic environments (Banik and Brady 2010; Robertson et al. 2004; Jackson et al. 2015).

Despite the above-cited successes, metagenomic strategies typically have low rates of target identification, and a number of issues need to be addressed in order to improve the screening efficiency of metagenomic libraries. The limits include (1) bias imposed by expression of genes in foreign host organisms, (2) low vector performance in particular hosts, and (3) absence of suitable screening strategies for

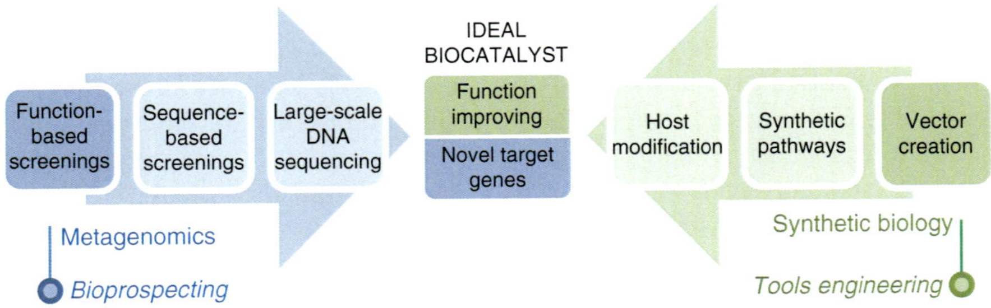

Fig. 5.1 Identification strategy of an ideal biocatalyst. The identification of enzymes from cultured microorganisms and metagenomics are the two principal approaches currently employed for recovering of genes encoding the desired enzymatic activity for industrial processes. Then, these genes may be cloned and expressed, and parameters such as activity, stability, specificity, and efficiency improved using protein rational design and in vitro evolution techniques. Metagenomics strategies are based on (1) function-based approaches, (2) sequence-based approaches, and (3) massive DNA sequencing methods. Synthetic biology can provide solutions to the current limitations in function-based metagenomic approaches, which involve the construction of expression libraries and its posterior activity screening. Development of methods for the engineering of new bacterial hosts and molecular biology tools promise to increase the efficiency of discovery of biotechnologically relevant enzymes

many target genes. As has already been demonstrated in several recent studies, all these issues may be addressed using synthetic biology approaches (Fig. 5.1) (Williamson et al. 2005; Uchiyama and Miyazaki 2010; Uchiyama and Watanabe 2008). In addition to the application of existing synthetic biology approaches, the development of new methodologies is imperative for the next generation of metagenomic studies that aim to recover ideal biocatalysts for given industrial processes or to identify novel genes involved in the synthesis of bioactive molecules. This chapter focuses on how the interplay between synthetic biology and functional metagenomics can yield novel strategies to obtain target genes. We also discuss how innovative synthetic biology applications could help to relieve current limitations in metagenomic screenings.

5.2 Metagenomes as Sources of Novel Genes with Potential Use in Biotechnology

5.2.1 State of the Art and Main Limitations

Although most environmental bacteria are refractory to cultivation, the biotechnological potential of uncultivated bacteria can be realized by directly cloning the DNA retrieved from the microbial community (Guazzaroni et al. 2010a). The construction and subsequent screening of metagenomic libraries allows identification of the targeted genes encoding the desired catalytic activities (Fig. 5.1). Accordingly, a well-planned strategy should take into consideration the vector to be used, the host organism for transformation and the screening strategy in order to maximize the rate of identification of the target activities. For example, if single genes or small operons are of interest, the best option is to use a small-insert metagenomic library instead of a large-insert library (Guazzaroni et al. 2010a). Small-insert expression libraries, especially those using lambda phage vectors and plasmids, are especially suitable for activity-based screening. The small size of the cloned fragments (up to 8 kb) means that most genes that are present in the appropriate orientation will be under the influence of strong promoters present at the cloning vector and thus have a good chance of being expressed and detected in activity screens (Ferrer et al. 2009). On the other hand, if the goal is biosynthetic pathway mining or functional expression of large multi-enzyme assemblies (e.g., in the case of gene clusters encoding polyketide synthases), the preferred option is library construction using cosmids or fosmids, which can harbor DNA inserts of up to 40 kb in size (Guazzaroni et al. 2010a).

It is important to appreciate that significant differences in expression modes exist between different taxonomic groups of prokaryotes, and that only 40% of enzymatic activities may be detected by random cloning in *E. coli* (Gabor et al. 2004). Therefore, it is also likely that performing metagenomic library screening in hosts other than *E. coli* will expand the range of detectable activities, although achieving this goal will require further optimization of the conditions for high transformation efficiency. Indeed, promising results of metagenomic library screening have been reported in *Streptomyces* spp. (Wang et al. 2000), *Rhizobium leguminosarum* (Wexler et al. 2005), and diverse *Proteobacteria* (Craig et al. 2010). For example, Wexler and collaborators constructed a library in the broad-host-range cosmid pLAFR3 using metagenomic DNA obtained from the microbial community of an anaerobic digester in a wastewater treatment plant (Wexler et al. 2005). After screening the metagenomic libraries in *R. leguminosarum*, a single cosmid that enabled *R. leguminosarum* to grow on ethanol as the sole carbon and energy source was recovered. Further analysis identified the presence of a gene encoding an atypical alcohol dehydrogenase that did not confer ethanol utilization ability to either *E. coli* or to *Pseudomonas aeruginosa*, even though the gene was transcribed in both hosts (Wexler et al. 2005). These results show that the use of broad-host-range vectors enhances the flexibility of metagenomic library screening. Furthermore, a recent functional metagenomic study showed that recovery of genes conferring acid resistance to *E. coli* and the subsequent transfer of some of these genes to *P. putida* and *B. subtilis* expanded the capabilities of these two bacteria to survive harsh acid conditions (Guazzaroni et al. 2013). However, in agreement with previous studies (Craig et al. 2010), variable gene doses were present due to the use of different cloning plasmids in each host organism, and no quantitative comparison could be made regarding gene expression or activity levels between the hosts. Thus, the developing of robust broad-host-range vectors capable of replication in several different hosts is one of the major challenges in metagenomics and one to which synthetic biology may make a significant contribution.

An additional bottleneck in metagenomic screening is related to the low frequency of positive clones that are typically recovered (Vieites et al. 2009). Common screening methods are based on the degradation of specific enzyme substrates that result either in the appearance of halos surrounding the positive clones or alternatively on the use of chromogenic substrates (Guazzaroni et al. 2010b). Depending on the type of substrate used and on the enzyme screened, detection of positive clones can be assayed directly in solid or in liquid media (Guazzaroni et al. 2010b), and the choice of medium will have consequences on the throughput level of the screening method. Similarly, adequate selection of specific activity-driven substrates plays a key role in the success of recovering of the desired enzymatic activity and decreasing the frequency of false positives.

5.2.2 *Screening Strategies*

There are three different strategies in functional metagenomics for the recovery of sequences encoding the desired enzymatic activity (Fig. 5.1). Firstly, activity-based approaches involve construction of small- to large-insert expression libraries that are suitable for direct activity screening, such as lambda phage, plasmid, cosmid, or copy-controlled fosmid vectors (Lorenz and Eck 2005). Once a library has been constructed, a critical step is the screening of a large number of clones, and in the case of activity-based screening, thousands of clones may be analyzed in a single screen, with the advantage that sequence information is not required. Therefore, this strategy has the potential to identify entirely novel classes of genes encoding known or novel functions (Handelsman 2004; Daniel 2005; Gloux et al. 2010; Bhat et al. 2013; Alcaide et al. 2013; Martinez-Martinez et al. 2013, 2014). Furthermore, activity-driven screening strategies can potentially provide the means to reveal undiscovered genes or gene families that cannot be detected by sequence-driven approaches.

However, sequence-based approaches, which is the second strategy used in functional metagenomics, have also led to the effective identification of genes relevant for biotechnology (Fig. 5.1). Several examples have been documented in the literature (Morimoto and Fujii 2009; Sul et al. 2009; Zaprasis et al. 2010; Varaljay et al. 2010; Gong et al. 2013; Jiang et al. 2013; Yan et al. 2013). In fact, sequence-based approaches involve the design of PCR primers for the target sequences that are derived from conserved regions of known protein families, and this dependence on prior knowledge limits the possibility for identifying new protein families (Ferrer et al. 2009). Finally, the large-scale sequencing of bulk DNA or metagenomic libraries through deep sequencing techniques provides the raw data for mining sequences encoding potentially useful enzymes (Fernandez-Arrojo et al. 2010). Since homology-based methods are effective only when the information regarding the reference sequences is accurate, a further disadvantage of this approach is its reliance both on existing genome annotations and on the quality and completeness of current databases (Hallin et al. 2008). It is worth a cautionary note, since a significant number of genomes in the current databases contain mis-annotations (Schnoes et al. 2009). Considering that the classification of protein families is based on amino acid similarity, novel enzyme families could not be detected by database searching with sequences from metagenomic sequencing or PCR-based detection methods and might be annotated as hypothetical proteins. A review of prokaryotic protein diversity in different shotgun metagenome studies indicated that 30–60% of the proteins could not be assigned known functions using current public databases (Vieites et al. 2009).

As previously mentioned, the maximum yield of industrial processes is achieved by optimization of physico-chemical parameters, and most currently available enzymes are incompatible with these conditions (Lorenz and Eck 2005). In this sense, the use of enzymes in industry requires the adaptation of processes to suit the optimal conditions of the applied enzyme, which can result in reduced production levels. Several examples have demonstrated that it is possible to recover enzymes displaying environment-compatible properties when metagenomic libraries derived from extreme habitats were screened. For example, solfataric hot springs (Rhee et al. 2005), Urania hypersaline basins (Ferrer et al. 2005), acid mine drainage biofilms (Guazzaroni et al. 2013), glacier soil (Yuhong et al. 2009), glacial ice (Simon et al. 2009), and Antarctic soil (Cieslinski et al. 2009; Heath et al. 2009) represent rich and largely unexploited reservoirs of novel genes with biotechnologically valuable properties. Although the diversity of microbial communities present in many extreme habitats is likely to be low, samples from these environments are still a valuable source of novel enzymes that are active under extreme conditions (Steele et al. 2009). Thus, as might be expected, the properties of the enzymes retrieved from extreme habitats are consistent with conditions of the source environments (Heath et al. 2009; Feng et al. 2007; Hu et al. 2012; Jiang et al. 2009).

5.2.3 New Screening Methods

Although several methods are available to the direct screening of many enzymes, there is an increasing need to expand the tools available for target gene detection. An alternative approach to this problem using synthetic biology is the design and implementation of in vivo biosensors capable of generate a detectable output in response to the degradation or production of a particular metabolite (Galvao and de Lorenzo 2006). In biosensor-based screening strategies, it is not the product of an enzyme or a pathway itself that results in a measurable property, but rather a genetically encoded reporter that provides a discriminating phenotype (Eggeling et al. 2015). As recently reviewed by Eggeling et al. (2015), there are basically three types of biosensors: (1) constitutively formed reporter proteins leading to signal generation due to interaction with the target product (small molecule), (2) RNA switches where an RNA aptamer controls expression of a reporter gene in response to binding of the product, and (3) systems based on transcriptional factors (TFs), where interaction of the small molecule with a regulatory protein controls

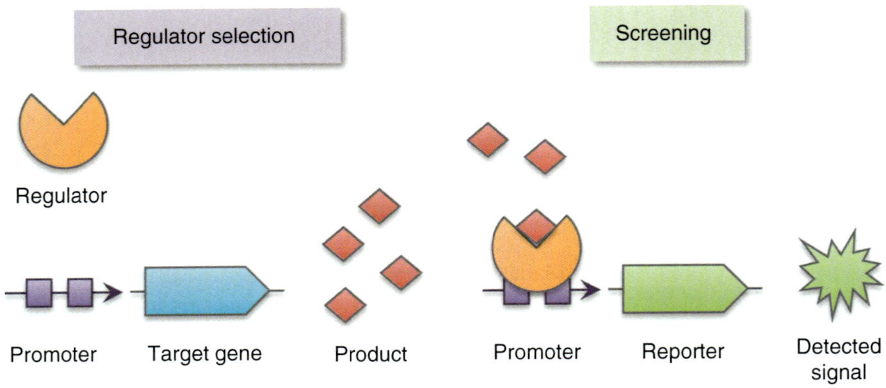

Fig. 5.2 Regulator-based sensors. In the biosensor approach, a target gene(s) is used to identify a transcriptional factor that responds to the molecules produced by the activity of a given enzyme(s). In this sense, metabolic products will act as effectors of the chosen regulator. Then, this gene is replaced by a reporter gene (such as *gfp, tetA, lacZ*) that in turn determines which screening or selection strategy is possible (FACS, agar plate, etc.). As a result, interaction of the produced molecules with the regulator controls expression of the reporter gene. Finally, signal emitted by the reporter gene such as fluorescence, antibiotic resistance, or color formation is screened instead of product formation

expression of the reporter (Fig. 5.2). Although each strategy presents advantages and disadvantages depending of the specific goal, the latter strategy, which involves screening methods using TF-based biosensors, has been the unique used in metagenomic screenings. In this context, substrate-induced gene expression (SIGEX) and product-induced gene expression (PIGEX) approaches have been developed and successfully applied to the detection of enzymatic activities associated with the metabolic modification of compounds of interest (Uchiyama et al. 2005; Uchiyama and Miyazaki 2010). In a broader approach, the concept of genetic traps has been used to guide the construction of synthetic circuits containing engineered regulators to control gene expression responses to metabolites generated by enzymatic activities present in cloned metagenomic fragments (Uchiyama and Watanabe 2008). In general, these strategies are particularly useful for screening of libraries where processing of small molecules is targeted (Williamson et al. 2005).

Strachan and co-workers (2014) have recently published another interesting example. In this study, authors develop a biosensor responsive to mono-aromatic lignin transformation products compatible with functional screening in *E. coli*. In particular, they used the EmrR regulator and its promoter as a versatile biosensor (P_{emrR}-GFP) in functional screens for lignin transformation, since this regulator is able to control a metabolic network responsive to mono-aromatic exposure in the environment (Strachan et al. 2014). In this sense, EmrR was used to retrieve metagenomic scaffolds sourced from coal bed bacterial communities conferring an array of lignin transformation phenotypes that synergize when combined. Additionally, studies of transposon mutagenesis and comparative sequence analysis of active clones identified genes encoding six functional classes mediating lignin transformation phenotypes (i.e., share common enzymatic, regulatory, and transport features) that appear to be re-arrayed in nature via horizontal gene transfer (HGT) (Strachan et al. 2014). Based on the foregoing observations, authors propose a model for lignin transformation in the environment in which reusable functional classes are shuffled through HGT to generate adaptive combinatorial arrays that promote niche adaptation based on a range of lignin substrates and transformation products. In this sense, this work showed that exploiting ecological design principles using naturally assembled genetic parts have a tremendous potential to build a new generation of biorefining microorganisms (Strachan et al. 2014).

5.3 Synthetic Biology Approaches to Improve Metagenomic Screening Strategies

The previous sections have highlighted a number of bottlenecks facing metagenomic screening that need to be resolved in order to improve the discovery rate of target genes (Fig. 5.3). These limitations can be grouped in three main categories. Firstly, there is a need for improvement of the host organism capabilities to improve the expression of the target enzymes. Secondly, the development of new genetic tools is necessary in order to improve the construction of metagenomic libraries suitable for screening in different hosts. Finally, continuation of ongoing search to elaborate novel screening strategies that enhance the discovery rate of the genes of interest is recommended. Synthetic biology could efficiently address all these limitations, and the potential of this rising field will be discussed below.

5.3.1 Fundaments of Synthetic Biology

The advances in synthetic biology over the past decade could provide the framework to address these constrains, and a particularly promising approach is the analysis of biological systems in an analogy to electronic devices, whereby cells can be reprogrammed to perform new tasks with high efficiency (Weber and Fussenegger 2010; Purnick and Weiss 2009). In fact, synthetic biology relies on a conceptual framework more closely related to engineering than biology, such as design, modeling, implementation, and debugging (Weber and Fussenegger 2010; Purnick and Weiss 2009; Eggeling et al. 2015). The design aspect focuses on the planning and construction of new gene circuits for the desired application (Eggeling et al. 2015). Modeling involves computational simulation of the proposed gene circuits in order to both evaluate performance and capabilities and to guide the selection of the suitable molecular components necessary for its construction (Koide et al. 2009). The implementation step encompasses the physical assembly of the DNA elements encoding the appropriate components (such as promoters, regulators, terminators, enzymes, transporters, etc.) and follows a specific assembly standard (Arkin 2008).

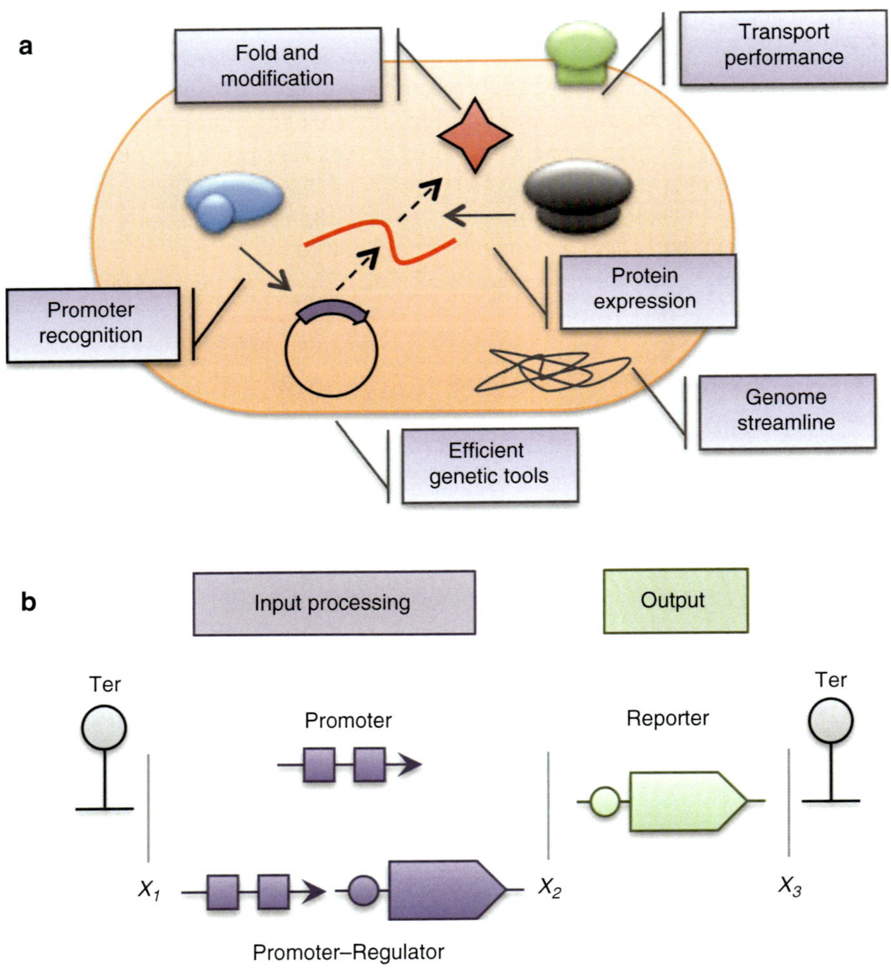

Fig. 5.3 Synthetic biology may overcome restrictive steps in activity-based metagenomic library screening. (**a**) Limitations in the host capabilities. In the case of the host, critical steps related to the recognition of transcriptional and translational signals, as well as the folding and modification of the expressed enzyme, need to be enhanced. Host performance might be improved by reducing the metabolic burden related to the expression of unnecessary genes and improving performance of membrane transporters involved in target substrate or product passage. In addition, the use of semisynthetic, high-efficiency genetic tools is essential for the construction of metagenomic libraries that can be maintained and screened in a wide number of microorganisms. (**b**) Availability of efficient screening methods. Genetic circuits constructed by combining input modules (e.g., promoters and regulators) and output devices (such as reporter proteins) assembled with a standard format that uses the same sets of restriction enzymes (represented by X_1, X_2, etc.). Such circuits facilitate the screening of target genes in metagenomic libraries. The standardization of the assembly process facilitates the combination of several independent modules to construct sophisticated activity-trigged biosensors

Finally, the debugging step requires the testing and validation of the circuit in vivo and includes the correction of undesirable traits that have their origin in the emergent properties of biology systems (Siuti et al. 2013; Moon et al. 2011; Gardner et al. 2000). Several examples of new biological circuits that have been successfully designed and implemented are currently available (Silva-Rocha and de Lorenzo 2011; Moon et al. 2011; Strachan et al. 2014; Otte and Hauer 2015; Gardner et al. 2000), and in recent years the field has developed at a remarkable speed (Siuti et al. 2013; Regot et al. 2011; Zhan et al. 2010; Weber and Fussenegger 2010).

5.3.2 Strategies to Improve the Level of Expression and Activity of Target Proteins

As mentioned in previous sections, failure of heterologous gene expression in host cells is among the main causes of the low recovery rates of enzymes of interest from metagenomic libraries (Handelsman 2004; Lorenz and Eck 2005). A combination of different strategies could be applied to optimize this critical event (Fig. 5.3a). In the initial transcription step, a reduced affinity of the RNA polymerase (RNAP) for intrinsic promoters derived from metagenomic fragments represents an important limitation for heterologous protein expression in the host organism. At the bottom level, promoter recognition in prokaryotes is strongly biased among the different phylogenetic groups (Gabor et al. 2004), and the ideal host for heterologous protein expression should be endowed with a transcriptional machinery with broad promoter recognition capability. Such a goal could be attained, for example, by co-expressing heterologous sigma factors with different promoter specificities (Osterberg et al. 2011; Rhodius et al. 2013), thereby allowing protein expression from promoters derived from different bacterial sources. In a more direct approach, the expression of foreign genes in the host organism could be driven by a high-efficiency expression system such as the T7 RNAP (Terron-Gonzalez et al. 2013). This has the advantage that genes controlled by complex signal transduction mechanisms could be easily expressed in response to a single inducer such as IPTG (Tabor 2001). The same line of reasoning may also be applied to mRNA translation, where poor recognition of the ribosome-binding site (RBS) can reduce protein expression levels (Zelcbuch et al. 2013). In this case, the co-expression of additional proteins related to the mRNA recognition step could expand the host capability for foreign gene expression (Uchiyama and Miyazaki 2009). In a very elegant work, the improvement of the expression of an entire metabolic pathway was achieved using a synthetic biology approach (Tang and Cirino 2011). In this study, the RBS was targeted using an AraC variant of *E. coli* responsive to mevalonate and an *E. coli* strain enabling mevalonate synthesis by three plasmid-encoded genes. Libraries of variant RBS in front of a hydroxymethylglutaryl-CoA reductase were successfully screened by colony color for increased mevalonate formation (Tang and Cirino 2011). In a similar way, a strategy could be developed for screening of entire pathways or genes acting in combination (such as multi-enzyme assemblies) in fosmids-/cosmids-based metagenomic libraries, considering that these vectors can harbor DNA inserts of high size (Guazzaroni et al. 2010a).

While strategies described above aim to improve the level of expression of target proteins, an additional factor is related to the activity levels of expressed enzymes. This is particularly true in the case of enzymes displaying either complex folding or requiring additional processing steps (such as cleavage, secretion, or peptide modification) (Bhat 2000; Ajikumar et al. 2008). For instance, in the case of protein folding, the co-expression of molecular chaperons (Leis et al. 2014) has been reported to enhance the expression of heterologous proteins in *E. coli* (Ferrer et al. 2004).

5.3.3 Strategies to Improve Host Organism Performance

Among the additional approaches to improve host organism capabilities, genome edition (or streamlining) deserves special attention. Since many of the genes in bacteria such as *E. coli* are not essential for growth under standard laboratory conditions (Medini et al. 2005), maintaining these genes and the consequent redundant expression of unnecessary proteins represents a significant energetic cost and a metabolic burden to the host cell (Posfai et al. 2006). The reduction of genome size in bacteria by the removal of nonessential genes has been shown to endow bacteria with renewed metabolic vigor that enhances the production level of heterologous proteins (Posfai et al. 2006; Martinez-Garcia et al. 2014). Due to their superior expression capabilities, bacterial strains with minimized genomes are therefore promising host organisms for screening of metagenomic libraries (Fig. 5.3a).

In addition to genetic engineering of the more common hosts such as *E. coli*, several attempts have focused on the use of alternative hosts to increase the rate of enzyme identification in metagenomic screening (Guazzaroni et al. 2013; Craig et al. 2010). Screening in alternative hosts requires either library construction in nonoptimal, broad-host-range vectors (Craig et al. 2010) or subcloning of target genes in appropriate vectors (Guazzaroni et al. 2013). Universal and standardized tools are required in order to facilitate this type of multi-organism approach for the screening of metagenomic libraries in a larger number of hosts, and synthetic biology approaches are particularly suited to the development and use of such tools for the construction of adequate gene circuits in an increasing number of host cell platforms (Arkin 2008; Shetty et al. 2008; Jackson et al. 2015). Among these new multi-host tools, the pSEVA vectors are of particular interest as synthetic broad-host-range vectors that are expected to work in about 100 different bacterial species (Silva-Rocha et al. 2013). Using this system as a starting point, new genetic tools could be developed for cloning and screening of environmental DNA from phylogenetically diverse bacteria.

Angelov and collaborators (2009) developed a new system for functional screening in extreme thermophiles. Authors described a two-host fosmid system, which allows the construction of large-insert fosmid libraries in *E. coli* and the transfer of the recombinant libraries to the extreme thermophile *Thermus thermophilus* via natural transformation (Angelov et al. 2009). Libraries are established in the thermophilic host by site-specific chromosomal insertion of the recombinant fosmids via single crossover or double crossover recombination at the *T. thermophilus pyr* locus. Furthermore, a recent study shows the success of the strategy when screenings were carried out using complex metagenomes isolated from thermophilic microbial communities (Leis et al. 2015). For the screenings in *T. thermophilus*, a multiple clean deletion mutant was used (termed BL03) which lacks several characterized extracellular and putative esterase encoding genes (Leis et al. 2014). The knockout strain had a substantially lower esterase activity than the wild type, and the deletions abolished its ability to grow on defined minimal medium supplemented with tributyrin as the sole carbon source. In this sense, this esterase-diminished strain was used as a host to screen for metagenomic DNA fragments that could complement growth on tributyrin. Several thousand single fosmid clones from thermophilic metagenomic libraries from heated compost and hot spring water samples were subjected to a comparative screening for esterase activity in both *T. thermophilus* strain BL03 and *E. coli* EPI300. Authors recovered a greater number of active esterase clones in the thermophilic bacterium than in the mesophilic *E. coli*, highlighting the benefit of using additional screening hosts other than *E. coli* for the identification of novel biocatalysts with industrial relevance (Leis et al. 2015).

5.3.4 Using Well-Characterized Parts of Regulatory Circuits for Metagenomic Screenings

Currently available methods that are used to screen for new catalytic activities in metagenomic libraries frequently rely either on the use of chromogenic enzyme substrates (Ko et al. 2013; Hu et al. 2012) or substrates that when degraded leave a clear halo around positive colonies (Robertson et al. 2004). More labor intensive screening procedures such as colony PCR are also used for enzyme discovery (Trincone 2011). The rapid expansion of synthetic biology has already resulted in the construction of regulatory circuits using well-characterized parts (Voigt 2006), and there is a tremendous potential to use this accumulated knowledge for further advances in the design of biosensors to screen for enzymatic activities (Williamson et al. 2005; Nasuno et al. 2012). The use of biosensors is an in vivo strategy that has been used to identify specific enzymatic activities using engineered regulatory circuits coupled to a reporter gene, such as *lacZ*, GFP, or luciferase (Mohn et al. 2006; Tang and Cirino 2011). This genetic trap approach (Uchiyama et al. 2005; Uchiyama and Miyazaki 2010; Uchiyama and Watanabe 2008) eliminates the necessity for extensive manipulation during screening yet allows the identification of positive clones in metagenomic libraries (Fig. 5.2). Synthetic biology approaches have also been used to modulate enzyme levels in biosynthetic pathways by combinatorial gene pairing with a defined set of ribosome-binding sites (Zelcbuch et al. 2013), resulting in the modulation of protein abundance by several orders of magnitude, which showed that engineering of metabolic pathways relies on precise control of enzyme levels (Zelcbuch et al. 2013). These examples demonstrate how synthetic biology approaches can improve the ability to interconnect regulatory components (e.g., promoters and regulators) to generate new circuits with reliable performance characteristics (Fig. 5.3b) (Schmidt and de Lorenzo 2012). The combination of available assembly platforms for circuit engineering (Nikel and de Lorenzo 2013; Zhan et al. 2010; Arkin 2008) together with approaches for the redesign of regulatory systems to recognize new molecules (Angelov et al. 2009) will allow the implementation of new genetic traps for the identification of novel biotechnologically relevant genes.

5.4 Concluding Remarks

The metagenomic approach embraces the idea that the ideal biocatalyst or metabolic pathway for synthesis of the desired bioactive compound may already exist in nature and that the exploration of the inherent diversity of natural environments is the solution. As a result, a broad repertoire of culture-independent techniques has been developed, and advances toward the identification of novel genes potentially useful for biotech applications have been made. However, the available metagenomic approaches require further refinement to achieve the goal of identifying biotechnological relevant genes. Existing limitations with respect to host expression, vector availability, and specific screening restrictions cannot be solved by using a single approach and require the synergic implementation of multiple methodologies (Fig. 5.3). One of the most elegant solutions that synthetic biology has already provided to overcome restrictions in metagenomic screenings is the use of engineered regulatory circuits coupled to a reporter gene, the so-called biosensors (Fig. 5.2). Biosensor-based screenings present the significant advantage that the catalytic activity or product of a metabolic pathway itself is not the measurable property but the phenotype produced by a reporter. In this sense, the detection of the output signal in the screening strategy, which is most of the times laborious, time-consuming, and partially efficient, is evaded. Recognition of the same output signal independently of the gene searched allows the use of always the same experimental conditions (such as background, sensitivity, measurement protocol), saving considerable efforts. Additionally, a growing number of studies have shown that other synthetic biology approaches may significantly improve metagenomic library screening and allow exploitation of the rich biochemical potential present in natural environments.

Acknowledgments The study was supported by the National Council of Technological and Scientific Development (CNPq 472893/2013-0 and 441833/2014-4) and by the Sao Paulo State Foundation (FAPESP, grant number 2015/04309-1 and 2012/21922-8). MEG was a beneficiary of a CNPq Young Talent Fellowship (CNPq 370630/2013-0). Authors have no conflict of interest to declare.

References

Ajikumar PK, Tyo K, Carlsen S, Mucha O, Phon TH, Stephanopoulos G (2008) Terpenoids: opportunities for biosynthesis of natural product drugs using engineered microorganisms. Mol Pharm 5(2):167–190. doi:10.1021/mp700151b

Alcaide M, Tornes J, Stogios PJ, Xu X, Gertler C, Di Leo R, Bargiela R, Lafraya A, Guazzaroni ME, Lopez-Cortes N, Chernikova TN, Golyshina OV, Nechitaylo TY, Plumeier I, Pieper DH, Yakimov MM, Savchenko A, Golyshin PN, Ferrer M (2013) Single residues dictate the co-evolution of dual esterases: MCP hydrolases from the alpha/beta hydrolase family. Biochem J 454(1):157–166

Angelov A, Mientus M, Liebl S, Liebl W (2009) A two-host fosmid system for functional screening of (meta)genomic libraries from extreme thermophiles. Syst Appl Microbiol 32(3):177–185. doi:10.1016/j.syapm.2008.01.003

Arkin A (2008) Setting the standard in synthetic biology. Nat Biotechnol 26(7):771–774. doi:10.1038/nbt0708-771

Banik JJ, Brady SF (2010) Recent application of metagenomic approaches toward the discovery of antimicrobials and other bioactive small molecules. Curr Opin Microbiol 13(5):603–609. doi:10.1016/j.mib.2010.08.012

Bharate SB, Sawant SD, Singh PP, Vishwakarma RA (2013) Kinase inhibitors of marine origin. Chem Rev 113(8):6761–6815. doi:10.1021/cr300410v

Bhat MK (2000) Cellulases and related enzymes in biotechnology. Biotechnol Adv 18(5):355–383

Bhat A, Riyaz-Ul-Hassan S, Ahmad N, Srivastava N, Johri S (2013) Isolation of cold-active, acidic endocellulase from Ladakh soil by functional metagenomics. Extremophiles 17(2):229–239

Bhatnagar I, Kim SK (2010) Immense essence of excellence: marine microbial bioactive compounds. Mar Drugs 8(10):2673–2701. doi:10.3390/md8102673

Blunt JW, Copp BR, Keyzers RA, Munro MH, Prinsep MR (2015) Marine natural products. Nat Prod Rep 32(2):116–211. doi:10.1039/c4np00144c

Brady SF, Simmons L, Kim JH, Schmidt EW (2009) Metagenomic approaches to natural products from free-living and symbiotic organisms. Nat Prod Rep 26(11):1488–1503. doi:10.1039/b817078a

Chen HL, Chen YC, Lu MY, Chang JJ, Wang HT, Ke HM, Wang TY, Ruan SK, Wang TY, Hung KY, Cho HY, Lin WT, Shih MC, Li WH (2012) A highly efficient beta-glucosidase from the buffalo rumen fungus Neocallimastix patriciarum W5. Biotechnol Biofuels 5(1):24

Cieslinski H, Bialkowskaa A, Tkaczuk K, Dlugolecka A, Kur J, Turkiewicz M (2009) Identification and molecular modeling of a novel lipase from an Antarctic soil metagenomic library. Pol J Microbiol 58(3):199–204

Craig JW, Chang FY, Kim JH, Obiajulu SC, Brady SF (2010) Expanding small-molecule functional metagenomics through parallel screening of broad-host-range cosmid environmental DNA libraries in diverse proteobacteria. Appl Environ Microbiol 76(5):1633–1641. doi:10.1128/AEM.02169-09. [pii].

Daniel R (2005) The metagenomics of soil. Nat Rev Microbiol 3(6):470–478

Dinsdale EA, Edwards RA, Hall D, Angly F, Breitbart M, Brulc JM, Furlan M, Desnues C, Haynes M, Li L, McDaniel L, Moran MA, Nelson KE, Nilsson C, Olson R, Paul J, Brito BR, Ruan Y, Swan BK,

Stevens R, Valentine DL, Thurber RV, Wegley L, White BA, Rohwer F (2008) Functional metagenomic profiling of nine biomes. Nature 452(7187):629–632
Eggeling L, Bott M, Marienhagen J (2015) Novel screening methods-biosensors. Curr Opin Biotechnol 35C:30–36. doi:10.1016/j.copbio.2014.12.021
Feng Y, Duan CJ, Pang H, Mo XC, Wu CF, Yu Y, Hu YL, Wei J, Tang JL, Feng JX (2007) Cloning and identification of novel cellulase genes from uncultured microorganisms in rabbit cecum and characterization of the expressed cellulases. Appl Microbiol Biotechnol 75(2):319–328
Feng Z, Kallifidas D, Brady SF (2011) Functional analysis of environmental DNA-derived type II polyketide synthases reveals structurally diverse secondary metabolites. Proc Natl Acad Sci U S A 108(31):12629–12634. doi:10.1073/pnas.1103921108
Fernandez-Arrojo L, Guazzaroni ME, Lopez-Cortes N, Beloqui A, Ferrer M (2010) Metagenomic era for biocatalyst identification. Curr Opin Biotechnol 21(6):725–733
Ferrer M, Chernikova TN, Timmis KN, Golyshin PN (2004) Expression of a temperature-sensitive esterase in a novel chaperone-based Escherichia coli strain. Appl Environ Microbiol 70(8):4499–4504. doi:10.1128/AEM.70.8.4499-4504.2004
Ferrer M, Golyshina OV, Chernikova TN, Khachane AN, Martins Dos Santos VA, Yakimov MM, Timmis KN, Golyshin PN (2005) Microbial enzymes mined from the Urania deep-sea hypersaline anoxic basin. Chem Biol 12(8):895–904
Ferrer M, Beloqui A, Timmis KN, Golyshin PN (2009) Metagenomics for mining new genetic resources of microbial communities. J Mol Microbiol Biotechnol 16(1–2):109–123
Gabor EM, Alkema WB, Janssen DB (2004) Quantifying the accessibility of the metagenome by random expression cloning techniques. Environ Microbiol 6(9):879–886. doi:10.1111/j.1462-2920.2004.00640.x
Galvao TC, de Lorenzo V (2006) Transcriptional regulators a la carte: engineering new effector specificities in bacterial regulatory proteins. Curr Opin Biotechnol 17(1):34–42
Gardner TS, Cantor CR, Collins JJ (2000) Construction of a genetic toggle switch in Escherichia coli. Nature 403(6767):339–342
Gloux K, Berteau O, El Oumami H, Beguet F, Leclerc M, Dore J (2010) A metagenomic beta-glucuronidase uncovers a core adaptive function of the human intestinal microbiome. Proc Natl Acad Sci U S A 108(Suppl 1):4539–4546
Gong JS, Lu ZM, Li H, Zhou ZM, Shi JS, Xu ZH (2013) Metagenomic technology and genome mining: emerging areas for exploring novel nitrilases. Appl Microbiol Biotechnol 97(15):6603–6611
Guazzaroni M-E, Golyshin PN, Ferrer M (2010a) Analysis of complex microbial communities through metagenomic survey. In: Marco D (ed) Metagenomics: theory, methods and applications. Caister Academic, Norfolk, pp 55–77
Guazzaroni ME, Beloqui A, Vieites JM, Al-ramahi Y, Cortes NL, Ghazi A, Golyshin PN, Ferrer M (2010b) Metagenomic mining of enzyme diversity. In: Timmis K (ed) Handbook of hydrocarbon and lipid microbiology. Springer, Berlin, pp 2911–2927. doi:10.1007/978-3-540-77587-4_216
Guazzaroni ME, Morgante V, Mirete S, Gonzalez-Pastor JE (2013) Novel acid resistance genes from the metagenome of the Tinto River, an extremely acidic environment. Environ Microbiol 15(4):1088–1102
Hallin PF, Binnewies TT, Ussery DW (2008) The genome BLASTatlas-a GeneWiz extension for visualization of whole-genome homology. Mol Biosyst 4(5):363–371
Handelsman J (2004) Metagenomics: application of genomics to uncultured microorganisms. Microbiol Mol Biol Rev 68(4):669–685
Heath C, Hu XP, Cary SC, Cowan D (2009) Identification of a novel alkaliphilic esterase active at low temperatures by screening a metagenomic library from Antarctic desert soil. Appl Environ Microbiol 75(13):4657–4659
Herrera S (2004) Industrial biotechnology-a chance at redemption. Nat Biotechnol 22(6):671–675
Homann MJ, Vail RB, Previte E, Tamarez M, Morgan B, Dodds DR, Zaks A (2004) Rapid identification of enantioselective ketone reductions using targeted microbial libraries, vol 60. Elsevier. First published

Houssen WE, Jaspars M (2012) Isolation of marine natural products. Methods Mol Biol 864:367–392. doi:10.1007/978-1-61779-624-1_14
Hu XP, Heath C, Taylor MP, Tuffin M, Cowan D (2012) A novel, extremely alkaliphilic and cold-active esterase from Antarctic desert soil. Extremophiles 16(1):79–86
Jackson SA, Borchert E, O'Gara F, Dobson AD (2015) Metagenomics for the discovery of novel biosurfactants of environmental interest from marine ecosystems. Curr Opin Biotechnol 33:176–182. doi:10.1016/j.copbio.2015.03.004
Jiang C, Ma G, Li S, Hu T, Che Z, Shen P, Yan B, Wu B (2009) Characterization of a novel beta-glucosidase-like activity from a soil metagenome. J Microbiol 47(5):542–548
Jiang C, Yin B, Tang M, Zhao G, He J, Shen P, Wu B (2013) Identification of a metagenome-derived prephenate dehydrogenase gene from an alkaline-polluted soil microorganism. Antonie Van Leeuwenhoek 103(6):1209–1219
Kirk O, Borchert TV, Fuglsang CC (2002) Industrial enzyme applications. Curr Opin Biotechnol 13(4):345–351
Ko KC, Han Y, Cheong DE, Choi JH, Song JJ (2013) Strategy for screening metagenomic resources for exocellulase activity using a robotic, high-throughput screening system. J Microbiol Methods 94(3):311–316
Koide T, Pang WL, Baliga NS (2009) The role of predictive modelling in rationally re-engineering biological systems. Nat Rev Microbiol 7(4):297–305. doi:10.1038/nrmicro2107
Krogh KB, Harris PV, Olsen CL, Johansen KS, Hojer-Pedersen J, Borjesson J, Olsson L (2010) Characterization and kinetic analysis of a thermostable GH3 beta-glucosidase from Penicillium brasilianum. Appl Microbiol Biotechnol 86(1):143–154
Leis B, Angelov A, Li H, Liebl W (2014) Genetic analysis of lipolytic activities in Thermus thermophilus HB27. J Biotechnol 191:150–157. doi:10.1016/j.jbiotec.2014.07.448
Leis B, Angelov A, Mientus M, Li H, Pham VT, Lauinger B, Bongen P, Pietruszka J, Goncalves LG, Santos H, Liebl W (2015) Identification of novel esterase-active enzymes from hot environments by use of the host bacterium Thermus thermophilus. Front Microbiol 6:275. doi:10.3389/fmicb.2015.00275
Lorenz P, Eck J (2005) Metagenomics and industrial applications. Nat Rev Microbiol 3(6):510–516
Lorenz P, Liebeton K, Niehaus F, Eck J (2002) Screening for novel enzymes for biocatalytic processes: accessing the metagenome as a resource of novel functional sequence space. Curr Opin Biotechnol 13(6):572–577
Lynd LR, Weimer PJ, van Zyl WH, Pretorius IS (2002) Microbial cellulose utilization: fundamentals and biotechnology. Microbiol Mol Biol Rev 66(3):506–577
Martinez-Garcia E, Nikel PI, Chavarria M, de Lorenzo V (2014) The metabolic cost of flagellar motion in Pseudomonas putida KT2440. Environ Microbiol 16(1):291–303. doi:10.1111/1462-2920.12309
Martinez-Martinez M, Alcaide M, Tchigvintsev A, Reva O, Polaina J, Bargiela R, Guazzaroni ME, Chicote A, Canet A, Valero F, Rico Eguizabal E, Guerrero Mdel C, Yakunin AF, Ferrer M (2013) Biochemical diversity of carboxyl esterases and lipases from Lake Arreo (Spain): a metagenomic approach. Appl Environ Microbiol 79(12):3553–3562
Martinez-Martinez M, Lores I, Pena-Garcia C, Bargiela R, Reyes-Duarte D, Guazzaroni ME, Pelaez AI, Sanchez J, Ferrer M (2014) Biochemical studies on a versatile esterase that is most catalytically active with polyaromatic esters. Microb Biotechnol 7(2):184–191
Maurer KH (2004) Detergent proteases. Curr Opin Biotechnol 15(4):330–334
Medini D, Donati C, Tettelin H, Masignani V, Rappuoli R (2005) The microbial pan-genome. Curr Opin Genet Dev 15(6):589–594. doi:10.1016/j.gde.2005.09.006
Mohn WW, Garmendia J, Galvao TC, de Lorenzo V (2006) Surveying biotransformations with a la carte genetic traps: translating dehydrochlorination of lindane (gamma-hexachlorocyclohexane) into lacZ-based phenotypes. Environ Microbiol 8(3):546–555. doi:10.1111/j.1462-2920.2006.00983.x
Moon TS, Clarke EJ, Groban ES, Tamsir A, Clark RM, Eames M, Kortemme T, Voigt CA (2011) Construction of a genetic multiplexer to toggle between chemosensory pathways in Escherichia coli. J Mol Biol 406(2):215–227. doi:10.1016/j.jmb.2010.12.019

Morimoto S, Fujii T (2009) A new approach to retrieve full lengths of functional genes from soil by PCR-DGGE and metagenome walking. Appl Microbiol Biotechnol 83(2):389–396

Nasuno E, Kimura N, Fujita MJ, Nakatsu CH, Kamagata Y, Hanada S (2012) Phylogenetically novel LuxI/LuxR-type quorum sensing systems isolated using a metagenomic approach. Appl Environ Microbiol 78(22):8067–8074

Nikel PI, de Lorenzo V (2013) Implantation of unmarked regulatory and metabolic modules in Gram-negative bacteria with specialised mini-transposon delivery vectors. J Biotechnol 163(2):143–154. doi:10.1016/j.jbiotec.2012.05.002

Osterberg S, del Peso-Santos T, Shingler V (2011) Regulation of alternative sigma factor use. Annu Rev Microbiol 65:37–55. doi:10.1146/annurev.micro.112408.134219

Otte KB, Hauer B (2015) Enzyme engineering in the context of novel pathways and products. Curr Opin Biotechnol 35C:16–22. doi:10.1016/j.copbio.2014.12.011

Percival Zhang YH, Himmel ME, Mielenz JR (2006) Outlook for cellulase improvement: screening and selection strategies. Biotechnol Adv 24(5):452–481

Posfai G, Plunkett G 3rd, Feher T, Frisch D, Keil GM, Umenhoffer K, Kolisnychenko V, Stahl B, Sharma SS, de Arruda M, Burland V, Harcum SW, Blattner FR (2006) Emergent properties of reduced-genome Escherichia coli. Science 312(5776):1044–1046. doi:10.1126/science.1126439

Purnick PE, Weiss R (2009) The second wave of synthetic biology: from modules to systems. Nat Rev Mol Cell Biol 10(6):410–422

Regot S, Macia J, Conde N, Furukawa K, Kjellen J, Peeters T, Hohmann S, de Nadal E, Posas F, Sole R (2011) Distributed biological computation with multicellular engineered networks. Nature 469(7329):207–211. doi:10.1038/nature09679

Rhee JK, Ahn DG, Kim YG, Oh JW (2005) New thermophilic and thermostable esterase with sequence similarity to the hormone-sensitive lipase family, cloned from a metagenomic library. Appl Environ Microbiol 71(2):817–825

Rhodius VA, Segall-Shapiro TH, Sharon BD, Ghodasara A, Orlova E, Tabakh H, Burkhardt DH, Clancy K, Peterson TC, Gross CA, Voigt CA (2013) Design of orthogonal genetic switches based on a crosstalk map of sigmas, anti-sigmas, and promoters. Mol Syst Biol 9:702. doi:10.1038/msb.2013.58

Robertson DE, Chaplin JA, DeSantis G, Podar M, Madden M, Chi E, Richardson T, Milan A, Miller M, Weiner DP, Wong K, McQuaid J, Farwell B, Preston LA, Tan X, Snead MA, Keller M, Mathur E, Kretz PL, Burk MJ, Short JM (2004) Exploring nitrilase sequence space for enantioselective catalysis. Appl Environ Microbiol 70(4):2429–2436

Rocha-Martin J, Harrington C, Dobson AD, O'Gara F (2014) Emerging strategies and integrated systems microbiology technologies for biodiscovery of marine bioactive compounds. Mar Drugs 12(6):3516–3559. doi:10.3390/md12063516

Schloss PD, Handelsman J (2003) Biotechnological prospects from metagenomics. Curr Opin Biotechnol 14(3):303–310

Schloss PD, Handelsman J (2006) Toward a census of bacteria in soil. PLoS Comput Biol 2(7):e92

Schmidt M, de Lorenzo V (2012) Synthetic constructs in/for the environment: managing the interplay between natural and engineered Biology. FEBS Lett 586(15):2199–2206. doi:10.1016/j.febslet.2012.02.022

Schnoes AM, Brown SD, Dodevski I, Babbitt PC (2009) Annotation error in public databases: misannotation of molecular function in enzyme superfamilies. PLoS Comput Biol 5(12):e1000605

Shetty RP, Endy D, Knight TF Jr (2008) Engineering BioBrick vectors from BioBrick parts. J Biol Eng 2:5. doi:10.1186/1754-1611-2-5

Silva-Rocha R, de Lorenzo V (2011) Implementing an OR-NOT (ORN) logic gate with components of the SOS regulatory network of Escherichia coli. Mol Biosyst 7(8):2389–2396. doi:10.1039/c1mb05094j

Silva-Rocha R, Martinez-Garcia E, Calles B, Chavarria M, Arce-Rodriguez A, de Las Heras A, Paez-Espino AD, Durante-Rodriguez G, Kim J, Nikel PI, Platero R, de Lorenzo V (2013) The Standard European Vector Architecture (SEVA): a coherent platform for the analysis and deployment of complex prokaryotic phenotypes. Nucleic Acids Res 41(Database issue):D666–D675. doi:10.1093/nar/gks1119

Simon C, Herath J, Rockstroh S, Daniel R (2009) Rapid identification of genes encoding DNA polymerases by function-based screening of metagenomic libraries derived from glacial ice. Appl Environ Microbiol 75(9):2964–2968

Singhania RR, Patel AK, Sukumaran RK, Larroche C, Pandey A (2013) Role and significance of beta-glucosidases in the hydrolysis of cellulose for bioethanol production. Bioresour Technol 127:500–507

Siuti P, Yazbek J, Lu TK (2013) Synthetic circuits integrating logic and memory in living cells. Nat Biotechnol 31(5):448–452. doi:10.1038/nbt.2510

Sleator RD, Shortall C, Hill C (2008) Metagenomics. Lett Appl Microbiol 47(5):361–366

Steele HL, Jaeger KE, Daniel R, Streit WR (2009) Advances in recovery of novel biocatalysts from metagenomes. J Mol Microbiol Biotechnol 16(1–2):25–37

Strachan CR, Singh R, VanInsberghe D, Ievdokymenko K, Budwill K, Mohn WW, Eltis LD, Hallam SJ (2014) Metagenomic scaffolds enable combinatorial lignin transformation. Proc Natl Acad Sci U S A 111(28):10143–10148. doi:10.1073/pnas.1401631111

Sul WJ, Park J, Quensen JF 3rd, Rodrigues JL, Seliger L, Tsoi TV, Zylstra GJ, Tiedje JM (2009) DNA-stable isotope probing integrated with metagenomics for retrieval of biphenyl dioxygenase genes from polychlorinated biphenyl-contaminated river sediment. Appl Environ Microbiol 75(17):5501–5506

Tabor S (2001) Expression using the T7 RNA polymerase/promoter system. In: Current protocols in molecular biology. Wiley. doi:10.1002/0471142727.mb1602s11

Tang SY, Cirino PC (2011) Design and application of a mevalonate-responsive regulatory protein. Angew Chem Int Ed Engl 50(5):1084–1086. doi:10.1002/anie.201006083

Terron-Gonzalez L, Medina C, Limon-Mortes MC, Santero E (2013) Heterologous viral expression systems in fosmid vectors increase the functional analysis potential of metagenomic libraries. Sci Rep 3:1107. doi:10.1038/srep01107

Trincone A (2011) Marine biocatalysts: enzymatic features and applications. Mar Drugs 9(4):478–499. doi:10.3390/md9040478

Uchiyama T, Miyazaki K (2009) Functional metagenomics for enzyme discovery: challenges to efficient screening. Curr Opin Biotechnol 20(6):616–622. doi:10.1016/j.copbio.2009.09.010

Uchiyama T, Miyazaki K (2010) Product-induced gene expression, a product-responsive reporter assay used to screen metagenomic libraries for enzyme-encoding genes. Appl Environ Microbiol 76(21):7029–7035

Uchiyama T, Watanabe K (2008) Substrate-induced gene expression (SIGEX) screening of metagenome libraries. Nat Protoc 3(7):1202–1212

Uchiyama T, Abe T, Ikemura T, Watanabe K (2005) Substrate-induced gene-expression screening of environmental metagenome libraries for isolation of catabolic genes. Nat Biotechnol 23(1):88–93

Varaljay VA, Howard EC, Sun S, Moran MA (2010) Deep sequencing of a dimethylsulfoniopropionate-degrading gene (dmdA) by using PCR primer pairs designed on the basis of marine metagenomic data. Appl Environ Microbiol 76(2):609–617

Vieites JM, Guazzaroni ME, Beloqui A, Golyshin PN, Ferrer M (2009) Metagenomics approaches in systems microbiology. FEMS Microbiol Rev 33(1):236–255

Voigt CA (2006) Genetic parts to program bacteria. Curr Opin Biotechnol 17(5):548–557. doi:10.1016/j.copbio.2006.09.001

Wang GY, Graziani E, Waters B, Pan W, Li X, McDermott J, Meurer G, Saxena G, Andersen RJ, Davies J (2000) Novel natural products from soil DNA libraries in a streptomycete host. Org Lett 2(16):2401–2404

Ward R (2011) Cellulase engineering for biomass saccharification. In: Buckeridge MS, Goldman GH (eds) Routes to cellulosic ethanol. Springer, New York, pp 135–151. doi:10.1007/978-0-387-92740-4_9

Warnecke F, Hess M (2009) A perspective: metatranscriptomics as a tool for the discovery of novel biocatalysts. J Biotechnol 142(1):91–95

Weber W, Fussenegger M (2010) Synthetic gene networks in mammalian cells. Curr Opin Biotechnol 21(5):690–696

Wexler M, Bond PL, Richardson DJ, Johnston AW (2005) A wide host-range metagenomic library from a waste water treatment plant yields a novel alcohol/aldehyde dehydrogenase. Environ Microbiol 7(12):1917–1926

Whitman WB, Coleman DC, Wiebe WJ (1998) Prokaryotes: the unseen majority. Proc Natl Acad Sci U S A 95(12):6578–6583

Williamson LL, Borlee BR, Schloss PD, Guan C, Allen HK, Handelsman J (2005) Intracellular screen to identify metagenomic clones that induce or inhibit a quorum-sensing biosensor. Appl Environ Microbiol 71(10):6335–6344

Yan X, Geng A, Zhang J, Wei Y, Zhang L, Qian C, Wang Q, Wang S, Zhou Z (2013) Discovery of (hemi-) cellulase genes in a metagenomic library from a biogas digester using 454 pyrosequencing. Appl Microbiol Biotechnol 97(18):8173–8182

Yuhong Z, Shi P, Liu W, Meng K, Bai Y, Wang G, Zhan Z, Yao B (2009) Lipase diversity in glacier soil based on analysis of metagenomic DNA fragments and cell culture. J Microbiol Biotechnol 19(9):888–897

Zaprasis A, Liu YJ, Liu SJ, Drake HL, Horn MA (2010) Abundance of novel and diverse tfdA-like genes, encoding putative phenoxyalkanoic acid herbicide-degrading dioxygenases, in soil. Appl Environ Microbiol 76(1):119–128

Zelcbuch L, Antonovsky N, Bar-Even A, Levin-Karp A, Barenholz U, Dayagi M, Liebermeister W, Flamholz A, Noor E, Amram S, Brandis A, Bareia T, Yofe I, Jubran H, Milo R (2013) Spanning high-dimensional expression space using ribosome-binding site combinatorics. Nucleic Acids Res 41(9):e98

Zhan J, Ding B, Ma R, Ma X, Su X, Zhao Y, Liu Z, Wu J, Liu H (2010) Develop reusable and combinable designs for transcriptional logic gates. Mol Syst Biol 6:388. doi:10.1038/msb.2010.42

Chapter 6
Metagenomics for the Discovery of Novel Biosurfactants

Wesley Williams and Marla Trindade

Abstract Biosurfactants offer a range of diverse applications in practically all areas of biotechnology (medical, industrial, environmental, and marine) and as such draw upon a number of overlapping fields. Due to their immense structural diversity, complexity, and biochemical properties, research is focused on the isolation of novel biosurfactants to replace chemically synthesized counterparts. Most of the described biosurfactants are of microbial origin, isolated through traditional culturing techniques. However, given that the vast majority of bacteria have yet to be cultured, metagenomics provides the potential to explore novel biosurfactants from bacteria which are recalcitrant to culturing and from exotic and unexplored environments. Despite the availability of numerous screening assays suited for high-throughput functional-based screening, there are very few examples of metagenomically derived biosurfactants. In this chapter we explore specific obstacles which have led to their underrepresentation in metagenomic screening studies and highlight the most recent successes which can inform future screening strategies.

6.1 Introduction

Surface-active compounds (surfactants) have both hydrophilic and hydrophobic structural units on one molecule that allow them to preferentially associate and alter the conditions of the interface/surface of the phase (Rosen 1989). Surfactants impact many areas of chemical and food processing technologies as emulsifiers, dispersants, and foaming and wetting agents, where the variety of applications has built a global market for surfactants of over $27 billion (Geys et al. 2014). Documented soap manufacture is dated to 2800 BCE and has served as the main surfactant used by cultures the world over, where the alkali salts of fatty acid soaps are produced by

W. Williams • M. Trindade (✉)
Department of Biotechnology, Institute for Microbial Biotechnology and Metagenomics, University of the Western Cape, Bellville 7535, South Africa
e-mail: ituffin@uwc.ac.za

T.C. Charles et al. (eds.), *Functional Metagenomics: Tools and Applications*,
DOI 10.1007/978-3-319-61510-3_6

saponification of animal or plant fats. The first large-scale application of non-soap synthetic surfactants was during World War I, developed in Germany to overcome the constrained supply of animal and plant products. Once in general use, synthetic surfactants came to hold significant market share of the global surfactant consumption estimated to be approximately 60% (Myers 2010). As the market for synthetic surfactants grew, negative ecological effects began to manifest due to surfactants' degradation resistance. The branched structures of the most widely used synthetic surfactant at the time, alkylbenzene sulfonates (ABS), were resistant to biodegradation leading to their accumulation in the environment. The more amenable to biodegradation, linear alkylbenzene sulfonates (LABS), were adopted voluntarily and by government regulation in response to the environmental damage caused by ABS. The bacterial communities of wastewater plants were able to degrade the linear structures of LABS more effectively than the branched structures of ABS. LABS remain resistant to anaerobic degradation; thus, the ecological concerns of synthetic surfactants have not completely dissipated. The shift in consumer demand for more sustainably sourced and environmentally compatible surfactants has led to an increased interest in biological surfactants (biosurfactants). The small-molecule secondary metabolite surfactant activity of microorganisms has shown surfactant performance equivalent to synthetic surfactants (Müller et al. 2012).

Biosurfactants are produced by microbial fermentation or from renewable resources (Müller et al. 2012). Since biosurfactants are biological in origin, they are biodegradable with the added ability to perform well over a wide pH and temperature range. The variety of structures and surfactant performance indicate an equally varied function for biosurfactants, though many of the functions remain unclear (Pamp and Tolker-Nielsen 2007; Ron and Rosenberg 2001). The diversity of biosurfactant molecules originates from the hydrophilic moiety, which could either be carbohydrates, carboxylic acids, phosphates, amino acids, cyclic peptides, or alcohols (Desai and Banat 1997). The less structurally diverse hydrophobic moiety comprises of varying chain lengths of long-chain fatty acids, hydroxy fatty acids, and α-alkyl or β-hydroxy fatty acids. Though many microorganisms are able to synthesize biosurfactants, most reports focus on glycolipids and lipopeptides such as rhamnolipids, sophorolipids, and surfactin. Glycolipids and lipopeptides feature prominently due to the potential of glycolipids to replace chemical surfactants (Henkel et al. 2012; Müller et al. 2012) and lipopeptide's success as antibacterial molecules in medicine (Cochrane and Vederas 2014). Several promising avenues for biosurfactant adoption in industry have been suggested for the bioremediation of oil-contaminated soil and water. Due to their biological compatibility, they are also suitable additives in food, cosmetic, and pharmaceutical formulations and as sustainably produced detergents. In order to increase the applicability of biosurfactants to as wide a range of functions as possible, the number of structures with varying surface-active properties needs to be expanded, because small changes in surfactant structure can significantly affect its performance in different solvent systems.

Microbial biosurfactant producers have been identified from bacteria, archaea, and eukaryotes, which were all isolated from pure cultures. Accessing unsampled sites that might contain an enriched population of novel biosurfactant-producing microorganisms is a strategy to circumvent the repeated isolation of already known producers, sites such as sponge-associated bacterial populations (Rizzo et al. 2013) or hypersaline environments (Kebbouche-Gana et al. 2013). However, molecular microbial ecology studies suggest that up to 99% of bacteria in any given environmental sample are not amenable to culture (Handelsman et al. 1998; Rappé and Giovannoni 2003). Invariably, limiting biosurfactant discovery to standard culturing techniques will likely result in the continued re-isolation of already known biosurfactant producers, no matter which environment is sampled. Circumventing the need to create ideal culture conditions for difficult to culture microorganisms, the total environmental DNA present in the sample, called the metagenome, can be extracted and analyzed for biosurfactant production potential (Handelsman et al. 2002; Kennedy et al. 2011; Jackson et al. 2015). Metagenomics has been an underutilized tool for the discovery of new biosurfactant synthetic pathways. This chapter introduces biosurfactants and the application of metagenomics as a tool for their discovery from uncultured bacteria.

6.2 Classes of Biosurfactants

Analyzing the known and well-studied biosurfactants provides important insight into the challenges that could face a metagenomic approach. Biosurfactants are grouped according to the composition of the hydrophilic head group and separated into distinct classes of biosurfactants, namely, glycolipids, lipopeptides, lipoamino acids, polymeric biosurfactants, and fatty acids (Desai and Banat 1997; Müller et al. 2012). The biosynthetic pathways and regulatory mechanisms for biosurfactants are as varied as their structures; therefore, the only overarching property that can be used to detect biosurfactants is the effect on surfaces of liquids (Table 6.1).

Table 6.1 The structural classification of biosurfactants is the most popular method for classifying the heterogeneous biosurfactant compounds (Desai and Banat 1997)

Structural class	Examples	Microorganism	References
Glycolipids	Rhamnolipids	*Pseudomonas aeruginosa*	Jarvis and Johnson (1949)
	Sophorolipids	*Starmerella bombicola*	Gorin et al. (1961)
	Mannosylerythritol lipids	*Pseudozyma* sp.	Kitamoto et al. (1990)
	Trehalose lipids	*Rhodococcus* sp.	Singer and Finnerty (1990)

(continued)

Table 6.1 (continued)

Structural class	Examples	Microorganism	References
Lipopeptides	Surfactin	*Bacillus subtilis*	Arima et al. (1968)
	Serrawettin	*Serratia marcescens*	Wasserman et al. (1962)
	Bacillomycin	*Bacillus subtilis*	Peypoux et al. (1984)
	Iturins	*Bacillus subtilis*	Maget-Dana and Peypoux (1994)
Lipoamino acids	*N*-acyl tyrosine	Soil metagenome	Brady and Clardy (2000)
	N-acyl asparagine	*Bacillus pumilus*	Peypoux et al. (2004)
	Ornithine lipid	*Sinorhizobium meliloti*	Gao et al. (2004)
Polymeric	Emulsan	*Acinetobacter calcoaceticus RAG1*	Rosenberg et al. (1979)
	Liposan	*Candida lipolytica*	Cirigliano and Carman (1985)
Fatty acids/ membrane associated	Phospholipids	*Acinetobacter* sp.	Kappeli and Finnerty (1979)
	Glycerolipids	*Microbacterium* sp.	Palme et al. (2010)

6.2.1 Glycolipids

Glycolipids contain a carbohydrate hydrophilic head group attached to a hydrophobic fatty acid tail via a glycosidic bond and vary according to which carbohydrate is attached. The most studied glycolipids belong to the rhamnolipid (RL), sophorolipid (SL), and mannosylerythritol lipid (MEL) classes due to their high production titers and potential applications. Alkyl glycosides are glycolipids synthesized in vitro by enzymatic or chemical catalysis. Though these molecules are not fermented products of bacteria, metagenomics can play a role in identifying suitable biocatalysts for their synthesis.

SL and MEL biosurfactants are produced by the yeast *Starmerella bombicola* (previously known as *Candida bombicola*) and *Pseudozyma* sp., respectively. Since metagenomics targets the prokaryotic fraction of the microorganisms in the sample, it is unlikely that these biosurfactants would feature as potential targets of functional or sequence-based metagenomic screening, unless an as yet unknown bacterial biosynthetic pathway for these biosurfactants exists. Rhamnolipids produced by members of the *Pseudomonas* genus are the most studied bacterial glycolipid, with its complex regulatory system being well elucidated (Müller and Hausmann 2011; Reis et al. 2011). Other less studied glycolipids have metabolic roles in cells; however, their properties in the field of biosurfactants appear unexplored, possibly due to their low production amounts.

Rhamnolipids are anionic glycolipids composed of one or two L-rhamnose molecules linked via a β-glycosidic bond to one or two hydrophobic β-hydroxy fatty acids (Jarvis and Johnson 1949). It is a secondary metabolite produced as a virulence factor

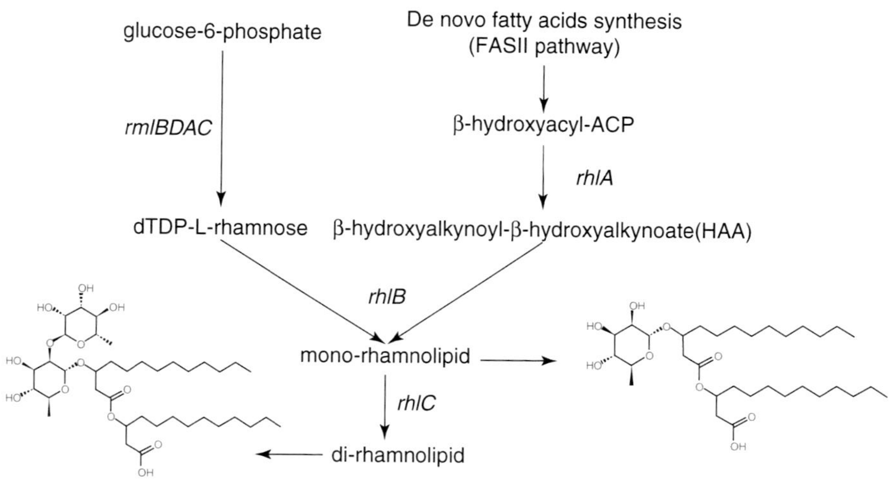

Fig. 6.1 Rhamnolipid biosynthesis pathway of *Pseudomonas aeruginosa* (Soberón-Chávez et al. 2005)

in *P. aeruginosa* lung infections. The rhamnose is synthesized (Fig. 6.1) from glucose-6-phosphate by the five-enzyme *rmlDBAC* biosynthetic pathway (Aguirre-Ramírez et al. 2012), and the β-hydroxy fatty acid is siphoned off the de novo fatty acid synthesis pathway (fatty acid synthesis type II (FASII)) (Zhu and Rock 2008). Both of these pathways are common among Gram-negative bacteria with the rhamnose pathway providing the substrate for lipopolysaccharide synthesis for outer membranes. Biosynthesis of rhamnolipids follows three enzymatic reactions to form either mono-rhamnolipids or di-rhamnolipids (Fig. 6.1). The first step entails the linking of two 3-hydroxyacyl-ACP molecules by the acyltransferase, RhlA, to form a dimer 3-(3-hydroxyalkanoyloxy) alkanoate (HAA) (Ochsner et al. 1994; Deziel 2003; Zhu and Rock 2008). The second step entails the condensation of HAA with dTDP-L-rhamnose by rhamnosyltransferase I (RhlB) to form mono-rhamnolipid. A third optional reaction adds a second rhamnose moiety by rhamnosyltransferase II (RhlC) to form di-rhamnolipids (Rahim et al. 2001). Rhamnolipid biosynthesis is tightly controlled by quorum sensing signaling (Reis et al. 2011).

6.2.2 Lipopeptides

Lipopeptides are linear or cyclic polypeptide chains attached to β-hydroxy fatty acid nonpolar tails of varying lengths. Lipopeptides feature strongly in antibiotic research due to their powerful cell membrane disruption properties. The most studied of the lipopeptide producers are *Bacillus* and *Pseudomonas* species, with *Bacillus subtilis* the producer of the most potent lipopeptide known, surfactin (Peypoux et al. 1999) (Fig. 6.2). Lipopeptides are a subset of natural products synthesized by large multifunctional enzyme clusters called non-ribosomal peptide synthases (NRPS) (Finking and Marahiel 2004). The enzymes form an assembly line made up of one or more

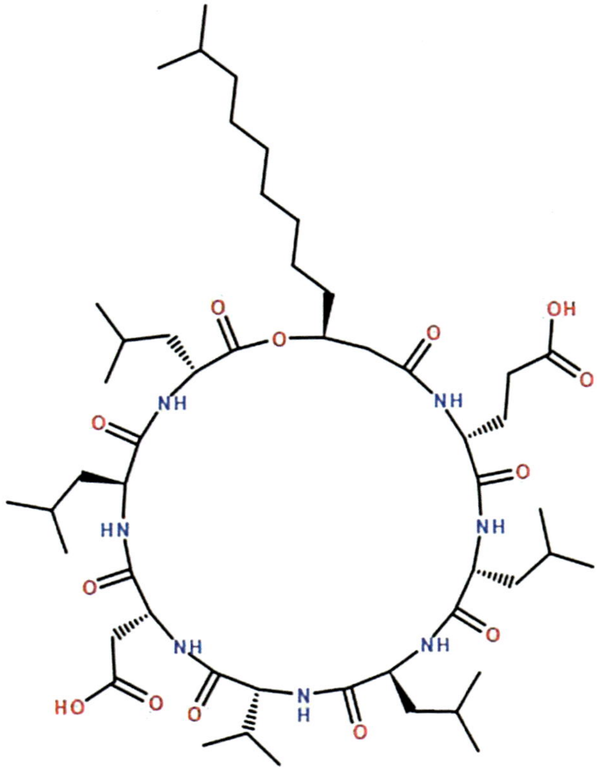

Fig. 6.2 Surfactin is a heptapeptide chain acylated with a fatty acid chain

clusters with each cluster arranged in modules. The modules are arranged in a canonical conformation consisting of a condensation (C) domain, adenylation (A) domain, and peptidyl carrier protein (PCP) domain, with a terminal thioesterase (TE) domain for peptide release and optional peptide chain cyclization at the end of the assembly line. Modules are subdivided into initiation and elongation modules (Finking and Marahiel 2004). Initiation modules in non-LP NRPS consist of an A-domain that selects and activates the amino acid residue and PCP domain, with no C-domains. In lipopeptides, the initiation domain contains a C-domain, which is necessary for attachment of the fatty acid nonpolar tails to the peptide molecule due to its function in lipoiniation (Kraas et al. 2010). The presence of a C-domain in the initiation module is an important feature that can be used to discriminate between lipopeptide- and non-lipopeptide-producing NRPS.

6.2.3 Lipoamino Acids

Lipoamino acids are amino acid residues attached by an amide bond to long-chain fatty acids. Despite consisting of a single amino acid, lipoamino acids still retain their biosurfactant activity (Peypoux et al. 2004). The family of lipoamino acids (Fig. 6.3)

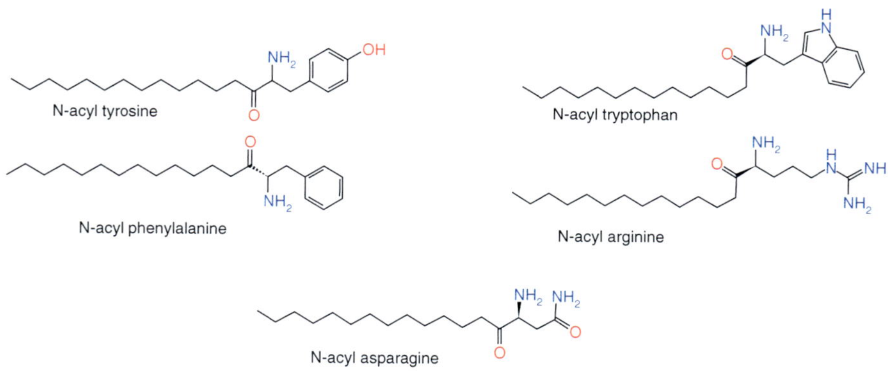

Fig. 6.3 The long-chain *N*-acyl amino acids, *N*-acyl tyrosine, tryptophan, phenylalanine, and arginine were identified by functional screening of metagenomic soil libraries (Brady et al. 2004; Brady and Clardy 2005a). The structures show the typical surfactant structural units, namely, a hydrophilic head, an amino acid residue, and a hydrophobic fatty acid tail. *N*-acyl asparagine was identified from cultured *Bacillus pumilus* (Peypoux et al. 2004), and its biosurfactant activity was determined confirming that these molecules are capable biosurfactants

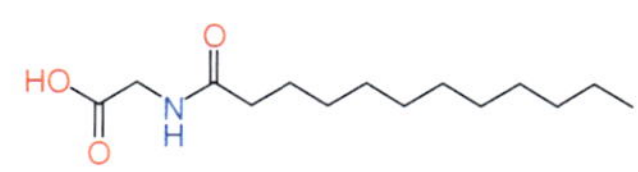

Fig. 6.4 The synthetic *N*-acyl sarcosinate that is structurally similar to *N*-acyl amino acids. This molecule has been used as an emulsifier in commercial formulations of toothpaste and cosmetics, perhaps indicating the potential uses of metagenomic-derived *N*-acyl amino acids

includes tyrosine, asparagine, tryptophan, phenylalanine, and arginine attached to a single fatty acid chain, whose production is catalyzed by *N*-acyl amino acid synthase (NAS) (Brady and Clardy 2000; Peypoux et al. 2004; Brady and Clardy 2005a). While ornithine lipids are structurally related to the *N*-acyl amino acids, they are synthesized by the unrelated enzymes, OlsA, and in a two-step pathway that adds a second fatty acid moiety to the amino lipid (Vences-Guzmán et al. 2012). While lipoamino acids do not appear to have significant interest in the biosurfactant field, similar molecules that have been chemically synthesized, such as *N*-acyl sarcosine, have long been used as emulsifiers in cosmetics and toothpaste (Fig. 6.4). Therefore, there is potential for these structures to be adopted for commercialization. For example, fatty-acyl-glutamate (FA-Glu), a surfactin variant, has been successfully produced using a genetically engineered strain of *Bacillus subtilis* by using a module from the NRPS biosynthetic gene cluster for surfactin (Marti et al. 2015). Similar to surfactin, it has fatty acid chains of 12–17 carbons; however, instead of the heptapeptide moiety, it has only a single glutamic acid residue. FA-Glu reduces the surface tension of water, and although it is less toxic than surfactin and has a lower dispersion-to-oil ratio, its lower toxicity makes it a candidate for use in oil spill remediation (Marti et al. 2015).

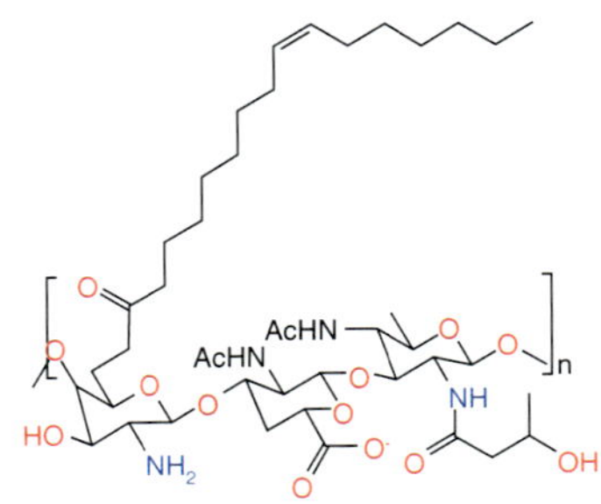

Fig. 6.5 The structure of the polymer biosurfactant commercially adopted, emulsan. The structure consists of the three-unit polysaccharide backbone with fatty acids attached to both amide and ester bonds

6.2.4 *Polymeric Biosurfactants*

Emulsan is an amphipathic, polyanionic, high molecular mass bioemulsifier that stabilizes oil in water emulsions. It consists of a three-unit polysaccharide backbone of *N*-acyl D-galactosamine, *N*-acyl L-galactosamine uronic acid, and 2,4-diamino-6-deoxy-D-glucosamine where the fatty acids can form both amide and ester bonds along the backbone to form an amphipathic compound (Fig. 6.5). Emulsan is produced by *Acinetobacter* species, most prominently *Acinetobacter lwoffii* RAG1. The polymer accumulates on the cell surface of RAG1 to form a mini-capsule during exponential phase. During the stationary phase of growth, an exocellular esterase releases the protein polysaccharide complex into the medium (Sen 2010). The 27 kb gene cluster for the biosynthesis of emulsan consists of 20 open reading frames (ORFs) known as the *wee* regulon. It encodes proteins whose functions include production of nucleotide amino sugar precursors, transglycosylation, transacetylation, polymerization, and transport (Nakar 2001). The *wee* regulon genes have closely related relatives that do not have any relation to biosurfactant synthesis.

6.2.5 *Alkyl Glycosides and Alkyl Polyglycosides*

Alkyl glycosides (AGs) and polyglycosides (APGs) are nonionic surfactants (Fig. 6.6), whose biodegradability and skin compatibility have led them to find commercial applications in detergents, cosmetics, and pesticide formulations (Van Rantwijk et al. 1999). Chemical production of APGs occurs by the Fischer glycosylation reaction, either through direct synthesis or transacetylation (Rather and Mishra 2013). The Fischer reaction results in multiple species of APGs with either more than one sugar group on the hydrophilic end or alkyl chains of different lengths and amounts attached to the glycosyl moiety (Rather and Mishra 2013). The chemical route to produce anomerically pure APGs involves extreme temperatures and pressures and use of toxic catalyst with multiple protection, deprotection, and activation steps making the process difficult and costly (Van Rantwijk et al. 1999; Rather and Mishra 2013). Enzymatic synthesis of AGs results in a single

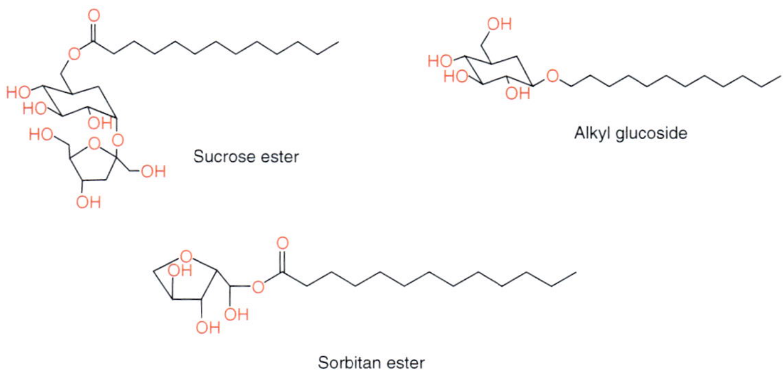

Fig. 6.6 Commercially available alkyl glycosides (von Rybinski and Hill 1998)

species and thus does not require the processes to produce anomerically pure compounds.

Alkyl glycoside synthesis is catalyzed by glycosidases under low water activity (a_w) conditions where the hydrolytic action of the glycosidases is reversed, to form rather than break glycosidic bonds. The glycosidic bonds can either form between mono- and oligosaccharides producing polysaccharides or between monosaccharides and fatty alcohols producing alkyl glycoside glycolipid surfactants. Enzymatic synthesis of AGs proceeds like chemical synthesis by two reaction mechanisms, reverse hydrolysis or transglycosylation (Van Rantwijk et al. 1999; Rather and Mishra 2013).

6.3 Metagenomics for the Discovery of Novel Biosurfactant Molecules

The majority of microorganisms in the environment have specific culturing requirements that are not met by standard culturing techniques. An estimated 95–99% of the bacterial population remains uncultured, representing the bulk of bacterial genetic and chemical diversity (Rondon et al. 2000; Handelsman et al. 2002; Rappé and Giovannoni 2003). Metagenomics allows access to the uncultured majority of microorganisms by extracting total environmental DNA (eDNA) directly from the sample, which could originate, for example, from soil (Zhou et al. 1996), sponge bacterial endosymbionts (Gurgui and Piel 2010), or gut microflora (Gloux et al. 2007). Given the number and diversity of biosurfactant-producing microorganisms among cultured isolates, it is expected that exploring the even larger uncultured fraction of the microbial population by using metagenomics can contribute significantly to new biosurfactant discoveries.

There are two general approaches to analyzing and exploiting metagenomic DNA, sequence-based and functional activity screening. Sequence-based metagenomics analyzes the genes or gene clusters either by direct sequencing using next-generation sequencing techniques or by PCR amplification that identifies regions with significant homology to known genes or biosynthetic pathways or conforms to predetermined patterns that would indicate that the sequence is a biosynthetic pathway. Analysis of genomes uploaded to online databases is also considered sequence homology metagenomics, where biosynthetic pathways of these genomes could be analyzed by online programs such as antiSMASH (Weber et al. 2015). In functional metagenomics, randomly sheared environmental DNA is cloned into a vector that is propagated in a bacterial host such as *Escherichia coli.* Using the native promoters within the eDNA insert, a clone producing an activity is detected by a reporter assay or physical effect. Many well-established surfactant screening methods are amenable to high-throughput screening (Walter et al. 2010) and can be adapted to perform functional screening on large environmental clone libraries.

6.3.1 Functional Metagenomics

Functional metagenomics identifies genes or pathways in an environmental clone library by their functional activity and offers three major advantages over sequence-based screening, mostly due to the fact that no prior knowledge of the gene sequence for the target activity is required (Tuffin et al. 2009; Suenaga 2012): (1) Given that biosurfactants span several different classes structurally and genetically, screening for biochemical properties might be expected to more easily yield new structures. (2) Furthermore, it is expected that a functional-based approach will increase the potential of identifying entirely new classes of biosurfactants. (3) Identification of an activity represents guaranteed success of expression in the chosen heterologous host, which will facilitate downstream production and analyses. Several well-established screening methods that are adapted to high-throughput biosurfactant activity discovery are available to identify a biosurfactant-producing clone.

6.3.1.1 Methods of Functional Screening for Biosurfactant Activity

The diversity of biosurfactant properties does not allow the direct detection of the molecules themselves. Rather the physical changes surfactants have on liquid media are the only general indication for the production of biosurfactants.

Atomized Spray Method

The atomized spray method (Burch et al. 2010) takes advantage of a surfactant's effect on droplets of liquid, such as paraffin. A thin mist of paraffin is sprayed over bacterial colonies arrayed on agar plates. The presence of surfactant is detected by the formation of a droplet halo around the colony (Fig. 6.7). The atomized paraffin spray is superior to the drop collapse method as it can detect biosurfactant production at higher sensitivities as well as hydrophobic biosurfactants (Burch et al. 2011).

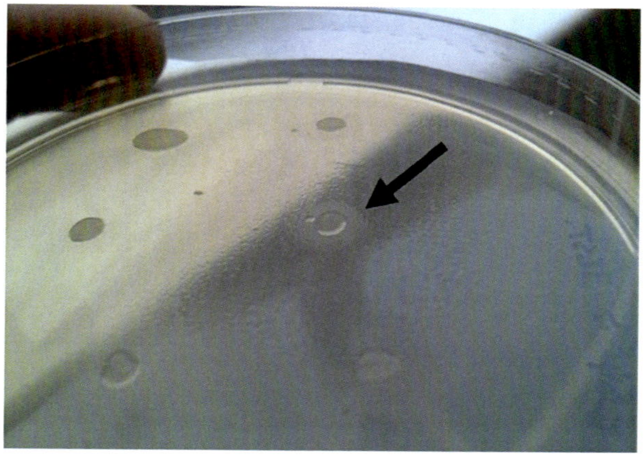

Fig. 6.7 A biosurfactant-producing *P. putida* colony from a metagenomic clone library displaying the halo (*arrow*) after being sprayed with paraffin (unpublished data)

By simply spraying arrayed colonies on an agar plate, the atomized spray method is ideally suited to identify biosurfactant production in metagenomic clone libraries that could comprise hundreds of thousands of clones.

The Drop Collapse Test

The drop collapse test determines a reduction in surface tension by testing the stability of a droplet on an oil-covered surface. Droplets containing surfactants collapse due to the lowering of the surface tension while non-surfactant-containing droplets remain beaded (Tugrul and Cansunar 2005). An important caveat is not all surfactants reduce surface tension that results in droplet collapse; the low solubility of hydrophobic biosurfactants cannot be detected in liquid media (Burch et al. 2011). Nevertheless, the drop collapse method is a high-throughput method that can be applied at large scales on the culture supernatant of metagenomic clone libraries to detect biosurfactants capable of reducing surface tension.

Microwell Plate Assay

This is a patented method (Vaux and Cottingham 2007) that qualitatively measures culture supernatant for surface activity. Pure water in a hydrophobic microwell has a flat surface. The edges of wells with surfactant-containing solutions will be wetted resulting in a concave surface that acts as a divergent lens (Walter et al. 2010). The plate is viewed using a backing sheet of paper, and an optical distortion is an indication of the presence of surfactants (Fig. 6.8). This method is amenable for efficient high-throughput screening for biosurfactants (Chen et al. 2007) therefore suitable as a functional metagenomic screening method.

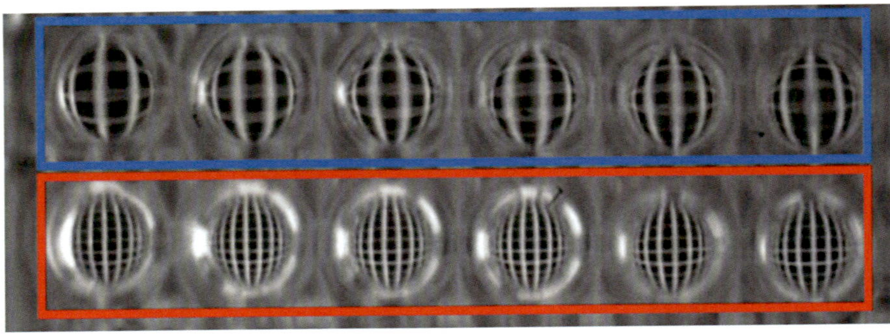

Fig. 6.8 The microwell plate assay where culture supernatant of the biosurfactant-producing microorganism (*red*) produces a distorted pattern when viewed with a grid backing sheet versus a non-biosurfactant-producing microorganism (*blue*) (Vaux and Cottingham 2007; Chen et al. 2007)

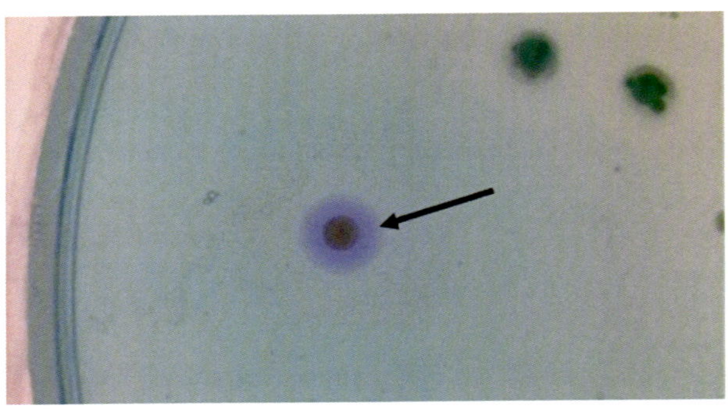

Fig. 6.9 The blue halo (indicated by the *arrow*) surrounds the biosurfactant-producing *P. aeruginosa* on CTAB-methylene blue agar indicating the production of rhamnolipids, an anionic biosurfactant (Pinzon and Ju 2009)

CTAB-Methylene Blue

The CTAB-methylene blue agar method (Siegmund and Wagner 1991; Pinzon and Ju 2009) detects the production of anionic surfactants that form an insoluble ion pair with the cationic surfactant (CTAB) and the dye (methylene blue) (Walter et al. 2010). The presence of an anionic surfactant is detected by a blue halo surrounding a bacterial colony (Fig. 6.9). The CTAB is toxic to many bacteria including *E. coli*, which limits its application to libraries prepared in tolerant hosts such as *P. putida* or *Streptomyces*.

6.3.1.2 Successful Identification of Biosurfactant-Like Molecules by Functional Metagenomics

Very recently it was reported that there is a surprising dearth of biosurfactant molecules discovered through metagenomic-based approaches and in only one occurrence has functional screening of a shotgun metagenomic clone library yielded a biosurfactant-producing clone (Jackson et al. 2015). However, we present here several examples of molecules discovered and described as antimicrobial compounds which bear biosurfactant-like properties. The *N*-acyl amino acids (Brady and Clardy 2000; Brady and Clardy 2005a) and palmitoyl putrescine (Brady and Clardy 2004) were isolated from soil metagenomic clone libraries screened in *E. coli* using antibacterial screening assays. Long-chain *N*-acyl amino acids (Fig. 6.3) are synthetized by the condensation of an ACP-activated fatty acid to the primary amine of an amino acid catalyzed by *N*-acyl amino acid synthase (NAS) (Brady et al. 2004). Additional members of the NAS family could not be identified by sequence-based strategies because they showed little overall similarity. The only feasible option was using functional activity to identify members from metagenomic libraries.

Therefore, the identification of *N*-acyl amino acids and palmitoyl putrescine could be considered as "coincidental" biosurfactant success stories and demonstrates the powerful advantage of employing the functional metagenomic approach to identify novel biosynthetic pathways and activities. Antibacterial activity is not a specific nor a universal test for biosurfactant activity as not all biosurfactants are antibacterial (Peypoux et al. 2004; Maneerat et al. 2006). In several functional metagenomic studies, several non-biosurfactant antimicrobial and pigment molecules (Fig. 6.9) were also identified through screening for antibacterial activity (Wang et al. 2000; Brady et al. 2001; Lim et al. 2005; Brady and Clardy 2005b; Fujita et al. 2011; He et al. 2012) (Fig. 6.10).

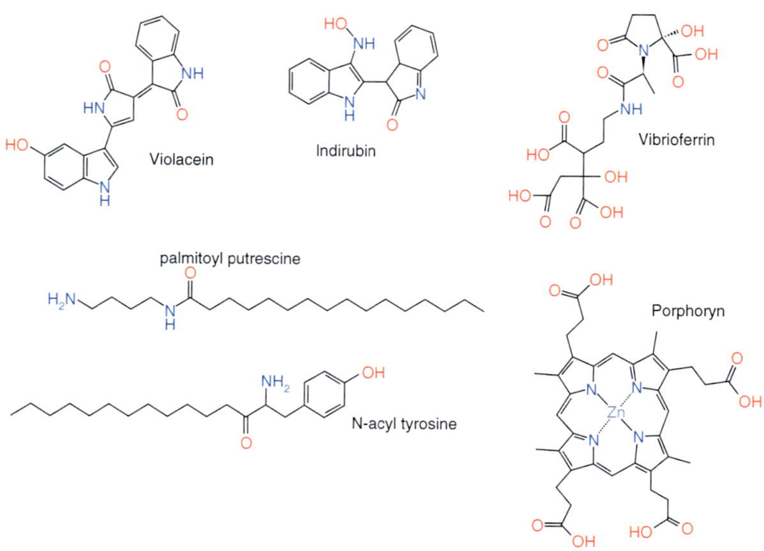

Fig. 6.10 Secondary metabolites identified by functional metagenomic screening. The molecules were either identified by antibacterial activity or by the production of pigment. Their biosynthetic pathways were elucidated by sequencing the environmental DNA insert of their respective clones (Brady and Clardy 2005a, b; Brady et al. 2001; Fujita et al. 2011; Lim et al. 2005; Wang et al. 2000)

6.3.1.3 Challenges Associated with Functional Metagenomics

A major technical challenge in functional metagenomics is that most genes in the coding metagenomic DNA cannot be accessed by the library host's transcriptional machinery. It is possible that the host cannot provide the substrates required to produce the molecule by that biosynthetic pathway; or it is unable to correctly fold the proteins of that pathway. Furthermore, the entire biosynthetic pathway comprising of multiple genes must be situated in a single metagenomic clone if the activity is to be detected. As the insert sizes increase, the probability that the entire pathway can be captured decreases. In addition, the number of clones present within a library should be large enough to adequately represent the total metagenome; depending on the nature of the community and the library, a representative library could be on the order of 10^6 clones. Libraries generally vastly underrepresent the actual diversity due to the difficulty in generating and screening such library sizes (Freeman et al. 2012; Trindade et al. 2015). Below are some considerations specific to biosurfactant discovery which can be employed in order to overcome some of the limitations.

Parallel Screening in Multiple Metagenomic Library Hosts

As the biosynthesis of rhamnolipids indicates, a complex regulatory signaling system can control the expression of the biosynthetic operon that is only activated at specific cellular conditions. In addition a supply of substrate is also required for the expressed enzyme complex to use to synthesize the rhamnolipid molecule. Therefore, for an activity to be detected, it is critical that all the genes responsible for synthesis are found within one clone and a requirement that the host provide building blocks for that biosynthetic pathway. Without the appropriate cellular environment, transcriptional machinery, and metabolic potential, the pathway will remain silent. Most functional metagenomic studies are conducted in *E. coli* as the heterologous host of the environmental library. Together with an estimated 40% of genes which are able to be heterologously expressed in *E. coli* (Gabor et al. 2004), the range of the environmental pathways potentially being recognized by the *E. coli's* transcriptional machinery is limited. Thus, for a rhamnolipid to be identified through screening a hypothetical metagenomic library in *E. coli*, a constitutive promoter for the *rhlAB* operon is required, as well as the addition of the *rmlDBAC* pathway to provide the dTDP-L-rhamnosyl moiety as the substrate for RhlB, the rhamnosyltransferase (Cabrera-Valladares et al. 2006). In short, it is not possible to detect rhamnolipid production in metagenomic libraries constructed solely in *E. coli* and, by extension, other unknown biosynthetic pathways.

In this regard, the multiple host screening of metagenomic libraries is a necessary strategy to identify pathways that would otherwise go undetected. *Streptomyces lividans*, *Pseudomonas putida*, and *Bacillus subtilis* are potential host candidates. *Bacillus subtilis* and *Streptomyces* are known NRPS biosurfactant producers. *P. putida* KT2440 is not a known biosurfactant producer, but other strains of *P. putida* are, and it belongs to the genus of prolific producers such as *P. aeruginosa*, thus making it a suitable candidate for the production of biosurfactants heterologously

(Wittgens et al. 2011). In addition, *P. putida* was found to be resistant to the antimicrobial effect of high concentrations of biosurfactants. Conveniently, shuttle vectors for the transfer and parallel screening of metagenomic libraries in all of these hosts have been developed (Martinez et al. 2004; Troeschel et al. 2012). While *P. putida* KT2440 does not have background biosurfactant activity, both *Streptomyces lividans* and *Bacillus subtilis* would require pathway silencing to remove the background biosurfactant activity that will confound any screen for biosurfactants. This approach has had success in identifying biosurfactant activity in a metagenomic library that was parallel screened in *E. coli* and *P. putida*, but activity was only identified in the latter host (Jackson et al. 2015).

Improvements in Large Insert Size Cloning Will Result in Greater Variety of Biosynthetic Pathways

The library size has implications for biosurfactant screening in two different ways. Cosmid and fosmid libraries with insert sizes of an average of 30 kb can be achieved relatively easily and successfully screened. However, in order to represent the total metagenome, millions of clones are required, and this is rarely obtained. The second consideration is the insert size. The larger the pathway, the less likely a randomly cloned environmental DNA insert will contain the entire pathway (Gabor et al. 2004). Many biosynthetic pathways can be exceedingly large, ranging above 40 kb in size comprising of multiple genes in a cluster (Walton 2000; Miao et al. 2005; Kim et al. 2010; Laureti et al. 2011). The lipopeptide antibiotic, for example, friulimicin's biosynthetic pathway, is 70 kb and required three overlapping cosmids to reconstruct (Müller et al. 2007). Due to the size limitation, the potentially novel biosurfactant pathway could go undetected using functional metagenomics.

Since the entire pathway needs to be cloned into a single metagenomic clone to detect functional activity, the large pathway size can significantly increase the number of clones required to find that pathway. Large insert libraries of up to 100 kb are thus required, which can be achieved using bacterial artificial chromosomes and would also reduce the number of clones to be screened. However, technical difficulties when working with high molecular weight DNA and cloning into BACs limit the creation of BAC libraries (Handelsman et al. 1998; Kakirde et al. 2010). Compared to natural product discovery of single-gene enzymes, where the rate of discovery is about one active clone per 1000 clones screened (Kakirde et al. 2010; Ferrer et al. 2015), the number of clones required to find a hit for large biosynthetic pathways can be from 1 in 10,000 to 20,000 clones (Brady et al. 2004).

6.3.2 Sequence-Based Metagenomics

Sequence-based approaches rely on shotgun sequencing or PCR to parse metagenomic DNA sequences for biosynthetic content, after which complete biosynthetic pathways can be recovered from the library of clones and expressed in a heterologous host (Kim et al. 2010; Owen et al. 2013; Charlop-Powers et al. 2014). A major advantage of this approach is that the entire pathway is not required to be identified during a screen for biosurfactant synthetic pathways. Fragments of the pathway can be identified in disparate clones, and overlapping regions can be reconstructed from the fragments into one functional unit (Kim et al. 2010; Loeschcke et al. 2013; Montiel et al. 2015). As a result the sequence-based approach has been the predominant method for identifying large secondary metabolite pathways from metagenomic DNA. Exceedingly large biosynthetic clusters would be better suited to sequence guided metagenomics rather than functional metagenomics as the number of clones that would need to be screened to properly ensure success might be out of the capabilities of the laboratory. However, given the structural and biosynthetic diversity of biosurfactants, there is no universal sequence-based screen to detect biosurfactants, excluding lipopeptides. Where the biosynthetic pathways are modular, the identification of conserved regions by PCR is especially useful. This is the case for lipopeptide biosurfactants as they are encoded by NRPS gene clusters, which are modularly arranged. Thus, PCR-based screening enables the identification of a diverse group of secondary metabolites while only targeting a few conserved regions within the gene cluster. Typically the adenylation domain of the synthetase gene is targeted (Owen et al. 2013), but this would detect both LP and non-LP biosynthetic clusters (Kim et al. 2010; Owen et al. 2013) and therefore will lead to false positives (Owen et al. 2013). Considering the differences between LP and non-LP lipopeptides, where the initiation domains are distinct, it should enable specific targeting of only LP biosynthetic pathways by a sequence-based screening method. LP biosynthetic clusters were identified using the degenerate PCR approach in *Pseudomonas* species (Rokni-Zadeh et al. 2011). While only tested on cultured bacteria, the method is ideal for the detection of LPs in environmental DNA clone libraries.

For biosynthetic pathways that are not modular, such as the glycolipid pathways, a sequence-based strategy is not viable, since there isn't a single universal conserved domain that can be targeted for screening. In glycolipid biosynthesis, for example, glycosyltransferases (GTs) are central for the transfer of glycosyl moieties onto their fatty acid hydrophobic structural units. GTs themselves have multiple functions in bacteria that include cell wall composition, energy storage, signaling, protein glycosylation (Coutinho et al. 2003), or flavonoid synthesis (Rabausch et al. 2013). Thus, targeting these genes might result in false positives of biosurfactant biosynthesis ability (discussed further in Sect. 6.3.3).

6.3.2.1 Biosynthetic Pathway Reconstruction

Identifying fragments of unique biosynthetic pathways in clone libraries is not particularly informative of the chemical properties and activity of their product. Expression of the entire pathways is required and thus necessitates the identification of the overlapping fragments of the pathway, reconstructing as a contiguous clone and its expression in a suitable heterologous host. Retrieving fully intact biosynthetic pathways using PCR is possible as a library of sufficient size would contain the required redundancy for clones containing overlapping sequences to be regularly identified in a metagenomic library. When empirically determined, ten million unique clones were required for sufficient numbers of redundant type II PKS sequences to be identified for the reconstruction of a 40 kb pathway (Kim et al. 2010). However, the number of clones required might be dependent on the diversity of inserts in the library that would increase the number of clones required for redundancy. Once the pathway is reconstructed, a suitable host as well as the appropriate promoters is required to express clusters that would otherwise remain silent (Montiel et al. 2015). While the heterologous expression of genes involved in biosurfactant production has been achieved (Jackson et al. 2015), this is a rather "hit and miss" process which is subject to all the same limitations presented for functional-based screening strategies.

6.3.3 Alternative Methods for Glycolipid Synthesis Could Be Augmented by Metagenomic-Derived Biocatalyst

Biosynthetic pathways of glycolipids have glycosyltransferase as the central enzyme responsible for the condensation of the acyl and glycosyl moieties (Bourne and Henrissat 2001; Coutinho et al. 2003). Glycosyltransferases belong to the group called Leloir glycosyltransferases which are highly efficient natural glycosylation catalysts (De Bruyn et al. 2015). They catalyze the glycosylation of acyl moieties using nucleotide-activated sugar substrates (Desmet et al. 2012; Bissaro et al. 2015). As an in vitro biocatalyst for the formation of glycolipids, glycosyltransferases are not particularly useful as the nucleotide-activated sugar substrates are expensive to manufacture, which hampers their adoption in the industry (Desmet et al. 2012). Identification of individual glycosyltransferases would require the heterologous host to produce a pool of nucleotide-activated sugars.

An alternative to identifying glycosyltransferases to produce glycolipids is using β-glycosidases for the in vitro production of glycolipids by reverse hydrolysis or transglycosylation. Glycosidases are popular targets of functional metagenomic screening (Ferrer et al. 2015). At this moment there is no literature describing the use of metagenomic-derived glycosidases to produce AGs. An ideal catalyst for the synthesis of AGs should be thermostable, low a_w, and solvent tolerant. Monosaccharides are relatively insoluble at low temperatures in organic solvents; therefore, elevated reaction conditions are required to dissolve the substrates. A general rule for thermostable enzymes is that they are conveniently also solvent tolerant, thus would make ideal biocatalysts for reverse hydrolysis reactions. In the search for the ideal catalyst, metagenomic screening could tap into a wide diversity of unexplored glycosidases. As yet, metagenomic screening for the reverse hydrolysis or transglycosylation has not been reported and thus constitutes a novel screening approach. Furthermore, no metagenomic-derived glycosidase has been employed for the synthesis of AGs, despite the isolation of several thermostable β-glucosidases from a number of metagenomes (Wang 2012; Schröder 2014). Just as with glycosidases from cultured bacteria and other sources, such as almonds, there is potential for a wide variety of glycolipid molecules to be produced by metagenome-derived β-glycosidases (Rather and Mishra 2013).

6.4 Conclusion

Employing a metagenomic approach to identify biosurfactant secondary metabolites is still in its infancy. In a recent review, the authors proposed that the dearth of biosurfactant molecules discovered through metagenomic-based approaches is likely due to a lack of research interest. In this chapter we present a number of novel metagenomically discovered compounds which remained unknown as biosurfactants since they were originally described as antibacterial compounds. Thus, the tools, more so for functional-based screening than sequence-based approaches, are available for novel molecules with biosurfactant properties to be discovered using metagenomics. Nonetheless, there are major obstacles to metagenomic biosurfactant biodiscovery. The biosynthetic potential of the microorganisms selected for library screening and to heterologously produce the compound, the recognition of transcriptional elements by the host, and the ability to clone large insert sizes and the assay to detect the compound in a high-throughput system present a variety of limitations which need concerted consideration. Parallel screening in multiple bacterial hosts allows a diverse metabolic environment for the heterologous production of a secondary metabolite of a metagenomic clone library. BACs, cosmids, and fosmids enable large contiguous sections of environment DNA to be cloned, which might hold entire biosynthetic pathways. If pathways are too large, as is the case for modular gene clusters such as NRPS, sequence-based strategies such as targeting the C-domain in the initiation module of lipopeptides are available. Finally, already well-established assays to detect biosurfactant activity can be adapted to be high throughput and applicable for functional metagenomic screening for biosurfactant-producing clones. As yet, there are few biosurfactant molecules discovered by the metagenomic approach, and nearly all have been isolated from cultured microorganisms. Therefore, there is significant potential to discover new biosurfactant molecules from the large uncultured fraction of microorganisms.

References

Aguirre-Ramírez M, Medina G, González-Valdez A et al (2012) The Pseudomonas aeruginosa rmlBDAC operon, encoding dTDP-L-rhamnose biosynthetic enzymes, is regulated by the quorum-sensing transcriptional regulator RhlR and the alternative sigma factor σS. Microbiology 158:908–916

Arima K, Kakinuma A, Tamura G (1968) Surfactin, a crystalline peptidelipid surfactant produced by Bacillus subtilis: isolation, characterization and its inhibition of fibrin clot formation. Biochem Biophys Res Commun 31:488–494
Bissaro B, Monsan P, Faure R, O'Donohue M (2015) Glycosynthesis in a waterworld: new insights into the molecular basis of transglycosylation in retaining glycoside hydrolases. Biochem J 467:17–35
Bourne Y, Henrissat B (2001) Glycoside hydrolases and glycosyltransferases: families and functional modules. Curr Opin Struct Biol 11:593–600
Brady SF, Clardy J (2000) Long-chain N-acyl amino acid antibiotics isolated from heterologously expressed environmental DNA. J Am Chem Soc 122:12903–12904
Brady SF, Clardy J (2004) Palmitoylputrescine, an antibiotic isolated from the heterologous expression of DNA extracted from bromeliad tank water. J Nat Prod 67:1283–1286
Brady SF, Clardy J (2005a) N-acyl derivatives of arginine and tryptophan isolated from environmental DNA expressed in Escherichia coli. Org Lett 7:3613–3616
Brady SF, Clardy J (2005b) Cloning and heterologous expression of isocyanide biosynthetic genes from environmental DNA. Angew Chem Int Ed Engl 44:7063–7065
Brady SF, Chao CJ, Handelsman J, Clardy J (2001) Cloning and heterologous expression of a natural product biosynthetic gene cluster from eDNA. Org Lett 3:1981–1984
Brady SF, Chao CJ, Clardy J (2004) Long-chain N-acyltyrosine synthases from environmental DNA. Appl Environ Microbiol 70:6865–6870
Burch AY, Shimada BK, Browne PJ, Lindow SE (2010) Novel high-throughput detection method to assess bacterial surfactant production. Appl Environ Microbiol 76:5363–5372
Burch AY, Browne PJ, Dunlap CA et al (2011) Comparison of biosurfactant detection methods reveals hydrophobic surfactants and contact-regulated production. Environ Microbiol 13:2681–2691
Cabrera-Valladares N, Richardson A-P, Olvera C et al (2006) Monorhamnolipids and 3-(3-hydroxyalkanoyloxy)alkanoic acids (HAAs) production using Escherichia coli as a heterologous host. Appl Microbiol Biotechnol 73:187–194
Charlop-Powers Z, Milshteyn A, Brady SF (2014) Metagenomic small molecule discovery methods. Curr Opin Microbiol 19C:70–75
Chen C-Y, Baker SC, Darton RC (2007) The application of a high throughput analysis method for the screening of potential biosurfactants from natural sources. J Microbiol Methods 70:503–510
Cirigliano MC, Carman GM (1985) Purification and characterization of liposan, a bioemulsifier from Candida lipolytica. Appl Environ Microbiol 50:846–850
Cochrane SA, Vederas JC (2014) Lipopeptides from Bacillus and Paenibacillus spp.: a gold mine of antibiotic candidates. Med Res Rev 36(1):1292–1327
Coutinho PM, Deleury E, Davies GJ, Henrissat B (2003) An evolving hierarchical family classification for glycosyltransferases. J Mol Biol 328:307–317
De Bruyn F, Maertens J, Beauprez J et al (2015) Biotechnological advances in UDP-sugar based glycosylation of small molecules. Biotechnol Adv 33:288–302
Desai JD, Banat IM (1997) Microbial production of surfactants and their commercial potential. Microbiol Mol Biol Rev 61:47–64
Desmet T, Soetaert W, Bojarová P et al (2012) Enzymatic glycosylation of small molecules: challenging substrates require tailored catalysts. Chemistry 18:10786–10801
Deziel E (2003) rhlA is required for the production of a novel biosurfactant promoting swarming motility in Pseudomonas aeruginosa: 3-(3-hydroxyalkanoyloxy)alkanoic acids (HAAs), the precursors of rhamnolipids. Microbiology 149:2005–2013
Ferrer M, Martínez-Martínez M, Bargiela R et al (2015) Estimating the success of enzyme bioprospecting through metagenomics: current status and future trends. Microb Biotechnol. doi:10.1111/1751-7915.12309
Finking R, Marahiel MA (2004) Biosynthesis of nonribosomal peptides. Annu Rev Microbiol 58:453–488

Freeman MF, Gurgui C, Helf MJ et al (2012) Metagenome mining reveals polytheonamides as posttranslationally modified ribosomal peptides. Science 338:387–390. doi:10.1126/science.1226121

Fujita MJ, Kimura N, Sakai A et al (2011) Cloning and heterologous expression of the vibrioferrin biosynthetic gene cluster from a marine metagenomic library. Biosci Biotechnol Biochem 75:2283–2287

Gabor EM, Alkema WBL, Janssen DB (2004) Quantifying the accessibility of the metagenome by random expression cloning techniques. Environ Microbiol 6:879–886

Gao JL, Weissenmayer B, Taylor AM et al (2004) Identification of a gene required for the formation of lyso-ornithine lipid, an intermediate in the biosynthesis of ornithine-containing lipids. Mol Microbiol 53:1757–1770

Geys R, Soetaert W, Van Bogaert I (2014) Biotechnological opportunities in biosurfactant production. Curr Opin Biotechnol 30:66–72

Gloux K, Leclerc M, Iliozer H et al (2007) Development of high-throughput phenotyping of metagenomic clones from the human gut microbiome for modulation of eukaryotic cell growth. Appl Environ Microbiol 73:3734–3737

Gorin PAJ, Spencer JFT, Tulloch AP (1961) Hydroxy fatty acid glycosides of sophorose from Torulopsis magnoliae. Can J Chem 39:846–855

Gurgui C, Piel J (2010) Metagenomic approaches to identify and isolate bioactive natural products from microbiota of marine sponges. In: Streit WR, Daniel R (eds) Metagenomics: methods and protocols, methods in molecular biology. Springer Science + Business Media, Berlin, pp 247–263

Handelsman J, Rondon MR, Brady SF et al (1998) Molecular biological access to the chemistry of unknown soil microbes: a new frontier for natural products. Chem Biol 5:R245–R249

Handelsman J, Liles M, Mann D, Riesenfeld C (2002) Cloning the metagenome: culture-independent access to the diversity and functions of the uncultivated microbial world. Methods Microbiol 33:241–255

He R, Wakimoto T, Takeshige Y et al (2012) Porphyrins from a metagenomic library of the marine sponge Discodermia calyx. Mol Biosyst 8:2334–2338

Henkel M, Müller MM, Kügler JH et al (2012) Rhamnolipids as biosurfactants from renewable resources: concepts for next-generation rhamnolipid production. Process Biochem 47:1207–1219

Jackson SA, Borchert E, O'Gara F, Dobson AD (2015) Metagenomics for the discovery of novel biosurfactants of environmental interest from marine ecosystems. Curr Opin Biotechnol 33:176–182

Jarvis F, Johnson M (1949) A glycolipid produced by Pseudomonas aeruginosa. J Am Chem Soc 71:4124–4126

Kakirde KS, Parsley LC, Liles MR (2010) Size does matter: application-driven approaches for soil metagenomics. Soil Biol Biochem 42:1911–1923

Kappeli O, Finnerty W (1979) Partition of alkane by an extracellular vesicle derived from hexadecane-grown Acinetobacter. J Bacteriol 140:707–712

Kebbouche-Gana S, Gana ML, Ferrioune I et al (2013) Production of biosurfactant on crude date syrup under saline conditions by entrapped cells of Natrialba sp. strain E21, an extremely halophilic bacterium isolated from a solar saltern (Ain Salah, Algeria). Extremophiles 17:981–993

Kennedy J, O'Leary ND, Kiran GS et al (2011) Functional metagenomic strategies for the discovery of novel enzymes and biosurfactants with biotechnological applications from marine ecosystems. J Appl Microbiol 111:787–799

Kim JH, Feng Z, Bauer JD et al (2010) Cloning large natural product gene clusters from the environment: piecing environmental DNA gene clusters back together with TAR. Biopolymers 93:833–844

Kitamoto D, Akiba S, Hiok C, Tabuchi T (1990) Extracellular accumulation of mannosylerythritol lipids by a strain of Candida antarctica. Agric Biol Chem 54:31–36

Kraas FI, Helmetag V, Wittmann M et al (2010) Functional dissection of surfactin synthetase initiation module reveals insights into the mechanism of lipoinitiation. Chem Biol 17:872–880
Laureti L, Song L, Huang S et al (2011) Identification of a bioactive 51-membered macrolide complex by activation of a silent polyketide synthase in Streptomyces ambofaciens. Proc Natl Acad Sci U S A 108:6258–6263
Lim HK, Chung EJ, Kim J et al (2005) Characterization of a forest soil metagenome clone that confers indirubin and Indigo production on Escherichia coli. Appl Environ Microbiol 71:7768–7777
Loeschcke A, Markert A, Wilhelm S et al (2013) TREX: a universal tool for the transfer and expression of biosynthetic pathways in bacteria. ACS Synth Biol 2:22–33
Maget-Dana R, Peypoux F (1994) Iturins, a special class of pore-forming lipopeptides: biological and physicochemical properties. Toxicology 87:151–174
Maneerat S, Bamba T, Harada K et al (2006) A novel crude oil emulsifier excreted in the culture supernatant of a marine bacterium, Myroides sp. strain SM1. Appl Microbiol Biotechnol 70:254–259
Marti ME, Colonna WJ, Reznik G et al (2015) Production of fatty-acyl-glutamate biosurfactant by Bacillus subtilis on soybean co-products. Biochem Eng J 95:48–55
Martinez A, Kolvek SJ, Lai C et al (2004) Genetically modified bacterial strains and novel bacterial artificial chromosome shuttle vectors for constructing environmental libraries and detecting heterologous natural products in multiple expression host. Appl Environ Microbiol 70:2452–2463
Miao V, Coëffet-Legal M-F, Brian P et al (2005) Daptomycin biosynthesis in Streptomyces roseosporus: cloning and analysis of the gene cluster and revision of peptide stereochemistry. Microbiology 151:1507–1523
Montiel D, Kang H-S, Chang F-Y et al (2015) Yeast homologous recombination-based promoter engineering for the activation of silent natural product biosynthetic gene clusters. Proc Natl Acad Sci U S A 112(29):8953–8958. doi:10.1073/pnas.1507606112
Müller MM, Hausmann R (2011) Regulatory and metabolic network of rhamnolipid biosynthesis: traditional and advanced engineering towards biotechnological production. Appl Microbiol Biotechnol 91:251–264
Müller C, Nolden S, Gebhardt P et al (2007) Sequencing and analysis of the biosynthetic gene cluster of the lipopeptide antibiotic Friulimicin in Actinoplanes friuliensis. Antimicrob Agents Chemother 51:1028–1037
Müller MM, Kügler JH, Henkel M et al (2012) Rhamnolipids-next generation surfactants? J Biotechnol 161:366–380
Myers D (2010) Surfactant science and technology, 3rd edn. Wiley, Hoboken
Nakar D, Gutnick DL (2001) Analysis of the wee gene cluster responsible for the biosynthesis of the polymeric bioemulsifier from the oil-degrading strain Acinetobacter lwoffii RAG-1. Microbiology 147:1937–1946
Ochsner UA, Fiechter A, Reiser J (1994) Isolation, characterization, and expression in Escherichia coli of the Pseudomonas aeruginosa rhlAB genes encoding a rhamnosyltransferase involved in rhamnolipid biosurfactant synthesis. J Biol Chem 269:19787–19795
Owen JG, Reddy BVB, Ternei MA et al (2013) Mapping gene clusters within arrayed metagenomic libraries to expand the structural diversity of biomedically relevant natural products. Proc Natl Acad Sci U S A 110:11797–11802
Palme O, Moszyk A, Iphöfer D, Lang S (2010) Selected microbial glycolipids: production, modification and characterization. In: Sen R (ed) Biosurfactants. Springer, New York, pp 185–202
Pamp SJ, Tolker-Nielsen T (2007) Multiple roles of biosurfactants in structural biofilm development by Pseudomonas aeruginosa. J Bacteriol 189:2531–2539
Peypoux F, Pommier MT, Das BC et al (1984) Structures of bacillomycin D and bacillomycin L peptidolipid antibiotics from Bacillus subtilis. J Antibiot (Tokyo) 37:1600–1604

Peypoux F, Bonmatin JM, Wallach J (1999) Recent trends in the biochemistry of surfactin. Appl Microbiol Biotechnol 51:553–563

Peypoux F, Laprevote O, Pagadoux M, Wallach J (2004) N-acyl derivatives of Asn, new bacterial N-acyl D-amino acids with surfactant activity. Amino Acids 26:209–214

Pinzon NM, Ju L-K (2009) Improved detection of rhamnolipid production using agar plates containing methylene blue and cetyl trimethylammonium bromide. Biotechnol Lett 31:1583–1588

Rabausch U, Juergensen J, Ilmberger N et al (2013) Functional screening of metagenome and genome libraries for detection of novel flavonoid-modifying enzymes. Appl Environ Microbiol 79:4551–4563

Rahim R, Ochsner UA, Olvera C et al (2001) Cloning and functional characterization of the Pseudomonas aeruginosa rhlC gene that encodes rhamnosyltransferase 2, an enzyme responsible for di-rhamnolipid biosynthesis. Mol Microbiol 40:708–718

Rappé MS, Giovannoni SJ (2003) The uncultured microbial majority. Annu Rev Microbiol 57:369–394

Rather MY, Mishra S (2013) β-Glycosidases: an alternative enzyme based method for synthesis of alkyl-glycosides. Sustain Chem Process 1:7

Reis RS, Pereira AG, Neves BC, Freire DMG (2011) Gene regulation of rhamnolipid production in Pseudomonas aeruginosa: a review. Bioresour Technol 102:6377–6384

Rizzo C, Michaud L, Hörmann B et al (2013) Bacteria associated with sabellids (Polychaeta: Annelida) as a novel source of surface active compounds. Mar Pollut Bull 70:125–133

Rokni-Zadeh H, Mangas-Losada A, De Mot R (2011) PCR detection of novel non-ribosomal peptide synthetase genes in lipopeptide-producing Pseudomonas. Microb Ecol 62:941–947

Ron EZ, Rosenberg E (2001) Natural roles of biosurfactants. Environ Microbiol 3:229–236

Rondon MR, August PR, Bettermann AD et al (2000) Cloning the soil metagenome: a strategy for accessing the genetic and functional diversity of uncultured microorganisms. Appl Environ Microbiol 66:2541–2547

Rosen M (1989) Surfactants and interfacial phenomenon, Third. John Wiley and Sons, New Jersey

Rosenberg E, Zuckerberg A, Rubinovitz C, Gutnick DL (1979) Emulsifier of Arthrobacter RAG-1: isolation and emulsifying properties. Appl Environ Microbiol 37:402–408

Sen R (2010) Biosurfactants. Landes Bioscience and Springer Science, New York

Schröder C, Elleuche S, Blank S, Antranikian G (2014) Characterization of a heat-active archaeal β-glucosidase from a hydrothermal spring metagenome. Enzyme Microb Technol 57:48–54. doi:10.1016/j.enzmictec.2014.01.010

Siegmund I, Wagner F (1991) New method for detecting rhamnolipids excreted by Pseudomonas species during growth on mineral agar. Biotechnol Tech 5:265–268

Singer ME, Finnerty WR (1990) Physiology of biosurfactant synthesis by Rhodococcus species H13-A. Can J Microbiol 36:741–745

Soberón-Chávez G, Lépine F, Déziel E (2005) Production of rhamnolipids by Pseudomonas aeruginosa. Appl Microbiol Biotechnol 68:718–725

Suenaga H (2012) Targeted metagenomics: a high-resolution metagenomics approach for specific gene clusters in complex microbial communities. Environ Microbiol 14:13–22. doi:10.1111/j.1462-2920.2011.02438.x

Trindade M, van Zyl LJ, Navarro-Fernandez J, Elrazak AA (2015) Targeted metagenomics as a tool to tap into marine natural product diversity for the discovery and production of drug candidates. Front Microbiol 6:1–14

Troeschel SC, Thies S, Link O et al (2012) Novel broad host range shuttle vectors for expression in Escherichia coli, Bacillus subtilis and Pseudomonas putida. J Biotechnol 161:71–79

Tugrul T, Cansunar E (2005) Detecting surfactant-producing microorganisms by the drop-collapse test. World J Microbiol Biotechnol 21:851–853

Tuffin M, Anderson D, Heath C, Cowan D (2009) Metagenomic gene discovery: how far have we moved into novel sequence space? Biotechnol J 4:1671–1683

Van Rantwijk F, Woudenberg-Van Oosterom M, Sheldon RA (1999) Glycosidase-catalysed synthesis of alkyl glycosides. J Mol Catal B Enzym 6:511–532

Vaux DJ, Cottingham M (2007) Method and apparatus for measuring surface configuration. 2:1–13
Vences-Guzmán MÁ, Geiger O, Sohlenkamp C (2012) Ornithine lipids and their structural modifications: from A to E and beyond. FEMS Microbiol Lett 335:1–10
von Rybinski W, Hill K (1998) Alkyl polyglycosides—properties and applications of a new class of surfactants. Angew Chem Int Ed 37:1328–1345
Walter V, Syldatk C, Hausmann R (2010) Screening concepts for the isolation of biosurfactant producing microorganisms. In: Sen R (ed) Biosurfactants. Landes Bioscience and Springer Science, New York, pp 1–13
Walton JD (2000) Horizontal gene transfer and the evolution of secondary metabolite gene clusters in fungi: an hypothesis. Fungal Genet Biol 30:167–171
Wang GY, Graziani E, Waters B et al (2000) Novel natural products from soil DNA libraries in a streptomycete host. Org Lett 2:2401–2404
Wang Q, Qian C, Zhang XZ et al (2012) Characterization of a novel thermostable beta-glucosidase from a metagenomic library of termite gut. Enzyme Microb Technol 51:319–324. doi:10.1016/j.enzmictec.2012.07.015
Wasserman HH, Keggi JJ, McKeon JE (1962) The structure of Serratamolide. J Am Chem Soc 84:2978–2982
Weber T, Blin K, Duddela S et al (2015) antiSMASH 3.0--a comprehensive resource for the genome mining of biosynthetic gene clusters. Nucleic Acids Res 43:1–7
Wittgens A, Tiso T, Arndt TT et al (2011) Growth independent rhamnolipid production from glucose using the non-pathogenic Pseudomonas putida KT2440. Microb Cell Fact 10:80
Zhou J, Bruns M, Tiedje JM (1996) DNA recovery from soils of diverse composition. Appl Environ Microbiol 62:316–322
Zhu K, Rock CO (2008) RhlA converts beta-hydroxyacyl-acyl carrier protein intermediates in fatty acid synthesis to the beta-hydroxydecanoyl-beta-hydroxydecanoate component of rhamnolipids in Pseudomonas aeruginosa. J Bacteriol 190:3147–3154

Chapter 7
Challenges and Opportunities in Discovery of Secondary Metabolites Using a Functional Metagenomic Approach

Alinne L.R. Santana-Pereira and Mark R. Liles

Abstract Bioprospecting for natural products via a metagenomic approach has been highly successful for enzyme discovery, yet methodological challenges have inhibited discovery of diverse secondary metabolites from environmental metagenomes. In this chapter, we discuss metagenomic approaches to identify and/or express secondary metabolites encoded from environmental DNA. The application of next-generation sequencing techniques has generated enormous metagenomic sequence databases for polyketide synthases. Isolated biosynthetic pathways can be introduced into multiple heterologous hosts, with some hosts engineered for expression of specific pathways. The goal of tapping into the extant diversity of secondary metabolites encoded by environmental metagenomes is being enabled by a combination of approaches, including advances in NGS technology, cloning methods, high-throughput screening, development of improved heterologous hosts, and pathway engineering.

7.1 Introduction

Natural products have been an important source of bioactive compounds thorough the history of humanity. Mankind has always taken advantage of molecules synthesized by other forms of life, remarkably plants I order to treat diseases, obtain pigments, and other useful activities. In the modern days, natural products still play a role as important as they used to play back then, especially bioactive compounds of microbiological origin.

Since the discovery of penicillin in 1929, a myriad of drugs have been developed based solely or inspired by metabolites synthesized by microorganisms. All the main classes of antibiotics—tetracyclines, cephalosporins, aminoglycosides, and

A.L.R. Santana-Pereira • M.R. Liles (✉)
Department of Biological Sciences, Auburn University, Auburn, AL 36849, USA
e-mail: lilesma@auburn.edu

T.C. Charles et al. (eds.), *Functional Metagenomics: Tools and Applications*,
DOI 10.1007/978-3-319-61510-3_7

Table 7.1 Total number of drugs for some relevant medical applications and number of natural product derived drugs

Activity	Total number of compounds	Natural compounds
Antibacterial	118	78
Antifungal	29	4
Antiviral	110	51
Antiparasitic	14	9
Anticancer	128	87
Antidiabetic	37	25
Immunosuppressant	12	12
Hypocholesterolemic	13	11

macrolides—were isolated from microbes, mainly from actinomycetes (Berdy 2005). In addition, microbial metabolites have been shown to have all sorts of useful medical applications and have been used as anticancer drugs, antivirals, antiparasitic, immunosuppressive, lipid control agents, and antidiabetics, revolutionizing medicine (Li and Vederas 2009; Newman and Cragg 2012). The antibiotic activities also make microorganisms an important source of pesticides and herbicides and therefore are also important in agriculture (Berdy 2005). A list of the main uses of natural products can be seen in Table 7.1.

However the rate of discovery of novel bioactive compounds has decreased significantly in the past decade, mainly due to two reasons: the pharmaceutical environment lack of investment in natural product discovery and technical limitations in identifying new compounds. The lack of novelty in the compounds isolated due to high rates of re-isolation has led the pharmaceutical industry to turn to synthetic chemistry and drop its investments in screening endeavors (Li and Vederas 2009).

For a long time, natural product discovery has relied on culture-dependent methods for isolation of bioactive compounds; however, because of the high rate of redundant isolation, these approaches have been becoming less efficient as means of drug discovery (Tulp and Bohlin 2005). Less than 1% of soil microorganisms can be readily cultured in a laboratory using nutrient-rich media. The percentage of extant bacteria that can be recovered on media will vary from environment to environment, with higher culture counts achieved from eutrophic environments such as gut microbiomes. Novel culture methods that utilize lower nutrient levels and extracts from natural environments can be used to obtain cultures from bacterial phyla that are underrepresented in culture collections, particularly when culturing on solid media (Schoenborn et al. 2004). The development of in situ incubation culturing methods has allowed cultivation of diverse bacteria (Nichols et al. 2010), some of which produce novel secondary metabolites such as teixobactin capable of inhibiting the growth of multidrug-resistant pathogens (Ling et al. 2015). While cultivation methods are increasingly allowing access to a greater diversity of microbial genomes, metabolisms, and metabolites, this approach is inherently limited in the degree to which it can sample environmental microbial diversity. A large percentage of microorganism cannot be maintained in culture (Nichols et al. 2010), and these approaches are time-consuming and require focused attention on specific strains to optimize them for natural product discovery. These approaches continue to yield novel phylogenetic and chemical diversity and are complementary to culture-independent approaches.

In this scenario, metagenomic libraries combined with next-generation sequencing (NGS) and heterologous expression techniques are a powerful tool to access the microbial "dark matter" and exploit this reservoir of bioactive compounds (Cragg and Newman 2013) while having a unique set of biases and limitations. In this chapter, we will discuss the use of a metagenomic approach in mining and expressing secondary metabolites from environmental samples and how these inherent biases and limitations can be mitigated and overcome to provide access to novel natural products.

7.2 Secondary Metabolites

Microorganisms are the main source of novel bioactive compounds, and the majority of these compounds are classified as secondary metabolites (Helfrich et al. 2014). Those are produced at the stationary phase and share characteristics such as (O'Brien and Wright 2011):

- They are not essential for growth nor reproduction.
- Their formation is dependent on growth conditions.
- They are produced as a group of closely related compounds.

The characteristics of secondary metabolites and their overwhelming diversity and complexity point toward a system that has long been adapted to produce molecules that are able to stop or slow down competitors in complex environments where the producer is under constant stress (Challis and Hopwood 2003) or as mediators of cross-species mutualism with other microorganisms, plants, and animals (O'Brien and Wright 2011).

Secondary metabolites can be classified in many different ways, according to biological activity, chemical structure, or genetic organization (Berdy 2005). From a metagenomic point of view, a classification based on biosynthetic gene clusters (BGCs) is more appropriate, since secondary metabolites will be dealt with in a genetic level. Thus, this classification can give information on genetic organization and active sites and even predict final product structures that can after guide screening and heterologous expression projects (Cimermancic et al. 2014; Weber et al. 2015).

There are several families of BGCs that can often share genes or even biosynthetic pathways leading to a remarkably complex network (Cimermancic et al. 2014). Here we discuss the main families of secondary metabolite BGCs.

Table 7.2 Bacteriocin classification

Classification	Characteristics	Group name
Class I		Lantibiotics
IA	Linear, rigid	
IB	Globular, flexible	
IC	Multicomponent	
Class II		Non-lantibiotics
IIa	Pediocin-like	
IIb	Miscellaneous	
IIc	Multicomponent	
Class III		Large heat stable
IIIa	Bacteriolytic	
IIIb	Non-bacteriolytic	
Class IV		Circular

Adapted from (Snyder and Worobo 2014)

7.3 Bacteriocins

Bacteriocins are ribosomally synthesized peptides with bactericidal activity, produced by a wide range of organisms, most notably lactic acid bacteria (LAB). Due to their bactericidal properties, they are widely used as preservatives in food industry (Snyder and Worobo 2014). Bacteriocins differentiate from traditional antibiotics as they are also produced during log phase and usually are narrow spectrum being active only against closely related strains (Zacharof and Lovitt 2012), even though some have a wider spectrum such as circular bacteriocins (Gabrielsen et al. 2014).

Bacteriocins are divided in four classes as shown in Table 7.2. They are usually coded in regulons containing genes for the bacteriocin peptide, secretion machinery (either transmembrane proteins or transporters), peptidases for bacteriocin activation, and an immunity peptide usually coded downstream the bacteriocin gene (Snyder and Worobo 2014). Regulon conformation shows significant variation from species to species, but despite the limited homology between genes of the BGC, they share features that make their identification possible such as multiple putative membrane-spanning domains and ATPases in the case of circular bacteriocins (van Belkum et al. 2011).

7.4 Non-ribosomal Peptide (NRP)

Non-ribosomal peptides are usually 2–48 amino acids in length and derive their diversity from the incorporation of more than 300 different amino acids including unusual ones such as D-amino acids and *N*-methylated amino acids (Du and Lou 2010). They work as mega-synthases, composed of at least three modules: (1) adenylation domain, recognizes and activates the entering amino acid; (2) peptidyl carrier protein (PCP) domain, covalently bonds to the amino acid; and (3) condensation domain, catalyzes the elongation of the peptide; in addition, different domains may be present at the modules acting as tailoring enzymes (Rausch et al. 2007). The termination is performed by a thioesterase domain usually located at the end of the modular "assembly line"; however, complete NRP synthesis is not always linear and may involve intra- and intermolecular domain interactions (Sundlov et al. 2012).

7.5 Polyketides

Many secondary metabolites produced by microorganisms are classified as polyketides, compounds synthesized by polyketide synthases (PKSs) that show a remarkable diversity in structure and bioactivities. Currently, polyketide compounds have been developed for multiple applications such as chemotherapy (e.g., rapamycin), anticholesterol (e.g., lovastatin), and a number of antibiotics including erythromycin, rifamycin, and amphotericin B (Staunton and Weissman 2001). Recently, metagenomic studies have led to the characterization of polyketides with antibiotic activity against methicillin-resistant *Staphylococcus aureus* (MRSA) and vancomycin-resistant *Enterococcus faecalis* (Feng et al. 2011).

Polyketides are produced via successive decarboxylative Claisen thioester condensations of an activated acyl starter unit and extension by the addition of malonyl-CoA units or homologs (Hertweck 2009; Khosla et al. 2014). The elongation is performed by modules that are commonly composed of an acyl transferase (AT) that activates the acyl starter and loads it on the acyl carrier protein (ACP) and a β-ketoacylsynthase (KS) that catalyzes the carbon–carbon condensation between molecules in two adjacent modules. Modules can contain other optional sites like ketoreductases and/or dehydrases, while the newly synthesized compounds can be later processed by methyltransferases/esterases, for example, to further contribute to the diversity of the compounds produced (Hertweck 2009). PKSs can be divided into three main categories that differ significantly in the arrangement of synthetic modules. Type I PKSs use diverse modules in an assembly-line fashion to synthesize the final compound, and each module performs only one elongation step. Type II PKSs use the same ACP molecule throughout the entire elongation process, and the modules function in an iterative way to produce highly oxidized aromatic compounds (Hertweck 2009). Type III PKSs are homodimeric enzymes that act directly on the Acyl-CoA substrate in an iterative manner to catalyze the condensations without the need of an ACP module (Shen 2003).

PKS pathways can also be found associated with other biosynthetic pathways such as non-ribosomal peptide synthases (NRPS) and fatty acid synthases (FAS) (Ansari et al. 2004; Masschelein et al. 2013; Milano et al. 2013). Interestingly, all of these biosynthetic pathways share striking homology in their basic chemistry allowing their modules to interact to produce hybrid molecules that can incorporate unusual amino acids other than the 20 classically incorporated by ribosomes (Du et al. 2001).

Given the organization and nature of PKS pathways, they can be genetically engineered in the laboratory and serve as a tool for drug production and novel bioactive compound development (Siezen and Khayatt 2008). Nevertheless, the lack of knowledge of the extant diversity of PKS pathway diversity in environmental metagenomes has limited the polyketide diversity that can be obtained from pathway engineering approaches. There are many unknowns even for the existing polyketide synthases in understanding spatial interactions between enzymes, stereochemistry control, or the correct flux of reagents and products between the active sites (Khosla et al. 2014). Given these current constraints in our knowledge, the cloning and expression of environmental PKS pathways could provide insights into PKS/NRPS biochemistry and genetics as well as provide new bioactive compounds of medicinal value.

7.6 Shotgun Metagenomics

The development and lowering cost of next-generation sequencing (NGS) methods have enabled massively parallel sequencing and sampling of diverse microbial assemblages (Suenaga 2012). In a "shotgun metagenomic" approach, the isolated genomic DNA is sequenced directly, removing the need for library construction. The use of shotgun sequencing has the advantages of speed and lower cost compared to a cloning approach and has become a powerful tool in understanding and characterizing microbial assemblages in diverse environments (Suenaga 2012). The metagenomic sequences generated can provide information on taxonomic composition and functional capabilities for an entire microbiome.

However, the application of shotgun metagenomics to access biosynthetic pathways for secondary metabolite synthesis is limited due to short sequence read lengths and, in complex microbiomes, the inability to assemble large, contiguous genomic regions. While the use of shotgun sequencing has been effective in assembly of bacterial genomes in extreme environments, such as the Iron Mountain acidophilic biofilms (Tyson et al. 2004), the assembly of microbial genomes is very problematic in non-extreme habitats. For example, a massive degree of shotgun sequencing was used for a single sample of a native prairie soil from Iowa, resulting in 256 Gb of sequence data, but assembly de novo of this sequence dataset required the adoption of a digital normalization method, and the largest contig from this assembly was less than 10 kb (Howe et al. 2014). The assembly of NGS raw sequence data using de novo algorithms can also lead to the formation of chimeric contigs, especially from repetitive genomic regions (Treangen and Salzberg 2012). As an additional challenge, the degree of genome coverage for diverse microbial assemblages is not sufficient to fully assemble the number of genomes predicted in a given sample, especially for bacterial taxa that are at a low relative abundance (Ladoukakis et al. 2014).

Advances in NGS technology are addressing the challenges faced by shotgun metagenomics such as short read lengths and appropriate genome coverage.

Increasingly a significant bottleneck is the bioinformatics analyses of the sequence data obtained by this approach. While there are now powerful tools available for metagenomic analyses (Darling et al. 2014; Meyer et al. 2008), the in silico prediction of secondary metabolite biosynthetic pathways, even with an excellent bioinformatics tool like antiSMASH3.0 (Weber et al. 2015), will be ineffective without complete or nearly complete clusters. Despite these limitations, shotgun sequencing of metagenomes has provided interesting insights in microbiome taxonomical composition and dynamics, even though especial attention has to be given to biasing toward high abundance species (Brooks et al. 2015; Ladoukakis et al. 2014). Furthermore, targeting conserved domains of biosynthetic pathways, such as the B-ketosynthase (KS) domains of PKS pathways or the condensation (C) domains of NRPS pathways, using NGS sequencing of bar-coded domain amplicons or mining these from shotgun sequencing data, can provide information on the diversity of these pathways present in an environmental sample (Cacho et al. 2014; Foerstner et al. 2008; Wawrik et al. 2005). The further development of NGS technologies will undoubtedly one day afford the ability to generate more complete genomic sequences directly from environmental samples that can be mined for complete biosynthetic clusters.

7.7 Cloning Metagenomic DNA

The only current methodological route to accomplishing the goal of obtaining complete biosynthetic clusters from metagenomic sources, without a cultivation step, is in the direct cloning of genomic fragments. Prior to constructing a metagenomic library, it is important to first select the source DNA, isolation method, vector, insert size, screening method, and expression host so that the output of the screening match the desired project goals (Kakirde et al. 2010). In particular, the host and cloning vector are critical for success. Different hosts have different metabolic capabilities and can be more or less feasible to culture and transform. Whereas the *E. coli* strain DH10B is readily transformable by electroporation, its capacity for secondary metabolite synthesis is marginal compared to *Streptomyces coelicolor*, which would conversely make a very poor host for constructing libraries. Because of these host differences in the relative degree of transformability and metabolic capacity, the development of broad host range shuttle vectors permits the construction of a metagenomic library in *E. coli* followed by its transfer to another heterologous host (Aakvik et al. 2009; Courtois et al. 2003; Craig et al. 2010; Kakirde et al. 2010). The resultant libraries may be conjugally transferred into a recipient host and can replicate autonomously as a plasmid (Kakirde et al. 2010) or can be integrated into the recipient host chromosome (Heil et al. 2012).

Generally, when the insert a vector can accommodate is larger, this reduces transformation efficiency and makes library construction more difficult. A clear exception to this rule is in the use of fosmid libraries (see below) which enhances transformation efficiency using phage packaging. On the other hand, vectors with

Table 7.3 Average sizes of biosynthetic gene clusters based on the curated AntiSMASH database (Blin et al. 2016)

Metabolite Type	Average BGC size (kb)
Bacteriocin	14.2
Polyketides:	
PKS I	61.9
PKS II	49.3
PKS III	44.9
NRPS	63.3

high efficiency of transformation but smaller insert sizes will fail to include the contiguity of the targeted pathways and will therefore be limiting for novel pathway and secondary metabolite discovery (Table 7.3).

7.8 High Molecular Weight DNA Cloning

When recovering PKS pathways from the environment for future heterologous expression, insert size is a key variable (Kakirde et al. 2010). The characteristic modular architecture of Type I polyketide synthases and NRPS's pathway are often coded as large assembly line pathways more than 20 kb long with some exceeding 100 kb (Fu et al. 2012). Even though the cluster size is smaller for Type II PKSs, which are commonly polycyclic, a large insert is still desirable in order to accommodate genes for tailoring enzymes that are not present in the core PKS pathway (Hertweck 2009). Therefore, metagenomic libraries have to be able to overcome insert size constraints in order to successfully clone complete biosynthetic gene clusters. Common large-insert cloning strategies for the construction of metagenomic libraries include the use of cosmid, fosmid, or bacterial artificial chromosome (BAC) vectors.

Cosmid vectors are plasmids containing the *cos* site of λ phage and an origin of replication allowing ligated sequences to be packed in λ phages and transduced into the host where the insert is maintained as a plasmid (Haley 1988). Because of the small size of the plasmid and in vitro packaging restraints, this technique can only successfully transform the host with inserts of usually 40 kb. The lack of controlling mechanisms on plasmid copy number and recombination between plasmids often decrease the stability of cosmid vectors. Functional screening of a cosmid metagenomic library using six different *Proteobacteria* hosts showed that not all hosts performed equally well and that phenotypes were rarely observed twice in different hosts (Craig et al. 2010). Thus, their work stresses the need of a vector system that can be used along different surrogate hosts. Cosmid vectors perform well in a narrow range of hosts, but its range can be widened via genetic manipulation. Broad host-range cosmid vectors pJC8 and pJC24 were designed by cleverly adding the Gateway® homology site *attL1* and *attL2* to the cosmid along with the RK-2 *otiT* (Cheng et al. 2014). This cosmid vectors were able to transfer the insert to a Gateway® plasmid via recombination, and the RK-2 *oriT* allows the transfer of the plasmid via conjugation to a number of other hosts including Gram-positive bacteria and even yeasts like *Saccharomyces cerevisiae*. These enhances in cosmid vectors increase the efficiency of screening techniques and allow higher rates of positive hits through heterologous expression.

Fosmid vectors are based on the same principles of cosmid vectors but use an F-plasmid origin of replication. F-plasmids have advantages such as existing as single copy plasmids and having a tight fertility control which decreases host intolerance to repetitive sequences and also limits homologous recombination between copies. Therefore, they are more stable and desirable when constructing genomic libraries (Quail et al. 2011) and have been used for many studies of microbial ecology and natural product discovery (Felczykowska et al. 2014; Parsley et al. 2011). Nevertheless, fosmid vectors have the same restrictions on host range faced by cosmid vectors, but similar approaches can be used to create broad host-range fosmid vectors. Broad host-range fosmid vectors were designed by adding RK-2 *oriV* and an *oriT* to the commercial fosmid vector pCC1FOS (Aakvik et al. 2009). By doing so, the fosmid vector can be transferred to other hosts from *E. coli* via conjugation, and the additional *oriV* allows for plasmid copy number regulation in hosts other than *E. coli* that doesn't recognize the original *ori2* site.

Even though cosmid or fosmid libraries provide a more efficient method to generate large-insert metagenomic libraries, because of the necessity to use phage packaging, these vector systems are inherently limited in size and are ineffective at recovering larger biosynthetic clusters, such as Type I PKS pathways. In contrast, BAC vectors can be used to stably clone insert DNA fragments as large as 300 kbp (Shizuya et al. 1992), which is sufficient to contain the largest of known biosynthetic clusters. Recently, metagenomic libraries containing inserts exceeding 100 kbp on average have been reported (Monsma et al. manuscript in preparation). These vectors provide means to have inserts big enough to accommodate complete secondary metabolite pathways, but because the transformation efficiency is lower, they are more labor intensive and therefore are not a suitable and/or preferred vector for all applications (Quail et al. 2011).

The first BAC vectors, such as pBELOBAC11, were kept as single copy plasmids in the host cell, granting high stability but decreasing the amount of product obtained by heterologous expression. Therefore, larger scale *E. coli* cultures were necessary (Wild et al. 2002). To increase plasmid yield and also potentially benefit heterologous expression, BAC vectors were engineered to have an inducible copy system in which the BAC vector carries an additional origin of replication *oriV* from RK2 that is dependent upon the expression of the TrfA protein. By expressing the *trfA* gene under the control of an arabinose-dependent promoter, the BAC vector is then conditionally copy-induced only in the presence of arabinose, permitting stable maintenance of libraries in single copy and then using copy induction for screening or DNA isolation purposes (Kakirde et al. 2010; Wild et al. 2002). The broad host-range vector pGNS-BAC has been shown to replicate in diverse Gram-negative bacteria and can be conjugally transferred from *E. coli* (Kakirde et al. 2010). Therefore, this vector is well suited for metagenomic library construction and screening, especially if the DNA isolated from an environmental sample is biased toward Gram-negative bacteria (Liles et al. 2003). Other derivatives of this vector have been generated that have the capacity to be introduced and stably integrated into Gram-positive hosts as well (Monsma et al. manuscript in preparation).

7.9 Functional Screening of Metagenomic Libraries

The research groups of Prof. Jo Handelsman and Prof. Robert Goodman in their seminal paper on soil metagenomics demonstrated that clone libraries could be screened for desired bioactivities by expressing metagenomic clones within an *E. coli* host and resulted in discovery of various enzymatic activities (Rondon et al. 2000). Since this method had been proposed, different groups have turned to functional screening to identify clones exhibiting a desired activity. High-throughput function-based screenings often rely on indicators to rapidly identify the activity of interest, such as inhibition halos, color change in the media or colonies, and degradation of media (Coughlan et al. 2015).

A high-throughput screening method for antibiotic activity was developed based on a sequential two antibiotic selection schemes (Brady and Clardy 2000). First clones are selected for kanamycin resistance in media containing the antibiotic. Then antibiotic activity is verified by an overlay of kanamycin-resistant *Bacillus subtilis*. Positive hits are recovered directly from the assay plate and then streaked in media containing ampicillin killing *B. subtilis*. This screening method have led to the characterization of a number of compounds including several long-chain *N*-acyl amino acid antibiotics (Brady et al. 2004; Brady and Clardy 2000), isocyanide antibiotic (Brady and Clardy 2005), antibiotic polyketides (Craig et al. 2009), novel metalloproteases, serine proteases, and lipolytic compounds (Iqbal et al. 2014).

One interesting way to track activities of interest that may not be readily accessible is the use of reporter genes. *S*ubstrate-*i*nduced *g*ene *ex*pression (SIGEX) screening allows the screening of catabolic genes by using a green fluorescent protein (GFP) as a reporter. Since the expression of catabolic pathways is usually triggered by the presence of the substrate, it was rationalized that clones containing catabolic pathways would be activated in the presence of specific substrates (Uchiyama et al. 2005). The vector has *gfp* genes that are translationally dependent on the expression of the eDNA. Thus, if a catabolic pathway is activated in a clone, the colonies get green fluorescence that can be sorted by fluorescence-activated cell sorting for high-throughput screening.

However the technique is unable to identify constitutively expressed catabolic genes and fails to identify genes inserted in the opposite orientation of *gfp* (Yun and Ryu 2005). PIGEX is an improvement to SIGEX, where the expression of *gfr* is triggered by the presence of a particular catabolic product (Uchiyama and Miyazaki 2010). By doing so, PIGEX is not sensitive to insert orientation and is able to detect constitutively expressed catabolic pathways. In the same work, the authors reported the discovery of three novel amidases (Uchiyama and Miyazaki 2010), and diverse groups have also been successful in isolating catabolic genes of interest (Coughlan et al. 2015) (Fig. 7.1).

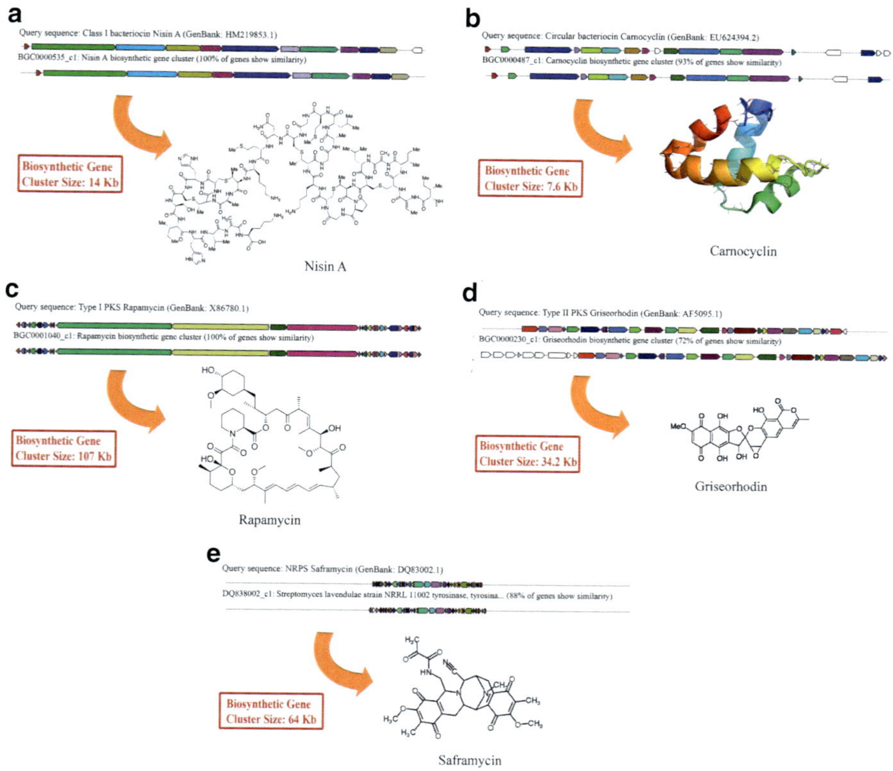

Fig. 7.1 Annotation of biosynthetic gene clusters (BCGs) of the main secondary metabolite groups using the program antiSMASH (Weber et al. 2015) using BGC sequences obtained from GenBank. Compounds from each group were chosen along with their chemical structures, illustrating the chemical diversity of secondary metabolites. (**a**) Nissin A (Class I bacteriocin). (**b**) Carnocyclin circular bacteriocin, (**c**) rapamycin (Type I PKS), (**d**) griseorhodin (Type II PKS), and (**e**) saframycin (NRPS peptide)

An interesting way to enrich clones containing not only PKS but also NRPS pathways is to utilize the biochemical characteristics of these pathways (Charlop-Powers et al. 2013). The ability of ACP/PCP to bind to the growing polymer chain relies on a phosphopantetheine prostetic group added by PPTases as part of the post-translational modification of the carrier proteins. In the absence of phosphopantetheine, neither ACP nor PCP is functional. Iron uptake is also dependent on NRPS-derived peptides, and in *E. coli* this function is performed by the siderophore enterobactin. The enterobactin gene cluster contains the PPTase EntD which is crucial to the peptide biosynthesis, thus *entD* mutants cannot grow in iron-limiting conditions. Therefore, *entD* mutants could have the phenotype reverted and be viable in low limiting conditions if the insert cloned coded for PPTase which is commonly associated with PKS and NRPS pathways. This approach provides a fast way to select promising clones within a metagenomic library in an easy, fast and low cost manner. However, there are still limitations intrinsic to heterologous expression, especially when using an *E. coli* host as proposed by Charlop-Powers et al. (2013).

Nevertheless, there are several limitations on heterologous expression of environmental DNA that can result in a subset of the cloned genes being transcribed and translated or their protein product(s) being modified and active (National Research Council (US) Committee on Metagenomics, 2007). There is also evidence for an under-abundance of strong rpoD consensus sequences within a human microbiome sample, suggesting that the loss of AT-rich sequences may be due to bias in favor or less actively transcribed genes (Lam and Charles 2015). Conversely, expression of a heterologous sigma factor, for example, from *Acidobacteria* or *Lactobacillus* taxa, has been shown to enhance transcription from metagenomic clones (Gaida et al. 2015; Sabree et al. 2006).

Therefore, because of these potential issues with transcription and translation of metagenome-derived genes in a particular heterologous host, there have been efforts to expand the range of heterologous hosts for metagenomic library expression (Courtois et al. 2003; Craig et al. 2010). With specific regard to secondary metabolite producers, hosts such as *Streptomyces* spp. and *Pseudomonas* spp. and other *Proteobacteria* taxa have been a particular emphasis. Development of new heterologous hosts that better reflect the phylogenetic origin of cloned biosynthetic clusters and further engineering of existing hosts will be important for the field of functional metagenomics for the goal of novel secondary metabolite discovery.

Expression of eDNA can also be influenced by the primary metabolism of the host; thus, the existing host metabolic pathways and their interactions have to be taken in consideration when inducing expression of an environmental biosynthetic pathway. PKS pathways are a good example of these interactions. PKS pathways and fatty acid synthesis show remarkable similarities and are thought to have a common evolutionary origin. Both pathways can share the precursors acetyl-CoA; thus, high expression of one metabolic pathway could inhibit the other by depleting the pool of acetyl-CoA in the cell (Cronan and Thomas 2009). In fact, compounds that block fatty acid metabolism are shown to enhance pigment production in *Streptomyces coelicolor*, demonstrating how primary and secondary metabolisms are linked together and stressing the significance of competition over precursors between metabolic pathways (Craney et al. 2012).

To overcome this limitations in secondary metabolite expression, different strains of *S. coelicolor* have been genetically engineered, with the strain M145 having deletions in four endogenous antibiotic gene clusters, thereby removing competing sinks for carbon and nitrogen, increasing precursor availability, and enhancing heterologous expression (Gomez-Escribano et al. 2012). The pathway deletions also help by simplifying the metabolic profile of *S. coelicolor*, making mass spectrometry analysis easier with a lower metabolite background and facilitating the identification of compounds encoded by the exogenous insert.

Another path to enhancing secondary metabolite expression in a host is to manipulate its transcriptional apparatus, especially when screening for antibiotics. To increase *S. coelicolor* antibiotic expression, point mutations were introduced in the *rpoB* and *rpsL* genes that encode the RNA polymerase β-subunit and the ribosomal protein S12, respectively. The mutations conferred increased host resistance to rifampicin, streptomycin, and paromomycin and increased antibiotic production 30- to 40-fold when exogenous antibiotic genes were cloned into the modified strain (Gomez-Escribano and Bibb 2014; Hu et al. 2002).

7.10 Sequence-Based Screening of Metagenomic Libraries

Function-based screening has been an important way to mine metagenomic libraries for molecules and/or processes of interest; however, one of its main limitations is redundant isolation (Tulp and Bohlin 2005). Redundant isolation can be overcome by prior bioinformatics analyses of inserts selected by sequence-based screening, making the process more efficient (Chang and Brady 2013a). By first identifying a specific gene target, this method allows the identification and characterization of a pathway prior to functional screening, thereby reducing the number of clones subjected to functional screening, saving time and effort, and increasing the potential for heterologous expression (Culligan et al. 2014). Secondary metabolite pathways can be targeted by PCR-, hybridization-, or homology-based methods in order to identify pathway-containing clones, utilizing conserved domains and/or regions.

PKS clusters represent an ideal target for pathways encoding diverse secondary metabolites, given their well-characterized architecture and conserved domains and their presence in diverse marine and terrestrial organisms (Fieseler et al. 2007; Ginolhac et al. 2004; Muller et al. 2015). A minimal polyketide synthase module is composed of an acyl-transferase (AT), a ketosynthase (KS), and an acyl-carrying protein, with the ketosynthase being the most conserved element of the module since it catalyzes the carbon condensation in polyketide synthesis (Khosla et al. 2014). Fortunately, there are conserved sequences within KS domains, and their DNA sequence is an ideal target for PCR amplification (Wawrik et al. 2005). Therefore, clones containing polyketide pathway can be identified by homology-based approaches by targeting KS and other conserved domains (Banik and Brady 2010; Chang and Brady 2013b). Nevertheless, there is great sequence diversity among KS domain sequences, and it is probable that there are KS domains present within environmental metagenomes that would not be amplified using the available degenerate primer sets, resulting in a biased set of sequences most similar to those from previously characterized pathways.

Another powerful approach is the utilization of next-generation sequence and bioinformatics approaches to mine for conserved motifs. In this approach, all of the clones in the library are sequenced, and computational analysis tools are used to identify clones containing the target sequences. The availability of a complete PKS pathway sheds light on modular organization of the pathway, and this can be used for in silico prediction of polyketide structure, even though there are limitations in predicting the final structure based only on sequence data (Hertweck 2009; Zerikly and Challis 2009). In addition, PKS domain sequences cluster together in functional groups, and therefore, a molecular phylogenetic analysis can be used as a guide to identify pathways that encode structurally related polyketide products (Metsa-Ketela et al. 2002). The use of NGS as a means of identifying NRPS, PKS, and other clusters has now been demonstrated using a large-insert soil metagenomic BAC library using a pooled strategy (rows, columns, and plates) and resulting in the discovery of diverse and novel pathways (Monsma et al. manuscript in preparation).

7.11 Future Perspectives

Secondary metabolites produced by as-yet-uncultured microorganisms could be a rich resource for bioactive compounds. Metagenomic approaches are one of the powerful tools that can be used to unravel the diversity of secondary metabolites present in nature, and the combination of tools now available is finally starting to bear fruit.

The rapid development of NGS platforms provides a means to sequence large amounts of DNA with increasing throughput, accuracy, and longer read lengths. Despite the rapid technological evolution of NGS platforms, read lengths currently are not sufficient to allow shotgun metagenomics to access larger biosynthetic pathways, and there is increasing computational demands to handle this deluge of sequencing data. The combination of new developments in NGS technology and bioinformatics approaches promises to make it one day possible to access more complete genomes and pathways directly from an environmental sample to mine these for their secondary metabolite encoding potential.

The advances in NGS technology and data processing have also revolutionized the way that screening is performed. The use of PCR to perform gene-targeted screens is still in use, but increasingly, NGS allows the discovery of a greater diversity of gene sequences and avoids the primary disadvantage of PCR in that it is more likely to identify genes that are most similar to previously known sequences. Thus, NGS is a much more desirable approach for gene targeting as bioinformatics analyses of the sequences may reveal novel genes and/or pathways configuration that were cryptic to PCR techniques. As NGS becomes cheaper and easier to process, more and more research projects are expected to shift from PCR-based screening to sequencing-based screening. Furthermore, the use of NGS as a first step in screening a library can avoid the significant biases inherent in functional screening, so that identified pathways of interest can then be selectively targeted for engineering toward enhanced expression.

When expressing secondary metabolite pathways for desired bioactivities, it is very important that environmental genes, operons, and/or pathways have their integrity preserved when cloning. For many biosynthetic clusters, such as Type I PKS, NRPS, or hybrid pathways, the use of BAC vectors constitutes the best option. Not only can BAC vectors carry inserts of greater than 100 kb, there are new derivatives engineered for copy induction, inducible expression of cloned inserts, and introduction into broad hosts. These features can be important in reducing host metabolic stress and allowing the cloning of potentially toxic pathways. These advances have set the stage for the isolation and expression of entire biosynthetic pathways that have been previously inaccessible, now permitting the mining of previously unexplored reservoirs of bioactive compounds. Complete pathway cloning is an important advance in the bioprospecting of active compounds via metagenomics, combined together with pathway discovery via NGS (Monsma et al. manuscript in preparation).

Despite these advances in cloning and NGS screening methods, the heterologous expression of cloned pathways has serious limitations related to the ability of the host to correctly and efficiently express the foreign DNA inserted. Some hosts have stricter promoter recognition and deviant codon usage or lack the metabolic ability to synthesize a particular product. Heterologous hosts can be successfully genetically engineered to increase their efficiency for secondary metabolite pathway expression, and there is value in attaining a deeper understanding of host physiology in order to rationally address expression limitations to facilitate better expression systems.

Given the vast metagenomic diversity represented in clone libraries, there is also a benefit in utilizing multiple heterologous hosts, including both Gram-negative hosts such as *E. coli* and other bacteria phylogenetically related to the origin of cloned genes and selected to have desired characteristics for expression and metabolic support. There is, therefore, great value in continued pursuit of novel microbial cultures for their metabolites and their use as hosts, which are representative of the microbial "dark matter" previously tapped into purely by culture-independent methods (Rinke et al. 2013).

Realizing the promise of functional metagenomics for secondary metabolite discovery, as articulated 17 years ago by Jo Handelsman and colleagues (Handelsman et al. 1998), has required concomitant development of many tool sets. Only through the wise combination of methods to clone, identify, and express the diverse pathways encoded by environmental microorganisms can this dream be fulfilled.

References

Aakvik T, Degnes KF, Dahlsrud R, Schmidt F, Dam R, Yu L et al (2009) A plasmid RK2-based broad-host-range cloning vector useful for transfer of metagenomic libraries to a variety of bacterial species. FEMS Microbiol Lett 296(2):149–158. doi:10.1111/j.1574-6968.2009.01639.x

Ansari MZ, Yadav G, Gokhale RS, Mohanty D (2004) NRPS-PKS: a knowledge-based resource for analysis of NRPS/PKS megasynthases. Nucleic Acids Res 32(Web Server Issue):W405–W413. doi:10.1093/nar/gkh359

Banik JJ, Brady SF (2010) Recent application of metagenomic approaches toward the discovery of antimicrobials and other bioactive small molecules. Curr Opin Microbiol 13(5):603–609. doi:10.1016/j.mib.2010.08.012

Berdy J (2005) Bioactive microbial metabolites. J Antibiot (Tokyo) 58(1):1–26. doi:10.1038/ja.2005.1

Blin K, Medema MH, Kottmann R, Lee SY, Weber T (2016). The antiSMASH database, a comprehensive database of microbial secondary metabolite biosynthetic gene clusters. Nucleic acids research, gkw960. doi:10.1093/nar/gkw960

Brady SF, Clardy J (2000) Long-chain N-acyl amino acid antibiotics isolated from heterologously expressed environmental DNA. J Am Chem Soc 122(51):12903–12904. doi:10.1021/ja002990u

Brady SF, Clardy J (2005) Cloning and heterologous expression of isocyanide biosynthetic genes from environmental DNA. Angew Chem Int Ed Engl 44(43):7063–7065. doi:10.1002/anie.200501941

Brady SF, Chao CJ, Clardy J (2004) Long-chain N-acyltyrosine synthases from environmental DNA. Appl Environ Microbiol 70(11):6865–6870. doi:10.1128/AEM.70.11.6865-6870.2004

Brooks JP, Edwards DJ, Harwich MD Jr, Rivera MC, Fettweis JM, Serrano MG et al (2015) The truth about metagenomics: quantifying and counteracting bias in 16S rRNA studies. BMC Microbiol 15(1):66. doi:10.1186/s12866-015-0351-6

Cacho RA, Tang Y, Chooi YH (2014) Next-generation sequencing approach for connecting secondary metabolites to biosynthetic gene clusters in fungi. Front Microbiol 5:774. doi:10.3389/fmicb.2014.00774

Challis GL, Hopwood DA (2003) Synergy and contingency as driving forces for the evolution of multiple secondary metabolite production by Streptomyces species. Proc Natl Acad Sci U S A 100(Suppl 2):14555–14561. doi:10.1073/pnas.1934677100

Chang F-YY, Brady SF (2013a) Discovery of indolotryptoline antiproliferative agents by homology-guided metagenomic screening. Proc Natl Acad Sci U S A 110(7):2478–2483. doi:10.1073/pnas.1218073110

Chang FY, Brady SF (2013b) Discovery of indolotryptoline antiproliferative agents by homology-guided metagenomic screening. Proc Natl Acad Sci U S A 110(7):2478–2483. doi:10.1073/pnas.1218073110

Charlop-Powers Z, Banik JJ, Owen JG, Craig JW, Brady SF (2013) Selective enrichment of environmental DNA libraries for genes encoding nonribosomal peptides and polyketides by phosphopantetheine transferase-dependent complementation of siderophore biosynthesis. ACS Chem Biol 8(1):138–143. doi:10.1021/cb3004918

Chatterjee C, Paul M, Xie L, van der Donk WA (2005) Biosynthesis and mode of action of lantibiotics. Chem Rev 105(2):633–684. doi:10.1021/cr030105v

Cheng J, Pinnell L, Engel K, Neufeld JD, Charles TC (2014) Versatile broad-host-range cosmids for construction of high quality metagenomic libraries. J Microbiol Methods 99:27–34. doi:10.1016/j.mimet.2014.01.015

Cimermancic P, Medema MH, Claesen J, Kurita K, Wieland Brown LC, Mavrommatis K et al (2014) Insights into secondary metabolism from a global analysis of prokaryotic biosynthetic gene clusters. Cell 158(2):412–421. doi:10.1016/j.cell.2014.06.034

Coughlan LM, Cotter PD, Hill C (2015) Biotechnological applications of functional metagenomics in the food and pharmaceutical industries. Front Microbiol 6:672. doi:10.3389/fmicb.2015.00672

Courtois S, Cappellano CM, Ball M, Francou FX, Normand P, Helynck G et al (2003) Recombinant environmental libraries provide access to microbial diversity for drug discovery from natural products. Appl Environ Microbiol 69(1):49–55. doi:10.1128/AEM.69.1.49-55.2003

Cragg GM, Newman DJ (2013) Natural products: a continuing source of novel drug leads. Biochim Biophys Acta 1830(6):3670–3695. doi:10.1016/j.bbagen.2013.02.008

Craig JW, Chang F-Y, Brady SF (2009) Natural products from environmental DNA hosted in Ralstonia metallidurans. ACS Chem Biol 4(1):23–28. doi:10.1021/cb8002754

Craig JW, Chang FY, Kim JH, Obiajulu SC, Brady SF (2010) Expanding small-molecule functional metagenomics through parallel screening of broad-host-range cosmid environmental DNA libraries in diverse proteobacteria. Appl Environ Microbiol 76(5):1633–1641. doi:10.1128/AEM.02169-09

Craney A, Ozimok C, Pimentel-Elardo SM, Capretta A, Nodwell JR (2012) Chemical perturbation of secondary metabolism demonstrates important links to primary metabolism. Chem Biol 19(8):1020–1027. doi:10.1016/j.chembiol.2012.06.013

Cronan JE, Thomas J (2009) Bacterial fatty acid synthesis and its relationships with polyketide synthetic pathways. Methods Enzymol 459:395–433. doi:10.1016/S0076-6879(09)04617-5

Culligan EP, Sleator RD, Marchesi JR, Hill C (2014) Metagenomics and novel gene discovery: promise and potential for novel therapeutics. Virulence 5(3):399–412. doi:10.4161/viru.27208
Darling AE, Jospin G, Lowe E, Matsen FA 4th, Bik HM, Eisen JA (2014) PhyloSift: phylogenetic analysis of genomes and metagenomes. PeerJ 2:e243. doi:10.7717/peerj.243
Du L, Lou L (2010) PKS and NRPS release mechanisms. Nat Prod Rep 27(2):255–278. doi:10.1039/b912037h
Du L, Sanchez C, Shen B (2001) Hybrid peptide-polyketide natural products: biosynthesis and prospects toward engineering novel molecules. Metab Eng 3(1):78–95. doi:10.1006/mben.2000.0171
Ennahar S, Sashihara T, Sonomoto K, Ishizaki A (2000) Class IIa bacteriocins: biosynthesis, structure and activity. FEMS Microbiol Rev 24(1):85–106. doi:10.1016/S0168-6445(99)00031-5
Felczykowska A, Dydecka A, Bohdanowicz M, Gasior T, Sobon M, Kobos J et al (2014) The use of fosmid metagenomic libraries in preliminary screening for various biological activities. Microb Cell Fact 13(1):105. doi:10.1186/s12934-014-0105-4
Feng Z, Kallifidas D, Brady SF (2011) Functional analysis of environmental DNA-derived type II polyketide synthases reveals structurally diverse secondary metabolites. Proc Natl Acad Sci U S A 108(31):12629–12634. doi:10.1073/pnas.1103921108
Fieseler L, Hentschel U, Grozdanov L, Schirmer A, Wen G, Platzer M et al (2007) Widespread occurrence and genomic context of unusually small polyketide synthase genes in microbial consortia associated with marine sponges. Appl Environ Microbiol 73(7):2144–2155. doi:10.1128/AEM.02260-06
Foerstner KU, Doerks T, Creevey CJ, Doerks A, Bork P (2008) A computational screen for type I polyketide synthases in metagenomics shotgun data. PLoS One 3(10):e3515. doi:10.1371/journal.pone.0003515
Foulston LC, Bibb MJ (2010) Microbisporicin gene cluster reveals unusual features of lantibiotic biosynthesis in actinomycetes. Proc Natl Acad Sci U S A 107(30):13461–13466. doi:10.1073/pnas.1008285107
Fu J, Bian X, Hu S, Wang H, Huang F, Seibert PM et al (2012) Full-length RecE enhances linear-linear homologous recombination and facilitates direct cloning for bioprospecting. Nat Biotechnol 30(5):440–446. doi:10.1038/nbt.2183
Gabrielsen C, Brede DA, Nes IF, Diep DB (2014) Circular bacteriocins: biosynthesis and mode of action. Appl Environ Microbiol 80(22):6854–6862. doi:10.1128/AEM.02284-14
Gaida SM, Sandoval NR, Nicolaou SA, Chen Y, Venkataramanan KP, Papoutsakis ET (2015) Expression of heterologous sigma factors enables functional screening of metagenomic and heterologous genomic libraries. Nat Commun 6:7045. doi:10.1038/ncomms8045
Ginolhac A, Jarrin C, Gillet B, Robe P, Pujic P, Tuphile K et al (2004) Phylogenetic analysis of polyketide synthase I domains from soil metagenomic libraries allows selection of promising clones. Appl Environ Microbiol 70(9):5522–5527. doi:10.1128/AEM.70.9.5522-5527.2004
Gomez-Escribano JP, Bibb MJ (2014) Heterologous expression of natural product biosynthetic gene clusters in Streptomyces coelicolor: from genome mining to manipulation of biosynthetic pathways. J Ind Microbiol Biotechnol 41(2):425–431. doi:10.1007/s10295-013-1348-5
Gomez-Escribano JP, Song LJ, Fox DJ, Yeo V, Bibb MJ, Challis GL (2012) Structure and biosynthesis of the unusual polyketide alkaloid coelimycin P1, a metabolic product of the cpk gene cluster of Streptomyces coelicolor M145. Chem Sci 3(9):2716–2720. doi:10.1039/c2sc20410j
Haley JD (1988) Cosmid library construction. Methods Mol Biol 4:257–283. doi:10.1385/0-89603-127-6:257
Handelsman J, Rondon MR, Brady SF, Clardy J, Goodman RM (1998) Molecular biological access to the chemistry of unknown soil microbes: a new frontier for natural products. Chem Biol 5(10):R245–R249. doi:10.1016/S1074-5521(98)90108-9
Heil JR, Cheng J, Charles TC (2012) Site-specific bacterial chromosome engineering: PhiC31 integrase mediated cassette exchange (IMCE). J Vis Exp (61). doi:10.3791/3698
Helfrich EJ, Reiter S, Piel J (2014) Recent advances in genome-based polyketide discovery. Curr Opin Biotechnol 29:107–115. doi:10.1016/j.copbio.2014.03.004

Hertweck C (2009) The biosynthetic logic of polyketide diversity. Angew Chem Int Ed Engl 48(26):4688–4716. doi:10.1002/anie.200806121

Hill AM (2006) The biosynthesis, molecular genetics and enzymology of the polyketide-derived metabolites. Nat Prod Rep 23(2):256–320. doi:10.1039/b301028g

Hopwood DA (1997) Genetic contributions to understanding polyketide synthases. Chem Rev 97(7):2465–2498. doi:10.1021/cr960034i

Howe AC, Jansson JK, Malfatti SA, Tringe SG, Tiedje JM, Brown CT (2014) Tackling soil diversity with the assembly of large, complex metagenomes. Proc Natl Acad Sci U S A 111(13):4904–4909. doi:10.1073/pnas.1402564111

Hu H, Zhang Q, Ochi K (2002) Activation of antibiotic biosynthesis by specified mutations in the rpoB gene (encoding the RNA polymerase beta subunit) of Streptomyces lividans. J Bacteriol 184(14):3984–3991. doi:10.1128/JB.184.14.3984-3991.2002

Iqbal HA, Craig JW, Brady SF (2014) Antibacterial enzymes from the functional screening of metagenomic libraries hosted in Ralstonia metallidurans. FEMS Microbiol Lett 354(1):19–26. doi:10.1111/1574-6968.12431

Kakirde KS, Parsley LC, Liles MR (2010) Size does matter: application-driven approaches for soil metagenomics. Soil Biol Biochem 42(11):1911–1923. doi:10.1016/j.soilbio.2010.07.021

Khosla C, Herschlag D, Cane DE, Walsh CT (2014) Assembly line polyketide synthases: mechanistic insights and unsolved problems. Biochemistry 53(18):2875–2883. doi:10.1021/bi500290t

Ladoukakis E, Kolisis FN, Chatziioannou AA (2014) Integrative workflows for metagenomic analysis. Front Cell Dev Biol 2:70. doi:10.3389/fcell.2014.00070

Lal R, Kumari R, Kaur H, Khanna R, Dhingra N, Tuteja D (2000) Regulation and manipulation of the gene clusters encoding type-I PKSs. Trends Biotechnol 18(6):264–274. doi:10.1016/S0167-7799(00)01443-8

Lam KN, Charles TC (2015) Strong spurious transcription likely contributes to DNA insert bias in typical metagenomic clone libraries. Microbiome 3:22. doi:10.1186/s40168-015-0086-5

Li JW, Vederas JC (2009) Drug discovery and natural products: end of an era or an endless frontier? Science 325(5937):161–165. doi:10.1126/science.1168243

Li L, Deng W, Song J, Ding W, Zhao QF, Peng C et al (2008) Characterization of the saframycin A gene cluster from Streptomyces lavendulae NRRL 11002 revealing a nonribosomal peptide synthetase system for assembling the unusual tetrapeptidyl skeleton in an iterative manner. J Bacteriol 190(1):251–263. doi:10.1128/JB.00826-07

Liles MR, Manske BF, Bintrim SB, Handelsman J, Goodman RM (2003) A census of rRNA genes and linked genomic sequences within a soil metagenomic library. Appl Environ Microbiol 69(5):2684–2691. doi:10.1128/Aem.69.5.2684-2691.2003

Ling LL, Schneider T, Peoples AJ, Spoering AL, Engels I, Conlon BP et al (2015) A new antibiotic kills pathogens without detectable resistance. Nature 517(7535):455–459. doi:10.1038/nature14098

Masschelein J, Mattheus W, Gao LJ, Moons P, Van Houdt R, Uytterhoeven B et al (2013) A PKS/NRPS/FAS hybrid gene cluster from Serratia plymuthica RVH1 encoding the biosynthesis of three broad spectrum, zeamine-related antibiotics. PLoS One 8(1):e54143. doi:10.1371/journal.pone.0054143

Metsa-Ketela M, Halo L, Munukka E, Hakala J, Mantsala P, Ylihonko K (2002) Molecular evolution of aromatic polyketides and comparative sequence analysis of polyketide ketosynthase and 16S ribosomal DNA genes from various streptomyces species. Appl Environ Microbiol 68(9):4472–4479. doi:10.1128/Aem.68.9.4472-4479.2002

Meyer F, Paarmann D, D'Souza M, Olson R, Glass EM, Kubal M et al (2008) The metagenomics RAST server—a public resource for the automatic phylogenetic and functional analysis of metagenomes. BMC Bioinformatics 9(1):386. doi:10.1186/1471-2105-9-386

Milano T, Paiardini A, Grgurina I, Pascarella S (2013) Type I pyridoxal 5′-phosphate dependent enzymatic domains embedded within multimodular nonribosomal peptide synthetase and polyketide synthase assembly lines. BMC Struct Biol 13(1):26. doi:10.1186/1472-6807-13-26

Muller CA, Oberauner-Wappis L, Peyman A, Amos GC, Wellington EM, Berg G (2015) Mining for nonribosomal peptide synthetase and polyketide synthase genes revealed a high level of diversity in the Sphagnum bog metagenome. Appl Environ Microbiol 81(15):5064–5072. doi:10.1128/AEM.00631-15

Newman DJ, Cragg GM (2012) Natural products as sources of new drugs over the 30 years from 1981 to 2010. J Nat Prod 75(3):311–335. doi:10.1021/np200906s

Nichols D, Cahoon N, Trakhtenberg EM, Pham L, Mehta A, Belanger A et al (2010) Use of ichip for high-throughput in situ cultivation of "uncultivable" microbial species. Appl Environ Microbiol 76(8):2445–2450. doi:10.1128/AEM.01754-09

O'Brien J, Wright GD (2011) An ecological perspective of microbial secondary metabolism. Curr Opin Biotechnol 22(4):552–558. doi:10.1016/j.copbio.2011.03.010

Parsley LC, Linneman J, Goode AM, Becklund K, George I, Goodman RM et al (2011) Polyketide synthase pathways identified from a metagenomic library are derived from soil Acidobacteria. FEMS Microbiol Ecol 78(1):176–187. doi:10.1111/j.1574-6941.2011.01122.x

Quail MA, Matthews L, Sims S, Lloyd C, Beasley H, Baxter SW (2011) Genomic libraries: I. Construction and screening of fosmid genomic libraries. Methods Mol Biol 772:37–58. doi:10.1007/978-1-61779-228-1_3

Rausch C, Hoof I, Weber T, Wohlleben W, Huson DH (2007) Phylogenetic analysis of condensation domains in NRPS sheds light on their functional evolution. BMC Evol Biol 7(1):78. doi:10.1186/1471-2148-7-78

Rinke C, Schwientek P, Sczyrba A, Ivanova NN, Anderson IJ, Cheng JF et al (2013) Insights into the phylogeny and coding potential of microbial dark matter. Nature 499(7459):431–437. doi:10.1038/nature12352

Rondon MR, August PR, Bettermann AD, Brady SF, Grossman TH, Liles MR et al (2000) Cloning the soil metagenome: a strategy for accessing the genetic and functional diversity of uncultured microorganisms. Appl Environ Microbiol 66(6):2541–2547

Sabree ZL, Bergendahl V, Liles MR, Burgess RR, Goodman RM, Handelsman J (2006) Identification and characterization of the gene encoding the Acidobacterium capsulatum major sigma factor. Gene 376(1):144–151. doi:10.1016/j.gene.2006.02.033

Schoenborn L, Yates PS, Grinton BE, Hugenholtz P, Janssen PH (2004) Liquid serial dilution is inferior to solid media for isolation of cultures representative of the phylum-level diversity of soil bacteria. Appl Environ Microbiol 70(7):4363–4366. doi:10.1128/AEM.70.7.4363-4366.2004

Shen B (2003) Polyketide biosynthesis beyond the type I, II and III polyketide synthase paradigms. Curr Opin Chem Biol 7(2):285–295. doi:10.1016/S1367-5931(03)00020-6

Shizuya H, Birren B, Kim UJ, Mancino V, Slepak T, Tachiiri Y, Simon M (1992) Cloning and stable maintenance of 300-Kilobase-pair fragments of human DNA in Escherichia-Coli using an F-factor-based vector. Proc Natl Acad Sci U S A 89(18):8794–8797. doi:10.1073/pnas.89.18.8794

Siezen RJ, Khayatt BI (2008) Natural products genomics. Microb Biotechnol 1(4):275–282. doi:10.1111/j.1751-7915.2008.00044.x

Snyder AB, Worobo RW (2014) Chemical and genetic characterization of bacteriocins: antimicrobial peptides for food safety. J Sci Food Agric 94(1):28–44. doi:10.1002/jsfa.6293

Staunton J, Weissman KJ (2001) Polyketide biosynthesis: a millennium review. Nat Prod Rep 18(4):380–416. doi:10.1039/a909079g

Suenaga H (2012) Targeted metagenomics: a high-resolution metagenomics approach for specific gene clusters in complex microbial communities. Environ Microbiol 14(1):13–22. doi:10.1111/j.1462-2920.2011.02438.x

Summers RG, Donadio S, Staver MJ, Wendt-Pienkowski E, Hutchinson CR, Katz L (1997) Sequencing and mutagenesis of genes from the erythromycin biosynthetic gene cluster of Saccharopolyspora erythraea that are involved in L-mycarose and D-desosamine production. Microbiology 143(10):3251–3262. doi:10.1099/00221287-143-10-3251

Sundlov JA, Shi C, Wilson DJ, Aldrich CC, Gulick AM (2012) Structural and functional investigation of the intermolecular interaction between NRPS adenylation and carrier protein domains. Chem Biol 19(2):188–198. doi:10.1016/j.chembiol.2011.11.013

Treangen TJ, Salzberg SL (2012) Repetitive DNA and next-generation sequencing: computational challenges and solutions. Nat Rev Genet 13(1):36–46. doi:10.1038/nrg3117

Tulp M, Bohlin L (2005) Rediscovery of known natural compounds: nuisance or goldmine? Bioorg Med Chem 13(17):5274–5282. doi:10.1016/j.bmc.2005.05.067

Tyson GW, Chapman J, Hugenholtz P, Allen EE, Ram RJ, Richardson PM et al (2004) Community structure and metabolism through reconstruction of microbial genomes from the environment. Nature 428(6978):37–43. doi:10.1038/nature02340

Uchiyama T, Miyazaki K (2010) Product-induced gene expression, a product-responsive reporter assay used to screen metagenomic libraries for enzyme-encoding genes. Appl Environ Microbiol 76(21):7029–7035. doi:10.1128/AEM.00464-10

Uchiyama T, Abe T, Ikemura T, Watanabe K (2005) Substrate-induced gene-expression screening of environmental metagenome libraries for isolation of catabolic genes. Nat Biotechnol 23(1):88–93. doi:10.1038/nbt1048

van Belkum MJ, Martin-Visscher LA, Vederas JC (2011) Structure and genetics of circular bacteriocins. Trends Microbiol 19(8):411–418. doi:10.1016/j.tim.2011.04.004

Wawrik B, Kerkhof L, Zylstra GJ, Kukor JJ (2005) Identification of unique type II polyketide synthase genes in soil. Appl Environ Microbiol 71(5):2232–2238. doi:10.1128/AEM.71.5.2232-2238.2005

Weber T, Blin K, Duddela S, Krug D, Kim HU, Bruccoleri R et al (2015) antiSMASH 3.0-a comprehensive resource for the genome mining of biosynthetic gene clusters. Nucleic Acids Res 43(W1):W237–W243. doi:10.1093/nar/gkv437

Wild J, Hradecna Z, Szybalski W (2002) Conditionally amplifiable BACs: switching from single-copy to high-copy vectors and genomic clones. Genome Res 12(9):1434–1444. doi:10.1101/gr.130502

Yun J, Ryu S (2005) Screening for novel enzymes from metagenome and SIGEX, as a way to improve it. Microb Cell Fact 4(1):8. doi:10.1186/1475-2859-4-8

Zacharof MP, Lovitt RW (2012) Bacteriocins produced by lactic acid bacteria a review article. APCBEE Procedia 2:50–56. doi:10.1016/j.apcbee.2012.06.010

Zerikly M, Challis GL (2009) Strategies for the discovery of new natural products by genome mining. ChemBioChem 10(4):625–633. doi:10.1002/cbic.200800389

Chapter 8
Enhancing Functional Metagenomics of Complex Microbial Communities Using Stable Isotopes

Marcela Hernández, Josh D. Neufeld, and Marc G. Dumont

Abstract Exploring the function of genes encoded by uncultivated microorganisms is one of the major challenges facing microbiologists. Functions can be predicted by sequence comparisons to known genes and proteins, but proof of function requires the analysis of gene products by in vitro or in vivo expression, which is referred to as functional metagenomics. Using this approach, genetic material is retrieved from the environment, cloned, and expressed under laboratory conditions in order to screen for specific biochemical activities. Stable-isotope probing (SIP) is an approach for capturing genetic material of active microorganisms in environmental samples. This method facilitates functional metagenomics by directing the search toward microorganisms that are likely to possess genes of relevance to a specific research objective. In this chapter, we discuss how combined DNA-SIP and metagenomic research has been used for enhancing functional screening efforts. In addition, we highlight emerging methods, such as mRNA-SIP and Raman microspectroscopy, that can help retrieve genetic material from targeted microbial groups for the discovery of novel functions.

M. Hernández
Max Planck Institute for Terrestrial Microbiology, Karl-von-Frisch-Strasse 10, Marburg 35043, Germany

Biological Sciences, University of Southampton, Southampton SO17 1BJ, UK
e-mail: m.t.hernandez-garcia@soton.co.uk

J.D. Neufeld
Department of Biology, University of Waterloo, Waterloo, ON, Canada, N2L 3G1
e-mail: jneufeld@uwaterloo.ca

M.G. Dumont (✉)
Biological Sciences, University of Southampton, Southampton SO17 1BJ, UK
e-mail: m.g.dumont@soton.ac.uk

T.C. Charles et al. (eds.), *Functional Metagenomics: Tools and Applications*,
DOI 10.1007/978-3-319-61510-3_8

8.1 Introduction

The microbial world encompasses an enormous diversity of organisms, which for billions of years have evolved to exploit habitats efficiently in order to best access limiting sources of energy and the molecules required for assimilation into cellular components. Today, microbial activity is critical for all life on our planet. For example, our microbiome is critical to our health; microorganisms are essential for soil formation and plant growth promotion and play an irreplaceable role in global biogeochemical cycles. Despite their enormous importance, the vast majority of species associated with Earth's microbial diversity has not yet been isolated and studied under laboratory conditions; the physiology of microbial "dark matter" remains largely uncharacterized (Rinke et al. 2013). Basic research on uncultivated microbial species sheds light on novel metabolism, physiology, and genetics of uncultivated microorganisms, and from the perspective of applied research, their genomes comprise a vast untapped resource of genes encoding proteins and enzymes with potential benefits to human society. These benefits include the potential for discovery of novel antibiotics, detergents, and biopolymers or the ability to synthesize fine chemicals (Wackett 2015). Mining this extensive microbial resource for valuable products, especially from complex microbial communities, offers strong potential for immediate benefits to human society.

Although the majority of microorganisms resist cultivation under conventional laboratory conditions (Torsvik and Øvreås 2002), the advent of cultivation-independent approaches has resulted in a massive increase in the number of recognized bacterial phyla, from 12 in 1987 to 52 by 2003 (Rappe and Giovannoni 2003), and this number continues to increase (Rinke et al. 2013; Hug et al. 2016). Circumventing the recalcitrance of most microbial taxa to laboratory cultivation, metagenomic approaches offer an alternative for investigating the diversity and genetic potential of microbial communities in environmental samples. Metagenomics, also known as ecological genomics, community genomics, or environmental genomics, is most commonly applied by sequencing extracted environmental DNA directly, then using in silico methods to analyze the resulting sequence data. Sequencing-based approaches are powerful tools for microbial ecologists, enabling the exploration of microbial community composition and comparison of multiple samples to one another (Riesenfeld et al. 2004; Gilbert and Dupont 2011; Simon and Daniel 2011; Weinstock 2012). Indeed, the advent of high-throughput sequencing technologies has helped establish metagenomics as a commonplace approach for microbial community analyses within microbial ecology.

In many instances, the specific function of open reading frames (ORFs) associated with metagenomes can be inferred by sequence comparisons to databases of known protein families, including KEGG (Kanehisa et al. 2012), COG (Tatusov et al. 2003), EggNOG (Powell et al. 2012), M5nr (Wilke et al. 2012), and MetaCyc (Caspi et al. 2012). Homology-based approaches for metagenome annotation enable sample-sample descriptions and comparisons of inferred functional profiles (e.g., Turnbaugh et al. 2009; Greenblum et al. 2011; Abubucker et al. 2012; Huttenhower et al. 2012; Yatsunenko et al. 2012). Such approaches can provide accurate functional predictions that represent a basis for hypothesis testing. Proposed functions of genes encoded by uncultivated organisms can also be used to design cultivation strategies to isolate or enrich uncultivated microorganisms in the laboratory. For example, Tank and Bryant (2015) provide a fascinating account of how metagenomics, in combination with other approaches, led to the isolation of *Chloracidobacterium thermophilum*, which is an unusual photosynthetic heterotroph present in hot spring microbial mats. Despite the successes and commonplace application of sequence-based metagenomic surveys, the majority of ORFs obtained from any given metagenome, or genome, cannot be annotated. This is a major limitation; we do not know what most genes of most microorganisms do. This is especially true for novel and uncultivated microorganisms. As a result, alternative approaches are essential for interrogating unknown ORF functions and the broader niches, including metabolism and ecological roles, of uncultivated microorganisms.

Several cultivation-independent approaches are available to identify the function of specific microorganisms in the environment. Microautoradiography (MAR), coupled with fluorescence in situ hybridization (FISH), can be used to identify microorganisms that incorporate a radiolabeled substrate (Kindaichi et al. 2004). Both the proportion of cells labeled in a sample and the extent of radiolabel incorporated can be determined. Although powerful in principle, this approach is limited by the inherent sensitivity of detection methods and the availability of suitable radioactive elements. An alternative approach is to detect the incorporation of stable isotopes (e.g., ^{3}H, ^{13}C, ^{15}N, ^{18}O, and ^{34}S) into biomass using nanoscale secondary ion mass spectrometry (NanoSIMS). Using this approach, Orphan et al. (2001) used the light $\delta^{13}C$ signature of methane to demonstrate that an archaeal partner in microbial consortia from anoxic marine sediment consumed methane. Although this study could take advantage of the unique natural $\delta^{13}C$ signature of methane, alternatively substrates can be enriched artificially with heavy isotopes and applied to samples to track their incorporation (Musat et al. 2012). Raman spectroscopy can also detect differences in isotope abundances in cells (Wagner 2009; Read and Whiteley 2011; Li et al. 2014). Although this approach is not as sensitive and precise as NanoSIMS, it has the advantage of being nondestructive; cells can subsequently be retrieved for additional analyses.

Stable-isotope probing (SIP) is an approach where samples are incubated with a labeled substrate and the labeled nucleic acids, either DNA, rRNA, or mRNA, are isolated by density-gradient ultracentrifugation. Of these nucleic acid-based approaches, DNA-SIP was first developed (Radajewski et al. 2000), followed by rRNA-SIP (Manefield et al. 2002) and then mRNA-SIP (Huang et al. 2009; Dumont et al. 2011). Most SIP studies have been performed with ^{13}C-labeled substrates, but the method can also monitor incorporation of other isotopes potentially incorporated into nucleic acids, such as ^{15}N, ^{18}O, and ^{2}H (Uhlik et al. 2013). Sequencing the 16S rRNA or functional genes from labeled DNA or RNA can help identify microorganisms that incorporated labeled substrate. Alternatively, coupling of SIP with metagenomics (Chen and Murrell 2010) or metatranscriptomics (Dumont et al. 2013) can be used to reveal genomic sequences or transcript profiles of labeled populations, respectively. Demonstrating the broad versatility of SIP, the approach can also be extended to labeling and analysis of other biomarkers, such as phospholipid fatty acids (Boschker et al. 1998; Tillmann et al. 2005) and proteins (Jehmlich et al. 2008a, b, 2009).

Despite the widespread use of metagenomics, and SIP, relatively few studies have leveraged these natural methodological companions for accessing functional profiles of active-yet-uncultivated microorganisms. Furthermore, this book highlights the critical importance of functional metagenomics, largely based on expressed activity, for discovery of novel genes. We specifically discuss studies that have combined SIP and functional metagenomics in order to better access novel functions encoded by genes of uncultivated microorganisms. In all studies using DNA-SIP, and metagenomics, microorganisms possessing genes of interest are outnumbered in the original microbial community, often by orders of magnitude. In the absence of suitable selections for a desired phenotype, combining DNA-SIP and metagenomics enables researchers to circumvent a needle-in-the-haystack situation that would otherwise require extensive effort, high costs, and relatively high risk of failure, in the search for novel functions or target sequences of interest. In addition to reviewing key literature relevant to DNA-SIP and functional metagenomics, we outline various experimental strategies and highlight advantages and disadvantages of SIP in relation to functional metagenomic analyses.

8.2 SIP and Metagenomics

SIP has been combined with metagenomics as an approach to link the identity of microorganisms and metabolic functions (reviewed by Friedrich 2006; Chen and Murrell 2010; Pinnell et al. 2011). The main benefits of combining DNA-SIP with metagenomics include the detection of low abundance species within metagenomic datasets and obtaining functional and genomic insights into specific microbial populations that assimilate specific labeled substrates. The first example of DNA-SIP coupled with functional metagenomics involved the enrichment of glycerol-fermenting microorganisms with ^{13}C-labeled substrate, followed by subsequent cloning of labeled DNA (Schwarz et al. 2006). The heavy DNA metagenomic library was screened for genes encoding coenzyme B_{12}-dependent glycerol dehydratases, demonstrating that these target genes occurred at higher proportions in the library constructed from labeled DNA compared to those derived from unlabeled DNA. Subsequently, Dumont et al. (2006) performed DNA-SIP with labeled methane and cloned the DNA into a BAC plasmid vector. They screened for methane monooxygenase genes and recovered a 15-kb genome fragment from an uncultivated *Methylocystis* species that contained genes encoding methane monooxygenase. A similar approach was used to clone ^{13}C-enriched DNA from river sediment that was incubated with ^{13}C-biphenyl (Sul et al. 2009). A 31.8-kb genomic fragment containing genes of biphenyl dioxygenase subunits was identified on a cosmid clone and sequenced. In subsequent experiments, multiple displacement amplification (MDA) was used to amplify small amounts of ^{13}C-DNA obtained from SIP incubations, which allowed for shorter incubations and lower concentrations of substrates to be used (Chen et al. 2008; Neufeld et al. 2008). Both of these studies used cloning and screening to select genomic fragments of interest. Kalyuzhnaya et al. (2008)

were the first to use a shotgun approach to sequence ^{13}C-labeled DNA randomly to then reconstruct a nearly complete genome of an uncultivated strain of *Methylotenera mobilis* from the environmental sample.

The coupling of SIP for metagenomic analysis has become more frequent as sequencing technology becomes less expensive and computational resources for metagenomic analyses become better established. Indeed, several studies have now combined DNA-SIP and metagenomics (Table 8.1), and a detailed protocol for performing metagenomics from labeled DNA has been published previously (Chen et al. 2010). Although there are still some obstacles to overcome in SIP-based metagenomic studies, such as minimizing the generation of sequence artifacts during MDA (Binga et al. 2008), direct sequencing of small quantities of heavy DNA is now possible with high-throughput sequencing (Pinnell et al. 2014).

Table 8.1 Studies prior to 2016 combining DNA-SIP with metagenomics

Habitat	Substrate	Vector	Target	References
Forest acidic soils	^{13}C-methane	Bacterial artificial chromosome (BAC)	Methanotrophs	Dumont et al. (2006)
Marine sediment	^{13}C-glycerol	Plasmid	Glycerol-fermenting microorganisms	Schwarz et al. (2006)
Lake sediment and water	^{13}C-substrates (methane, methanol, methylamine, formaldehyde, and formate)	Plasmid	Methylotrophs	Kalyuzhnaya et al. (2008)
Marine surface water	^{13}C-methanol	Fosmid	Methylotrophs	Neufeld et al. (2008)
Acidic peat soil	^{13}C-methane	Fosmid	Methanotrophs	Chen et al. (2008)
Contaminated river sediment	^{13}C-biphenyl	Cosmid	Biphenyl-degrading bacteria	Sul et al. (2009)
Surface water from Canadian tailings ponds	^{13}C-methane	No vector, whole genome amplification	Methanotrophs	Saidi-Mehrabad et al. (2013)
Arctic tundra from Canadian soils	^{13}C-carbohydrates (glucose and cellulose)	No vector, whole genome amplification	Glycoside hydrolases	Pinnell et al. (2014)
Canadian soils: tundra, temperate rainforest, and agricultural	^{13}C-carbohydrates (glucose, cellobiose, xylose, arabinose, and cellulose)	Cosmid	Glycoside hydrolases	Verastegui et al. (2014)
Agricultural soils and lake sediments	$^{13}C_2$-dimethylsulfide	No vector, whole genome amplification	Dimethylsulfide-degrading bacteria	Eyice et al. (2015)

8.3 Functional Metagenomics Using Isotopes

There are several approaches to obtaining metagenomic fragments that can be expressed and characterized in the laboratory in order to identify functions of genes from environmental microorganisms (Fig. 8.1). One possibility is to obtain metagenomic sequences from heavy DNA, using high-throughput sequencing, which are

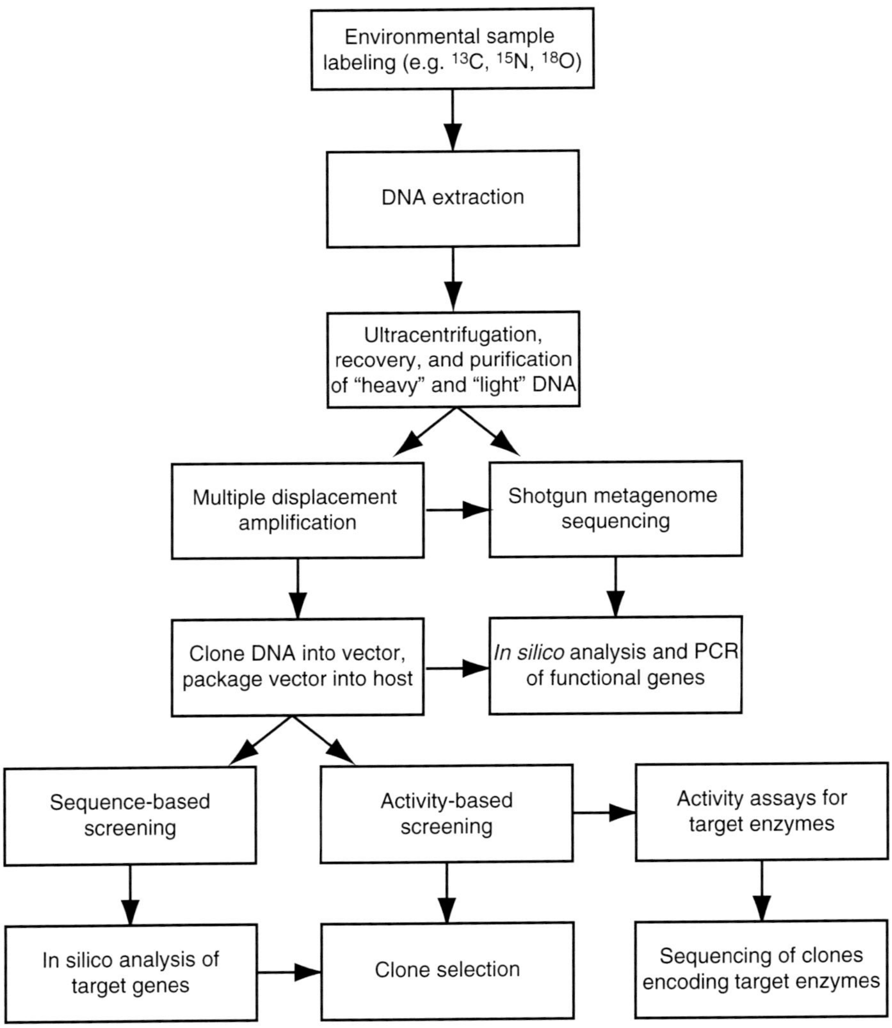

Fig. 8.1 Flow diagram of experimental design for the combination of DNA-SIP and metagenomics

then screened by in silico analyses for genes with potential functions of interest. These genes are then selectively amplified from the original sample using a specific PCR assay, cloned into an expression vector, and screened for activity using various assays. This was the general approach used by Wang et al. (2012) to obtain naphthalene degradation enzymes from contaminated groundwater. An alternative approach is to clone labeled DNA directly into an expression system, followed by screening for functions of interest (Verastegui et al. 2014). Whereas the PCR-based approach has the advantage of eliminating the need to screen clone libraries, the latter approach has a powerful advantage in being able to uncover novel enzymes with functions that could not be predicted based on sequence identity to characterized genes and proteins.

8.3.1 DNA-SIP Combined with Functional Metagenomics

In addition to the pioneering study of Schwarz et al. (2006), Verastegui et al.'s study (2014) is another example of DNA-SIP being used to generate DNA that is analyzed directly by functional metagenomics, involving screening of cloned inserts by function. Building on preliminary work testing sequence-based metagenomics for the recovery of glycoside hydrolases from Arctic tundra (Pinnell et al. 2014), the objective of this work was to recover glycoside hydrolases from diverse soils. The DNA-SIP incubations used common plant-associated carbohydrates (i.e., glucose, cellobiose, xylose, arabinose, and cellulose), and labeled DNA was cloned into a cosmid expression vector that was hosted in *E. coli*. The library was randomly screened for glycoside hydrolase activity using agar supplemented with carboxymethyl cellulose, which was post-stained with Congo red dye, in addition to screening libraries on a variety of 4-methylumbelliferone-labeled fluorogenic substrates for enzyme activity. From these functional screens, eight positive clones were obtained from a library of 2876 random clones, which is a detection rate much higher than previously obtained in similar screens from soils without using SIP enrichment (Kim et al. 2008; Wang et al. 2009; Liu et al. 2011; Nacke et al. 2012).

An alternative to direct cloning and screening of heavy DNA (Table 8.1) involves leveraging functional metagenomics by sequencing labeled DNA to identify putative functional genes, followed by amplification and cloning of the target genes for functional screening. For example, Wang et al. (2012) targeted naphthalene degraders using ^{13}C-naphthalene; the recovered heavy DNA was sequenced, and three naphthalene degradation operons were identified. One of the operons, *nagFCQED*, was amplified from the original sample by designing specific PCR primers. In their case, incorporation of a PCR step helped avoid possible MDA-associated errors that might have prevented correct expression of active enzyme. The authors cloned the operon into a salicylate biosensor system, where active enzyme converts 1,2-dihydroxynaphthalene (DHN) into salicylate, which activates a promoter and triggered the expression of lux-based bioluminescence. This study identified a novel naphthalene-degrading operon that was likely associated with *Acidovorax* sp. WH, the dominant bacterium assimilating naphthalene-derived carbon in their incubations.

8.3.2 Functional Metatranscriptomics and mRNA-SIP

Metatranscriptomics is the sequencing and analysis of gene transcripts obtained from a microbial community of interest. Such an approach, especially when combined with SIP (Fig. 8.1), enables a possible route toward accessing function by heterologous expression of environmental genetic material. In order to clone environmental mRNA and express it in a heterologous host, it must first be reverse transcribed to double-stranded DNA. This cDNA can then be cloned into a suitable expression vector and expressed in a heterologous host, screening for functions of interest. One drawback of targeting mRNA is that it is highly sensitive to enzymatic degradation and, as a result, is challenging to handle. However, methods of purifying mRNA from environmental samples have improved in recent years, and the analysis of mRNA from environmental samples is becoming commonplace.

The transcription of genes in bacteria is often upregulated in response to environmental conditions, which can potentially provide clues as to which genes are necessary for growth under those conditions. For example, if a researcher is interested in discovering novel enzymes for the metabolism of "compound X," then the addition of compound X to an environmental sample is likely to increase the abundance of relevant degradation genes in the metatranscriptome. Therefore, identifying upregulated genes in the metatranscriptome could help identify candidates encoding a novel enzyme of interest. These candidates could be targeted and amplified from the environmental sample by designing specific PCR primers based on the cDNA sequences, which could then be cloned and expressed in a heterologous system.

The combination of SIP and metatranscriptomics was first demonstrated by labeling methane oxidizers in lake sediment with ^{13}C-methane and separating labeled and unlabeled mRNA by ultracentrifugation, followed by RNA-Seq (Dumont et al. 2013). The approach successfully recovered the metatranscriptome from methanotrophs in the sample, indicating that the approach is feasible and that mRNA survives the ultracentrifugation treatment. Two specific advantages of mRNA-SIP compared with DNA-SIP are that mRNA is more rapidly labeled than DNA and the isopycnic density of mRNA is not affected by G + C content (Dumont et al. 2011). As a hypothetical example of a possible experimental approach for functional metatranscriptomics, coupled with SIP, consider a researcher interested in discovering novel genes for metabolism of "compound Y" by methanotrophs. Methanotrophs could be labeled with ^{13}C-methane and then briefly challenged with compound Y before extracting mRNA. Some of the transcripts with higher abundance with the addition of compound Y, compared to the methane addition alone, are likely to be involved in metabolism of the compound and would make ideal targets for further analysis.

8.3.3 *SIP Combined with Raman Microspectroscopy and Functional Metagenomics*

Raman microspectroscopy enables the nondestructive detection of stable isotope enrichment within individual cells. Because it is nondestructive, labeled cells can be recovered using optical tweezers, and their DNA can be amplified and sequenced. This offers a powerful alternative to the SIP approach based on nucleic acids because labeling would be faster and more sensitive.

Previous work by Berry et al. (2015) examined the activity of intestinal microorganisms using heavy water (D_2O) and Raman microspectroscopy. They identified cells with higher metabolic activity after the addition of selected carbohydrates. Active cells were captured using optical tweezers, and MDA was used to amplify the genomic DNA for sequencing. This approach has several advantages over traditional SIP. First, labeling is very quick, and microorganisms do not need to undergo cell division for labeling; DNA-SIP requires several cell divisions of active microorganisms (Neufeld et al. 2007). Another advantage of this approach is that labeled water was used rather than a ^{13}C-labeled growth substrate. By doing this, all active populations are labeled, rather than those involved in the metabolism and assimilation of a specific carbon source. Finally, the approach eliminates the need to separate labeled and unlabeled nucleic acids by ultracentrifugation, which is a step that requires relatively high proportions of isotope enrichment and can be complicated by variable inherent density differences of genomes by G + C content. Extending this method to functional metagenomics is an exciting prospect, but some technical obstacles would need to be overcome, such as potential errors and genetic rearrangements introduced by MDA (Binga et al. 2008; Neufeld et al. 2008) that would impede expression of functional proteins.

8.4 Conclusions and Perspectives

Although still nascent, we believe that functional metagenomics will be an important approach to discerning the function of ORFs encoded by uncultivated microorganisms and will lead scientists toward valuable discoveries. One major challenge is that functional screening is laborious. Nonetheless, SIP methods that are coupled with functional metagenomics can offer the means of focusing the effort and for designing intelligent strategies that suit the research objective.

References

Abubucker S, Segata N, Goll J et al (2012) Metabolic reconstruction for metagenomic data and its application to the human microbiome. PLoS Comput Biol 8:e1002358

Berry D, Mader E, Lee TK et al (2015) Tracking heavy water (D_2O) incorporation for identifying and sorting active microbial cells. Proc Natl Acad Sci U S A 112:194–203

Binga EK, Lasken RS, Neufeld JD (2008) Something from (almost) nothing: the impact of multiple displacement amplification on microbial ecology. ISME J 2:233–241

Boschker HTS, Nold SC, Wellsbury P et al (1998) Direct linking of microbial populations to specific biogeochemical processes by ^{13}C-labelling of biomarkers. Nature 392:801–805

Caspi R, Altman T, Dreher K et al (2012) The MetaCyc database of metabolic pathways and enzymes and the BioCyc collection of pathway/genome databases. Nucleic Acids Res 40:742–753

Chen Y, Murrell JC (2010) When metagenomics meets stable-isotope probing: progress and perspectives. Trends Microbiol 18:157–163

Chen Y, Dumont MG, Neufeld JD et al (2008) Revealing the uncultivated majority: combining DNA stable-isotope probing, multiple displacement amplification and metagenomic analyses of uncultivated *Methylocystis* in acidic peatlands. Environ Microbiol 10:2609–2622

Chen Y, Vohra J, Murrell JC (2010) Applications of DNA-stable isotope probing in bioremediation studies. Methods Mol Biol 599:129–139

Dumont MG, Radajewski SM, Miguez CB et al (2006) Identification of a complete methane monooxygenase operon from soil by combining stable isotope probing and metagenomic analysis. Environ Microbiol 8:1240–1250

Dumont MG, Pommerenke B, Casper P et al (2011) DNA-, rRNA- and mRNA-based stable isotope probing of aerobic methanotrophs in lake sediment. Environ Microbiol 13:1153–1167

Dumont MG, Pommerenke B, Casper P (2013) Using stable isotope probing to obtain a targeted metatranscriptome of aerobic methanotrophs in lake sediment. Environ Microbiol Rep 5:757–764

Eyice Ö, Namura M, Chen Y et al (2015) SIP metagenomics identifies uncultivated *Methylophilaceae* as dimethylsulphide degrading bacteria in soil and lake sediment. ISME J 9:2336–2348

Friedrich MW (2006) Stable-isotope probing of DNA: insights into the function of uncultivated microorganisms from isotopically labeled metagenomes. Curr Opin Biotechnol 17:59–66

Gilbert JA, Dupont CL (2011) Microbial metagenomics: beyond the genome. Annu Rev Mar Sci 3:347–371

Greenblum S, Turnbaugh PJ, Borenstein E (2011) Metagenomic systems biology of the human gut microbiome reveals topological shifts associated with obesity and inflammatory bowel disease. Proc Natl Acad Sci U S A 109:594–599

Huang WE, Ferguson A, Singer AC et al (2009) Resolving genetic functions within microbial populations: *in situ* analyses using rRNA and mRNA stable isotope probing coupled with single-cell Raman-fluorescence *in situ* hybridization. Appl Environ Microbiol 75:234–241

Hug LA, Baker BJ, Anantharaman K et al (2016) A new view of the tree of life. Nat Microbiol 1:16048

Huttenhower C, Gevers D, Knight R et al (2012) Structure, function and diversity of the healthy human microbiome. Nature 486:207–214

Jehmlich N, Schmidt F, Hartwich M et al (2008a) Incorporation of carbon and nitrogen atoms into proteins measured by protein-based stable isotope probing (protein-SIP). Rapid Commun Mass Spectrom 22:2889–2897

Jehmlich N, Schmidt F, von Bergen M et al (2008b) Protein-based stable isotope probing (protein-SIP) reveals active species within anoxic mixed cultures. ISME J 2:1122–1133

Jehmlich N, Schmidt F, Taubert M et al (2009) Comparison of methods for simultaneous identification of bacterial species and determination of metabolic activity by protein-based stable isotope probing (protein-SIP) experiments. Rapid Commun Mass Spectrom 23:1871–1878

Kalyuzhnaya MG, Lapidus A, Ivanova N et al (2008) High-resolution metagenomics targets specific functional types in complex microbial communities. Nat Biotechnol 26:1029–1034
Kanehisa M, Goto S, Sato Y et al (2012) KEGG for integration and interpretation of large-scale molecular data sets. Nucleic Acids Res 40:109–114
Kim SJ, Lee CM, Han BR et al (2008) Characterization of a gene encoding cellulase from uncultured soil bacteria. FEMS Microbiol Lett 282:44–51
Kindaichi T, Ito T, Okabe S (2004) Ecophysiological interaction between nitrifying bacteria and heterotrophic bacteria in autotrophic nitrifying biofilms as determined by microautoradiography-fluorescence in situ hybridization. Appl Environ Microbiol 70:1641–1650
Li M, Boardman DG, Ward A (2014) Single-cell Raman sorting. Methods Mol Biol 1096:147–153
Liu J, Liu WD, Zhao XL et al (2011) Cloning and functional characterization of a novel endo-beta-1,4-glucanase gene from a soil-derived metagenomic library. Appl Microbiol Biotechnol 89:1083–1092
Manefield M, Whiteley AS, Griffiths RI et al (2002) RNA stable isotope probing, a novel means of linking microbial community function to phylogeny. Appl Environ Microbiol 68:5367–5373
Musat N, Foster R, Vagner T, Adam B, Kuypers MM (2012) Detecting metabolic activities in single cells, with emphasis on nanoSIMS. FEMS Microbiol Rev 36:486–511
Nacke H, Engelhaupt M, Brady S et al (2012) Identification and characterization of novel cellulolytic and hemicellulolytic genes and enzymes derived from German grassland soil metagenomes. Biotechnol Lett 34:663–675
Neufeld JD, Dumont MG, Vohra J et al (2007) Methodological considerations for the use of stable isotope probing in microbial ecology. Microb Ecol 53:435–442
Neufeld JD, Chen Y, Dumont MG et al (2008) Marine methylotrophs revealed by stable-isotope probing, multiple displacement amplification and metagenomics. Environ Microbiol 10:1526–1535
Orphan VJ, House CH, Hinrichs KU et al (2001) Methane-consuming archaea revealed by directly coupled isotopic and phylogenetic analysis. Science 293:484–487
Pinnell LJ, Charles TC, Neufeld JD (2011) Stable-isotope probing and metagenomics. In: Murrell JC, Whiteley AS (eds) Stable isotopes in microbial molecular ecology. ASM Press, Washington, DC, pp 97–114
Pinnell LJ, Dunford E, Ronan P et al (2014) Recovering glycoside hydrolase genes from active tundra cellulolytic bacteria. Can J Microbiol 60:469–476
Powell S, Szklarczyk D, Trachana K et al (2012) eggNOG v3.0: orthologous groups covering 1133 organisms at 41 different taxonomic ranges. Nucleic Acids Res 40:284–289
Radajewski S, Ineson P, Parekh NR et al (2000) Stable-isotope probing as a tool in microbial ecology. Nature 403:646–649
Rappe MS, Giovannoni SJ (2003) The uncultured microbial majority. Annu Rev Microbiol 57:369–394
Read DS, Whiteley AS (2011) Identity and function of single microbial cells within a community by Raman Microspectroscopy and related single-cell techniques. In: Sen K, Ashbolt NJ (eds) Environmental microbiology: current technology and water applications. Horizon Press, Poole, pp 163–178
Riesenfeld CS, Schloss PD, Handelsman J (2004) Metagenomics: genomic analysis of microbial communities. Annu Rev Genet 38:525–552
Rinke C, Schwientek P, Sczyrba A et al (2013) Insights into the phylogeny and coding potential of microbial dark matter. Nature 499:431–437
Saidi-Mehrabad A, He Z, Tamas I et al (2013) Methanotrophic bacteria in oilsands tailings ponds of northern Alberta. ISME J 7:908–921
Schwarz S, Waschkowitz T, Daniel R (2006) Enhancement of gene detection frequencies by combining DNA-based stable-isotope probing with the construction of metagenomic DNA libraries. World J Microbiol Biotechnol 22:363–367
Simon C, Daniel R (2011) Metagenomic analyses: past and future trends. Appl Environ Microbiol 77:1153–1161

Sul WJ, Park J, Quensen JF 3rd et al (2009) DNA-stable isotope probing integrated with metagenomics for retrieval of biphenyl dioxygenase genes from polychlorinated biphenyl-contaminated river sediment. Appl Environ Microbiol 75:5501–5506
Tank M, Bryant DA (2015) Nutrient requirements and growth physiology of the heterotrophic Acidobacterium, *Chloracidobacterium thermophilum*. Front Microbiol 6:226
Tatusov RL, Fedorova ND, Jackson JD et al (2003) The COG database: an updated version includes eukaryotes. BMC Bioinformatics 4:41
Tillmann S, Strömpl C, Timmis KN et al (2005) Stable isotope probing reveals the dominant role of *Burkholderia* species in aerobic degradation of PCBs. FEMS Microbiol Ecol 52:207–217
Torsvik V, Øvreås L (2002) Microbial diversity and function in soil: from genes to ecosystems. Curr Opin Microbiol 5:240–245
Turnbaugh PJ, Hamady M, Yatsunenko T et al (2009) A core gut microbiome in obese and lean twins. Nature 457:480–484
Uhlik O, Leewis MC, Strejcek M et al (2013) Stable isotope probing in the metagenomics era: a bridge towards improved bioremediation. Biotechnol Adv 31:154–165
Verastegui Y, Cheng J, Engel K et al (2014) Multisubstrate isotope labeling and metagenomic analysis of active soil bacterial communities. mBio 5:e01157–e01114
Wackett LP (2015) Specialty chemicals from microbes: an annotated selection of World Wide Web sites relevant to the topics in microbial biotechnology. Microb Biotechnol 8:614–615
Wagner M (2009) Single-cell ecophysiology of microbes as revealed by Raman microspectroscopy or secondary ion mass spectrometry imaging. Annu Rev Microbiol 63:411–429
Wang F, Li F, Chen G et al (2009) Isolation and characterization of novel cellulase genes from uncultured microorganisms in different environmental niches. Microbiol Res 164:650–657
Wang Y, Chen Y, Zhou Q et al (2012) A culture-independent approach to unravel uncultured bacteria and functional genes in a complex microbial community. PLoS One 7:e47530
Weinstock GM (2012) Genomic approaches to studying the human microbiota. Nature 489:250–256
Wilke A, Harrison T, Wilkening J et al (2012) The M5nr: a novel non-redundant database containing protein sequences and annotations from multiple sources and associated tools. BMC Bioinformatics 13:141
Yatsunenko T, Rey FE, Manary MJ et al (2012) Human gut microbiome viewed across age and geography. Nature 486:222–227

Chapter 9
Metagenome Assembly and Functional Annotation

Adina Howe, Fan Yang, and Qingpeng Zhang

Abstracts This chapter reviews metagenome de novo assembly and currently available assembly algorithms and tools. Challenges and opportunities from metagenomic assembly are presented, with a summary of a typical metagenomic assembly workflow. Finally, approaches to reduce large datasets and identify the functions of assembled genes are also discussed.

Metagenomic assembly is the process of reconstructing multiple, possibly thousands of, genomes from sequencing reads. In contrast to single-genome assembly, metagenome assembly involves sequencing reads originating from *multiple* species that are present at *variable* abundances. Additionally, these assemblies often are associated with high volumes of data that may represent variable types of sequences (e.g., paired and unpaired ends or varying lengths and insert sizes). In this review, we discuss metagenome de novo assembly, as it is often the case that little is known about source genomes in environmental samples. There is no single assembly pipeline that can be applied to all metagenomic sequencing datasets. The choice of an assembly algorithm or tool depends on the question at hand and the characteristics of the sequencing data to be assembled (e.g., the complexity and coverage of the sample).

All assemblers use the assumption that highly similar sequencing reads originate from the same genome region and attempt to merge overlapping sequencing reads into a longer contiguous consensus sequence, most commonly called a contig. These contigs can subsequently be further assembled into scaffolds using information from the sequencing library preparation that inform contig order,

---

Adina Howe, Fan Yang and Qingpeng Zhang contributed equally to this work.

A. Howe (✉) • F. Yang
Department of Agricultural and Biosystems Engineering,
Iowa State University, Ames, IA, USA
e-mail: adina@iastate.edu

Q. Zhang
Department of Energy, Joint Genome Institute, Walnut Creek, CA, USA

T.C. Charles et al. (eds.), *Functional Metagenomics: Tools and Applications*,
DOI 10.1007/978-3-319-61510-3_9

orientation, and distance (e.g., insert sizes, paired-end information). The definition of a consensus sequence in a contig varies per assembler and can incorporate various heuristics including information about the frequency of alignment at each base pair (e.g., base pair sequencing coverage), the quality scores of aligned base pairs, or available paired-end information. Readers interested in more details describing specific assemblers and algorithms would benefit from the following reviews (Miller et al. 2010; Nagarajan and Pop 2013).

The main challenge for assembly arises when similar sequences do *not* as assumed originate from the same region of a genome. Examples of such cases are repetitive sequences and sequencing errors. Repetitive sequences challenge assemblers because of the difficulty in distinguishing where repetitive genomic regions originate, especially when repeats are longer than reads. Sequencing errors are also problematic for assembly and lead to erroneous sequence alignments. Given the genomic complexity associated with conserved repetitive gene regions between or within genomes and the presence of sequencing errors and artifacts, metagenomic assemblies often contain chimeric, or falsely joined, contigs. A solution for this problem is to combine sequencing reads from varying technologies, such as high volumes of short reads (e.g., Illumina) with long reads with variable insert sizes (e.g., PacBio), but this can significantly increase the budget needed for a project.

For computational memory efficiency, all assemblers convert sequencing reads into a set of strings and attempt to find the shortest common superstring of a set of sequences. These strings can be represented in a data structure called a graph, where nodes and edges of a graph are components of strings. Nodes and edges of assembly graphs often represent overlaps between sequences so that a path drawn through the graph represents consensus contiguous sequences (Fig. 9.1). For metagenomics,

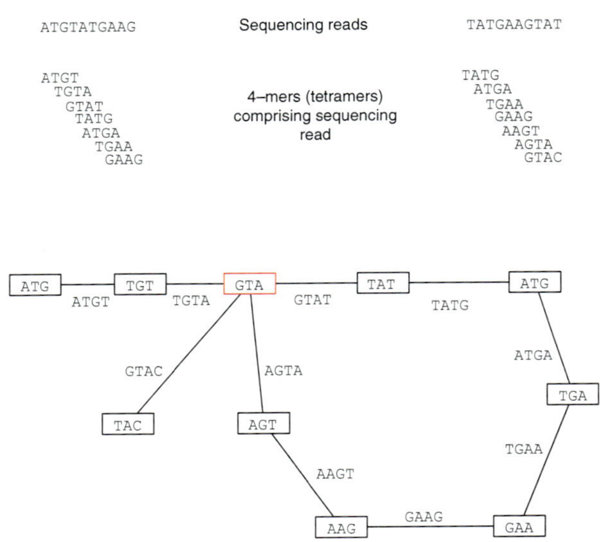

Fig. 9.1 An example of a representative graph constructed from two sequencing reads where the edges are representative of 4-mers. The *red box* is a repeat sequencing element shared between the two sequences

there are two main categories of assemblers, varying in how their graphs are constructed: the overlap/layout/consensus (OLC) and the de Bruijn graph (DBG) methods, reviewed in Li et al. (2012). Historically, OLC assemblers have been more effective for longer read lengths and required significant computational power because it computes all pairwise overlaps in a dataset. The development of the string graph assembler (Simpson and Durbin 2012) has reduced the computational requirements of the application of OLC leveraging efficient methods of string indexing. DBG assemblers do not explicitly calculate pairwise overlaps and thus require less computational power for large volumes of sequencing data. The majority of metagenomic assemblers use DBG methods, including Velvet (Zerbino and Birney 2008), SOAP (Li et al. 2008), and Megahit (Li et al. 2015).

Though the specific workflow for metagenomic assembly varies depending on the choice of assembler and algorithms applied, most analyses share common steps to identify a consensus assembled metagenomic reference (Box 9.1).

Box 9.1:Common Workflow for Metagenomic Assembly

1. Sequencing libraries are initially evaluated and trimmed by quality scores.
2. Data reduction or subdivision to reduce large volumes of complex data ((Brown et al. 2012), see Sect. 9.1).
3. Sequencing reads and associated library metadata (paired end, insert size) are provided to an assembler and merged into contigs.
4. Scaffolding software used to identify supercontigs or scaffolds.
5. Assembly quality is evaluated using read coverage and library preparation data.

9.1 Data Reduction: Partitioning, Binning, and Read Clustering

Various methods have been developed to address the complexity of analyzing metagenome datasets. Taxonomic binning is an approach that groups sequences into different clusters, or bins, representing different taxonomic units. The abundance of each taxonomic bin can be subsequently used to evaluate the taxonomic structure and diversity represented in a dataset. Most binning methods are applied to assembled contigs because these methods generally perform better on longer sequences. However, when assembled contigs are unavailable or difficult to resolve, binning unassembled reads directly into taxonomic units can offer valuable insights into the diversity (Mande et al. 2012). Additionally, directly binning reads can benefit other downstream analysis, including assembly. For example, individual bins can be assembled separately, often resulting in more efficient and possibly more effective assembly.

Generally, there are two types of sequence binning methods: taxonomy dependent and taxonomy independent. The "taxonomy-dependent" approach identifies the specific species or taxonomic unit of each read. This approach often involves supervised learning methods or classifiers. Alternately, reads can be clustered without assigning sequences into specific nodes of a taxonomic tree. Such "taxonomy-independent" methods often involve unsupervised learning, resulting in clustering sets of sequences by their sequence similarity.

Binning methods can employ similarity-based and/or composition-based approaches. For similarity-based methods, external reference sequences are used to guide binning. These reference databases can include known genomes or nucleotide/protein sequences of interesting marker genes. Reference and metagenomic sequences are often compared using sequence alignment or mapping tools. For example, a similarity-based binning program, MEGAN (Huson et al. 2007), utilizes BLAST (see Sect. 9.2) to calculate sequence similarity between metagenomic sequences and reference sequences originating from the National Center for Biotechnology Information (NCBI) reference database. MG-RAST (Meyer et al. 2008) employs a similar approach using a similar alignment tool, BLAT, to assign metagenomic sequences to different taxonomic units. Comparing metagenomic reads to reference models, such as Hidden Markov Models (HMMs), can also be done (Krause et al. 2008). A number of programs and software employ marker genes as references for binning including MetaPhlAn (Segata et al. 2012), MetaPhyler (Liu et al. 2011), PhyloSift (Darling et al. 2014), and AMPHORA (Wu and Eisen 2008). Composition-based methods use sequence characteristics, often tetramer or 4-mer frequency, to bin sequences. The distribution of these sequence characteristics can be evaluated and used to build classifier models that are used for binning. Tools that employ these methods include PhyloPythia (McHardy et al. 2007) and PhyloPythiaS (Patil et al. 2012), which employ support vector machine classifiers; Phymm (Brady and Salzberg 2009) and TETRA (Teeling et al. 2004), which use Markov models; NBC (Rosen et al. 2011) and RDP classifier (Wang et al. 2007), which use naïve Bayesian classifiers; and eSOMs (Dick et al. 2009) and TF-ESOMs (Wrighton et al. 2012), which use emergent self-organizing maps. An extensive review on these binning approaches is available (Mande et al. 2012).

New approaches for metagenomic sequence binning have recently been developed based on the abundance distribution of reads across different samples. This binning method is based on the assumption that sequences with similar abundance profiles across different samples are prone to originate from the same species (Albertsen et al. 2013; Karlsson et al. 2013; Sharon et al. 2013). Tools employing these methods include CONCOCT (Alneberg et al. 2014), GroopM (Imelfort et al. 2014), and MetaBAT (Kang et al. 2015). An important requirement of this approach is that data be available from multiple samples. Also, these tools are best applied on longer reads or assembled contigs. A notable new approach binning reads based on abundances has recently been developed (Cleary et al. 2015), indicating that the future is quite promising for binning tools.

9.2 Functional Annotation

Once metagenomic assembly or binning is complete, the next step of analysis is often the functional annotation of sequences. Microbial gene annotation uses the similarity of sequences to reference genes or proteins to identify potential gene functions. This process relies on the existence and quality of current reference datasets or databases.

Generalized databases are available and represent microbial functions that have been previously observed and deposited into searchable databases. These databases contain gene or protein sequences that represent functions that have been validated in the laboratory as well as sequences that have been observed but whose function is unknown (e.g., hypothetical proteins). The NCBI (http://www.ncbi.nlm.nih.gov/), Joint Genome Institute Integrated Microbial Genomes (JGI-IMG: https://img.jgi.doe.gov/cgi-bin/edu/main.cgi), and Universal Protein Resource (UniProt: http://www.uniprot.org/) are the most widely used generalized databases.

Full-genome databases are collections of annotated individual, often non-redundant, microorganisms. Besides gene functions, full-genome databases also include linked gene nucleotide and protein sequences, genome information, and organism classification. These databases are particularly useful for annotating the taxonomic origins of metagenomic reads. The NCBI RefSeq (Pruitt et al. 2007; Tatusova et al. 2014, 2015) and JGI-IMG (Nordberg et al. 2014) are the two major reference genome databases. To date (RefSeq Release 73), RefSeq contains 55,966 organisms, among which 3563 are fungi; 39,567 are bacteria; 919 are archaea; and 5027 are viruses. JGI-IMG contains 32,859 organisms, including 25,871 bacteria, 532 archaea, 3888 viruses, and 87 fungi. JGI-IMG is a smaller collection of genomes, mostly sequenced by JGI. This collection overlaps with RefSeq and can be linked through NCBI accession and JGI-IMG identifiers. A challenge for gene annotation is linking information and resolving inconsistencies across different databases, and caution must be exercised when comparing functional genes analyzed using different databases.

Many genes have been experimentally studied without having full genomes available and are deposited into gene collection databases. These gene collections, representing the largest databases of potential microbial functions, are often used for metagenomic functional annotation. The NCBI hosts the non-redundant nucleotide collection (nt) and non-redundant protein sequence collection (nr), where nt contains 33,791,422 nucleotide sequences and nr contains 77,881,237 protein sequences. Both nt and nr include microbial and nonmicrobial sequences. Additionally, protein sequences of microbial genes are available from the UniProt consortium (The UniProt Consortium 2014). These databases, including a manually curated literature-based protein sequence database (Swiss-Prot) and a machine-annotated unreviewed protein sequence database (TrEMBL), can be used for metagenomic protein annotations. To date, Swiss-Prot contains 550,116 protein sequences (19,370 archaeal; 332,327 bacterial; 31,732 fungal; and 16,605 viral) and

TrEMBL contains 55,270,679 protein sequences (1,079,783 archaeal; 33,700,407 bacterial; 5,554,136 fungal; and 2,447,876 viral).

Some databases also employ orthologous hierarchical method to classify genes, where genes are grouped based on similar functions. For example, genes encoding phosphoribulokinase and transketolase can be grouped together as part of the reductive pentose phosphate cycle, which is a subset of carbon fixation processes. Orthologous hierarchical grouping allows users to reduce annotation complexity according to functional relevance. Reference databases that employ such hierarchies include:

The Kyoto Encyclopedia of Genes and Genomes (KEGG: http://www.genome.jp/kegg/kegg1.html) (Kanehisa et al. 2004)
Evolutionary genealogy of genes: Non-supervised Orthologous Groups (eggNOG: http://eggnogdb.embl.de/#/app/home) (Jensen et al. 2008)
Clusters of Orthologous Groups/EuKaryotic Orthologous Groups (COG/KOG: http://genome.jgi.doe.gov/help/kogbrowser.jsf) (Tatusov et al. 2003)
Pfam (http://pfam.xfam.org/) (Finn et al. 2010)

While generalized databases are useful for broad annotation of sequencing reads, specialized databases focus on specific functions, which can specifically target annotation efforts. For example, the Carbohydrate-Active enZYmes Database (CAZY: http://www.cazy.org/) (Lombard et al. 2014) is a specialized collection of carbohydrate-active enzyme genes that perform roles in carbohydrate metabolisms and catabolism. Currently, CAZY contains genes from 3940 bacterial genomes, 220 archaeal genomes, and 277 virus genomes. These organisms are also linked to NCBI via taxonomy ID. Other examples of specialized databases include the RESFAMS antibiotic resistance database (http://www.dantaslab.org/resfams/) (Gibson et al. 2014) and the Comprehensive Antibiotic Resistance Database (CARD: http://arpcard.mcmaster.ca/) (McArthur et al. 2013). The Functional Gene Repository, hosted by the Ribosomal Database Project (RDP FunGene: http://fungene.cme.msu.edu/) (Fish et al. 2013), also contains a broad range of ecologically relevant gene databases.

A number of tools have been developed to compare metagenome sequences and gene reference databases, varying in their specificity, sensitivity, and speed. Developed by NCBI, BLAST+ (Camacho et al. 2009), a new version of BLAST (Basic Local Alignment Search Tools) (Altschul et al. 1990), performs a local alignment search between query sequences and sequences in a database. BLAST+ has a variety of available comparisons, including nucleotide sequence comparison (blastn), protein-to-protein sequence comparison (blastp), and nucleotide-to-protein sequence comparison (blastx). Similar to BLAST+, DIAMOND performs local alignments between query sequences and database sequences (Buchfink et al. 2014). DIAMOND can currently perform local alignments between nucleotide and protein sequences 20,000 times faster than blastx. This speed makes DIAMOND particularly suitable for the annotation of high volumes of metagenomic data. HMMER (http://hmmer.org/) is another metagenomic annotation tool that performs alignments and comparison using hidden Markov models (HMMs) (Finn et al. 2015).

The use of HMMs allows for comparisons that include the underlying structural profile of references, though at a cost of speed. The current version of HMMER (v3.1) is comparable to blastp in speed and is 5–10 times slower than blastn. Further, because HMMER is a profile-driven alignment search, it can only perform nucleotide-to-nucleotide sequence or protein-to-protein sequence comparisons.

With rapidly increasing volumes and types of sequencing datasets, we are facing many challenges in gene annotation, particularly to accurately and efficiently provide gene annotations. Importantly, errors in databases are present and will most certainly affect annotations. Further, sequence similarity does not necessarily imply functional similarity, and inferred gene functions should always be evaluated carefully. In current metagenomes, a large number of genes (over 30%) (Bench et al. 2007; Qin et al. 2010; Lamendella et al. 2011) do not share similarity with known genes, highlighting a critical need to continue building better gene references.

References

Albertsen M, Hugenholtz P, Skarshewski A et al (2013) Genome sequences of rare, uncultured bacteria obtained by differential coverage binning of multiple metagenomes. Nat Biotechnol 31:533–538. doi:10.1038/nbt.2579

Alneberg J, Bjarnason BS, de Bruijn I et al (2014) Binning metagenomic contigs by coverage and composition. Nat Methods 11:1144–1146. doi:10.1038/nmeth.3103

Altschul SF, Gish W, Miller W et al (1990) Basic local alignment search tool. J Mol Biol 215(3):403–410. doi:10.1016/S0022-2836(05)80360-2

Bench SR, Hanson TE, Williamson KE et al (2007) Metagenomic characterization of Chesapeake Bay virioplankton. Appl Environ Microbiol 73:7629–7641. doi:10.1128/AEM.00938-07

Brady A, Salzberg SL (2009) Phymm and PhymmBL: metagenomic phylogenetic classification with interpolated Markov models. Nat Methods 6:673–676. doi:10.1038/nmeth.1358

Brown CT, Howe A, Zhang Q, et al (2012) A reference-free algorithm for computational normalization of shotgun sequencing data. arXiv 1203.4802:1–18. doi: 10.1128/genomeA.00802-14. Copyright

Buchfink B, Xie C, Huson DH (2014) Fast and sensitive protein alignment using DIAMOND. Nat Methods 12:59–60. doi:10.1038/nmeth.3176

Camacho C, Coulouris G, Avagyan V et al (2009) BLAST+: architecture and applications. BMC Bioinformatics 10:421. doi:10.1186/1471-2105-10-421

Cleary B, Brito IL, Huang K et al (2015) Detection of low-abundance bacterial strains in metagenomic datasets by eigengenome partitioning. Nat Biotechnol 33(10):1053–1060

Darling AE, Jospin G, Lowe E et al (2014) PhyloSift: phylogenetic analysis of genomes and metagenomes. PeerJ 2:e243. doi:10.7717/peerj.243

Dick GJ, Andersson AF, Baker BJ et al (2009) Community-wide analysis of microbial genome sequence signatures. Genome Biol 10:R85. doi:10.1186/gb-2009-10-8-r85

Finn RD, Mistry J, Tate J et al (2010) The Pfam protein families database. Nucleic Acids Res 38:D211–D222. doi:10.1093/nar/gkm960

Finn RD, Clements J, Arndt W et al (2015) HMMER web server: 2015 update. Nucleic Acids Res 43:W30–W38. doi:10.1093/nar/gkv397

Fish JA, Chai B, Wang Q et al (2013) FunGene: the functional gene pipeline and repository. Front Microbiol 4:1–14. doi:10.3389/fmicb.2013.00291

Gibson MK, Forsberg KJ, Dantas G (2014) Improved annotation of antibiotic resistance determinants reveals microbial resistomes cluster by ecology. ISME J 9:1–10. doi:10.1038/ismej.2014.106

Huson DH, Auch AF, Qi J, Schuster SC (2007) MEGAN analysis of metagenomic data. Genome Res 17:377–386. doi:10.1101/gr.5969107
Imelfort M, Parks D, Woodcroft BJ et al (2014) GroopM: an automated tool for the recovery of population genomes from related metagenomes. PeerJ 2:e603. doi:10.7717/peerj.603
Jensen LJ, Julien P, Kuhn M et al (2008) eggNOG: automated construction and annotation of orthologous groups of genes. Nucleic Acids Res 36:250–254. doi:10.1093/nar/gkm796
Kanehisa M, Goto S, Kawashima S et al (2004) The KEGG resource for deciphering the genome. Nucleic Acids Res 32:D277–D280. doi:10.1093/nar/gkh063
Kang DD, Froula J, Egan R, Wang Z (2015) MetaBAT, an efficient tool for accurately reconstructing single genomes from complex microbial communities. PeerJ 3:e1165
Karlsson FH, Tremaroli V, Nookaew I et al (2013) Gut metagenome in European women with normal, impaired and diabetic glucose control. Nature 498:99–103. doi:10.1038/nature12198
Krause L, Diaz NN, Goesmann A et al (2008) Phylogenetic classification of short environmental DNA fragments. Nucleic Acids Res 36:2230–2239. doi:10.1093/nar/gkn038
Lamendella R, Domingo JWS, Ghosh S et al (2011) Comparative fecal metagenomics unveils unique functional capacity of the swine gut. BMC Microbiol 11:103. doi:10.1186/1471-2180-11-103
Li R, Li Y, Kristiansen K, Wang J (2008) SOAP: short oligonucleotide alignment program. Bioinformatics 24:713–714. doi:10.1093/bioinformatics/btn025
Li Z, Chen Y, Mu D et al (2012) Comparison of the two major classes of assembly algorithms: overlap-layout-consensus and de-bruijn-graph. Brief Funct Genomics 11:25–37
Li D, Liu C-M, Luo R et al (2015) MEGAHIT: an ultra-fast single-node solution for large and complex metagenomics assembly via succinct de Bruijn graph. Bioinformatics 31(10):1674–1676. doi:10.1093/bioinformatics/btv033
Liu B, Gibbons T, Ghodsi M et al (2011) Accurate and fast estimation of taxonomic profiles from metagenomic shotgun sequences. BMC Genomics 12:S4. doi:10.1186/1471-2164-12-S2-S4
Lombard V, Golaconda Ramulu H, Drula E et al (2014) The carbohydrate-active enzymes database (CAZy) in 2013. Nucleic Acids Res 42:490–495. doi:10.1093/nar/gkt1178
Mande SS, Mohammed MH, Ghosh TS (2012) Classification of metagenomic sequences: methods and challenges. Brief Bioinform 13:669–681. doi:10.1093/bib/bbs054
McArthur AG, Waglechner N, Nizam F et al (2013) The comprehensive antibiotic resistance database. Antimicrob Agents Chemother 57:3348–3357. doi:10.1128/AAC.00419-13
McHardy AC, Martín HG, Tsirigos A et al (2007) Accurate phylogenetic classification of variable-length DNA fragments. Nat Methods 4:63–72. doi:10.1038/nmeth976
Meyer F, Paarmann D, D'Souza M et al (2008) The metagenomics RAST server - a public resource for the automatic phylogenetic and functional analysis of metagenomes. BMC Bioinformatics 9:386. doi:10.1186/1471-2105-9-386
Miller JR, Koren S, Sutton G (2010) Assembly algorithms for next-generation sequencing data. Genomics 95:315–327. doi:10.1016/j.ygeno.2010.03.001
Nagarajan N, Pop M (2013) Sequence assembly demystified. Nat Rev Genet 14:157–167. doi:10.1038/nrg3367
Nordberg H, Cantor M, Dusheyko S et al (2014) The genome portal of the Department of Energy Joint Genome Institute: 2014 updates. Nucleic Acids Res 42:D26–D31. doi:10.1093/nar/gkt1069
Patil KR, Roune L, McHardy AC (2012) The PhyloPythiaS web server for taxonomic assignment of metagenome sequences. PLoS One 7:e38581. doi:10.1371/journal.pone.0038581
Pruitt KD, Tatusova T, Maglott DR (2007) NCBI reference sequences (RefSeq): a curated non-redundant sequence database of genomes, transcripts and proteins. Nucleic Acids Res 35:61–65. doi:10.1093/nar/gkl842
Qin J, Li R, Raes J et al (2010) A human gut microbial gene catalogue established by metagenomic sequencing. Nature 464:59–65
Rosen GL, Reichenberger ER, Rosenfeld AM (2011) NBC: the Naive Bayes Classification tool webserver for taxonomic classification of metagenomic reads. Bioinformatics 27:127–129. doi:10.1093/bioinformatics/btq619

Segata N, Waldron L, Ballarini A et al (2012) Metagenomic microbial community profiling using unique clade-specific marker genes. Nat Methods 9:811–814. doi:10.1038/nmeth.2066

Sharon I, Morowitz MJ, Thomas BC et al (2013) Time series community genomics analysis reveals rapid shifts in bacterial species, strains, and phage during infant gut colonization. Genome Res 23:111–120. doi:10.1101/gr.142315.112

Simpson JT, Durbin R (2012) Efficient de novo assembly of large genomes using compressed data structures. Genome Res 22:549–556

Tatusov RL, Fedorova ND, Jackson JD et al (2003) The COG database: an updated version includes eukaryotes. BMC Bioinformatics 4:41. doi:10.1186/1471-2105-4-41

Tatusova T, Ciufo S, Fedorov B et al (2014) RefSeq microbial genomes database: new representation and annotation strategy. Nucleic Acids Res 42:5000. doi:10.1093/nar/gkt1274

Tatusova T, Ciufo S, Federhen S et al (2015) Update on RefSeq microbial genomes resources. Nucleic Acids Res 43:D599–D605. doi:10.1093/nar/gku1062

Teeling H, Waldmann J, Lombardot T et al (2004) TETRA: a web-service and a stand-alone program for the analysis and comparison of tetranucleotide usage patterns in DNA sequences. BMC Bioinformatics 5:163

The UniProt Consortium (2014) UniProt: a hub for protein information. Nucleic Acids Res 43:D204–D212. doi:10.1093/nar/gku989

Wang Q, Garrity GM, Tiedje JM, Cole JR (2007) Naive Bayesian classifier for rapid assignment of rRNA sequences into the new bacterial taxonomy. Appl Environ Microbiol 73:5261–5267. doi:10.1128/AEM.00062-07

Wrighton KC, Thomas BC, Sharon I et al (2012) Fermentation, hydrogen, and sulfur metabolism in multiple uncultivated bacterial phyla. Science 337(6102):1661–1665. doi:10.1126/science.1224041

Wu M, Eisen J (2008) A simple, fast, and accurate method of phylogenomic inference. Genome Biol 9:R151. doi:10.1186/gb-2008-9-10-r151

Zerbino DR, Birney E (2008) Velvet: algorithms for de novo short read assembly using de Bruijn graphs. Genome Res 18:821–829. doi:10.1101/gr.074492.107

Chapter 10
Human Gut Metagenomics: Success and Limits of the Activity-Based Approaches

Alexandra S. Tauzin, Elisabeth Laville, Davide Cecchini, Hervé M. Blottière, Marion Leclerc, Joël Doré, and Gabrielle Potocki-Veronese

Abstract The symbiotic relationship between the host and its gut microbiota has evolved to benefit both parties, especially by maintaining the host health and providing nutrients to the microbiota. Until recently, the study of this complex relationship was limited to culture-based studies. For this purpose, functional metagenomics has proved to be a powerful technique for exploring the diversity of the microbiota, discovering new functions and signaling pathways and extending our knowledge in the cross talk between the host and its gut microbiota. By assigning functions to proteins without any a priori knowledge of their sequences, the activity-based metagenomic approach allows to rationalize the sequencing efforts, to boost enzyme discovery, and to obtain knowledge on their mode of action, from the molecular to the ecosystem scales. This review offers an overview of the recent results obtained by activity-based metagenomic approach on the exploration of the human gut microbiota and gives insights on future developments and promising discoveries in the field.

A.S. Tauzin • E. Laville • D. Cecchini • G. Potocki-Veronese (✉)
LISBP, Université de Toulouse, CNRS, INRA, INSA, Toulouse, France
e-mail: veronese@insa-toulouse.fr

H.M. Blottière • J. Doré
Micalis Institute, INRA, AgroParisTech, Université Paris-Saclay,
78352 Jouy-en-Josas, France

MGP MetaGenoPolis, INRA, Univsersité Paris-Saclay, 78352 Jouy en Josas, France

M. Leclerc
Micalis Institute, INRA, AgroParisTech, Université Paris-Saclay,
78352 Jouy-en-Josas, France

T.C. Charles et al. (eds.), *Functional Metagenomics: Tools and Applications*,
DOI 10.1007/978-3-319-61510-3_10

10.1 Introduction

Commonly referred to as an "organ" on its own, the human gut microbiota plays a crucial role in the health and well-being of the host by contributing, among other functions, to the digestion of several dietary constituents, especially fibers, and to the control of immune response against microbes and viruses. The development of sequence-based meta-omic approaches, including 16S rRNA gene sequencing (sometimes called "monogenic" or "targeted" metagenomics) and random and massive sequencing of functional genes ("sequence-based functional metagenomics") and their transcripts ("metatranscriptomics"), allowed in the last decade to access to the diversity and functional potential of the entire gut microbiota, while around 70% of dominant gut bacteria are still uncultured. An extensive gene catalog of the gut microbiome has been established (Qin et al. 2010) further completed by (Li et al. 2014), and the relationships between humans, their symbiotic microbiota, and diet have been investigated deeper and deeper (Wu et al. 2011; David et al. 2014). Comparative metagenomics thus revealed that dysbiosis, as well as low structural and functional diversity of the gut microbiota, is associated with many inflammatory and metabolic diseases, encompassing ulcerative colitis and Crohn's disease (Frank and Amand 2007), and also asthma (Abrahamsson et al. 2014), cardiovascular diseases (Karlsson et al. 2013), obesity (Turnbaugh et al. 2008; Le Chatelier et al. 2013), and even depression (Park et al. 2013).

If these huge sequence datasets and their statistical analyses are indispensable to demonstrate the link between human health and equilibrium/diversity of the gut microbiota, they still do not allow identification of the fine mechanisms which regulate the interactions between gut microbes, host, and food. Indeed, despite the huge sequencing effort, the up-to-date gene catalogs (Li et al. 2014) are composed of a large proportion of sequences that are of unknown function.

In order to explore their diversity, to characterize them at the molecular level, and even to discover novel functions of gut microbes which can be exploited in white and red biotechnologies, activity-based metagenomics represents the most powerful technique. As often described in review papers that are not particularly focused on the human gut ecosystem (Handelsman 2004; Ferrer et al. 2009; Steele et al. 2009; Ufarté et al. 2015), it consists of constructing libraries by cloning metagenomic DNA fragments in an expression vector, expressing the genes in a heterologous host (most of the time *Escherichia coli*) for screening the targeted functions (by clone cultivation on selective media, by using chromogenic substrates, or by employing a reporter system), and finally sequencing the hit clones to identify the genes responsible for the screened activity. The last decade has seen development of the functional screening on human gut metagenomic libraries with the first study proposed in 2007 by Gloux et al. (2007).

In this paper, we review the successes of the various activity-based functional metagenomic studies targeted on the human gut microbiota to better understand (a) the implication of the microbiota in the metabolization of food constituents,

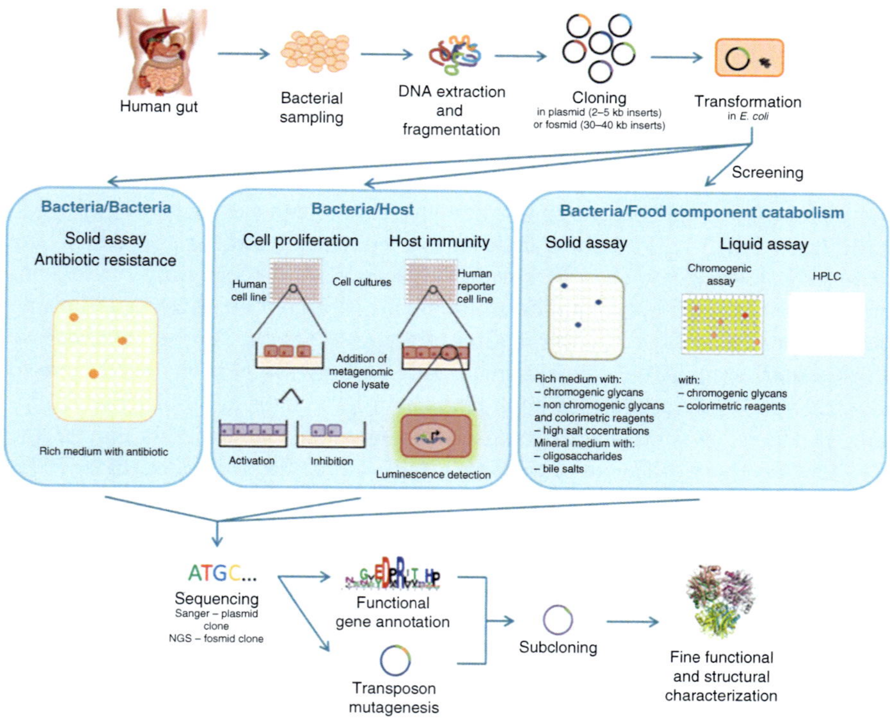

Fig. 10.1 Schematic overview of the activity-based metagenomic approaches developed to study the human gut microbiota. After isolation of fecal microbiota, DNA is extracted and sheared into fragments of appropriate size which are then ligated into vector (plasmid or fosmid) before *E. coli* transformation. The screening is performed depending on the activity of interest in high-throughput manner. The metagenomic clone with interesting characteristics is sequenced. The gene responsible for the activity is identified by annotation or by transposon mutagenesis to be subcloned for further structural and functional characterization. *NGS* next-generation sequencing technologies

(b) the cross talk between host cells and bacteria, and (c) the relationships between the different bacterial species fighting for an ecological edge in the gastrointestinal tract (overview presented in Fig. 10.1).

10.2 Food Metabolization by Gut Bacteria

Humans consume a broad range of foods rich in polysaccharides and oligosaccharides, which constitute the main carbon and energy source for the growth of intestinal microbes. Our gut microbiota is indeed exposed to a huge diversity of complex glycans since birth, in the form of lactose and human milk oligosaccharides of which the composition varies daily or of prebiotic oligosaccharides used to supplement baby formula milks. Further on in life, carbohydrate sources are even more diverse, as soon as our diet integrates cereals (especially whole grains), fruits, vegetables, legumes, and dietary fungi. For adults, plant cell wall polysaccharides represent the highest proportion of fibers, in addition to plant storage polysaccharides (resistant starch and inulin), lignin, and, to a minor extent, fungal glycans. These dietary constituents are recalcitrant to digestion in the upper part of the gut, since the human genome encodes only 97 glycoside hydrolases, the main carbohydrate-active enzymes (CAZymes) which are involved in the breakdown of glycosidic linkages. Among them, only 17 are demonstrated or thought to be implicated in dietary carbohydrate digestion, limiting our intrinsic ability for the degradation of a part of starch, lactose, and sucrose (El Kaoutari et al. 2013). Thus, humans rely entirely on their gut microbiota to digest dietary polysaccharides into short-chain fatty acids and other metabolizable products. Several sequence-based metagenomic studies highlighted the CAZyme diversity in the gut microbiome (Gill et al. 2006; Turnbaugh et al. 2009, 2010) revealing up to 156 CAZy families including 77 glycoside hydrolase, 21 carbohydrate-binding module, 35 glycosyltransferase, 12 polysaccharide lyase, and 11 carbohydrate esterase families (Turnbaugh et al. 2009). In contrast with the number of CAZymes of the human genome (less than 0.5% of the genes), the CAZymes represent on average 2.6% of the sequenced genes in the gut microbiome (Turnbaugh et al. 2009) bringing to the forefront the crucial role of gut bacteria in glycan metabolism.

In order to access the entire sequences of these enzymes, to explore their functional diversity, and to test their potential for plant biomass breakdown, an extensive activity-based metagenomic study was performed in 2010, targeting the fecal microbiome of a vegetarian subject (Tasse et al. 2010). The authors screened on solid plates a library of 156,000 clones, covering 5.4 Gbp of metagenomic DNA, for hydrolytic activities on plant cell wall (β-glucan, xylan, β-(1,4)-galactan, pectin) and energy storage polysaccharides (amylose) of various sizes, structures, and resistance to enzymatic degradation (Table 10.1). On the 0.85 Mbp of sequence (the equivalent of a small bacterial genome) obtained after functional screening, they identified 73 CAZy-encoding genes assigned to 35 known CAZy families but also several unclassified CAZymes and additional modules, highlighting the high potential of this approach for enzyme discovery and targeted metagenomic sequencing. Nearly all the CAZymes discovered in this study, assigned to the *Eubacterium*, *Bacteroides*, and *Bifidobacterium* genera or non-assignable, were found to be encoded, together with carbohydrate transporters and expression regulators, by multigenic clusters resembling the *Bacteroidetes* Polysaccharide Utilization Loci (PULs). PULs were first described in 2000 by the Salyers' team as allowing starch sensing, binding, transport, and hydrolysis (Anderson and Salyers 1989a, b) and, since 2010, extended to the harvesting of more complex polysaccharides such as porphyran (a red algal polysaccharide), xyloglucan, or mannan (Hehemann et al. 2010; Larsbrink et al. 2014; Cuskin et al. 2015). In addition, thanks to the sequencing depth of the cloned metagenomic DNA fragments (which prevents misassembling), Tasse et al. (2010) demonstrated for the

Table 10.1 Summary of the activity-based metagenomic studies targeted on the human gut microbiota

Environment	Target activity	Surrogate host, vector	Average size of the insert (kb)	# positives/# screened clones	Gbp DNA screened	Yield (‰)	Functional assay	Reference
Microbiota/host								
Healthy feces and CD remission ileum	Epithelial cell growth modulation	*E. coli*, fosmid	40	50/565 9/20160	0.8	88.5 0.5	Cell line growth—HT-29 and CV-1	Gloux et al. (2007)
CD feces	NF-κB modulation	*E. coli*, fosmid	40	171/2640	0.1	64.8	Reporter cell line—NF-κB-luciferase reporting HT-29 and Caco-2 cells	Lakhdari et al. (2010)
Healthy, CD, and ulcerative colitis feces	NF-κB modulation	*E. coli* cosmid	30	32/50000 17/12144 94/12859	2.3	0.6 1.4 7.3	Reporter cell line—NF-κB-GFP reporting HEK193-TN cells	Cohen et al. (2015)
Microbiota/food								
Feces	Bile salt hydrolase activity	*E. coli*, fosmid	25	142/89856	3.6	1.6	Solid assay—bile agar plate supplemented with taurodeoxycholic acid	Jones et al. (2008)
Feces	Dietary fiber-degrading activities	*E. coli*, fosmid	35	310/156000	5.4	2.0	Solid assay—agar plate supplemented with chromogenic (6) and non-chromogenic (14) substrates	Tasse et al. (2010)
Healthy and CD feces Colorectal cancer ileum	β-glucuronidase activity	*E. coli*, fosmid	40	19/6144	0.2	3.1	Liquid and solid assays—activity on *p*-nitrophenyl β-D-glucuronide and agar plate supplemented with 5-bromo-4-chloro-3-indolyl-β-D-glucuronide	Gloux et al. (2011)

(continued)

Table 10.1 (continued)

Environment	Target activity	Surrogate host, vector	Average size of the insert (kb)	# positives/# screened clones	Gbp DNA screened	Yield (‰)	Functional assay	Reference
Feces	Salt tolerance	*E. coli*, fosmid	40	53/23040	0.9	2.3	Solid assay—agar plate supplemented with NaCl	Culligan et al. (2012)
Feces and ileum	Prebiotic-degrading activities	*E. coli*, fosmid	35	11/20000 49/20000	1.4	0.6 2.5	Solid and liquid assays—agar plate supplemented and activity assay with prebiotic oligo- and polysaccharides (6) and HPAEC-PAD analysis	Cecchini et al. (2013)
Microbiota/microbe								
Feces	Antibiotic resistance	*E. coli*, fosmid	35	80/4000	0.1	20.0	Solid assay—agar plate supplemented with antibiotic (1)	Kazimierczak et al. (2008)
Feces	Antibiotic resistance	*E. coli*, plasmid	2	95/~4,650,000	9.3	0.02	Solid assay—agar plate supplemented with antibiotics (13)	Sommer et al. (2009)
Feces	Antibiotic resistance	*E. coli*, fosmid	30	17/415000	12.5	0.04	Solid assay—agar plate supplemented with antibiotics (7)	Cheng et al. (2012)
Feces	Antibiotic resistance	*E. coli*, plasmid	3.5	2489/~6,340,000	22.2	0.4	Solid assay—agar plate supplemented with antibiotics (18)	Moore et al. (2013)

CD Crohn's disease

first time by using metagenomic that these PUL-like systems are often transferred by horizontal gene transfer (HGT) between bacterial species which can be distantly related on a taxonomic point of view. Indeed, rupture of the synteny between the cloned metagenomic sequences and parts of gut bacterial genomes was observed around mobile elements, witnesses of HGTs between *Bacteroides* species but also between *Bacteroides* and *Roseburia* or yet unknown species. These data highlighted the plasticity of the human gut metagenomics, which is thus not restricted to that of *Bacteroidetes* genomes, even if, up to now, it is best described in the literature to explain the high adaptability of these bacteria to their environment and the niche specialization of different species (for review, see Thomas et al. 2011).

Further, other activity-based metagenomic approaches targeted CAZymes involved in the catabolism of specific dietary carbohydrates. Gloux et al. (2011) focused their functional metagenomic screen specifically on the β-D-glucuronidase activity (Gloux et al. 2011). They screened over 6144 clones from three different human samples by comparing the activity of the metagenomic clone on *p*-nitrophenyl-β-D-glucuronide substrate with basal activity of the host strain (here *E. coli* DH10B which has a constitutive β-D-glucuronidase activity), then by subcloning the positive clones into an *E. coli* strain deprived of β-D-glucuronidase activity (*E. coli* L90 Δ*uidA*), and by testing the activity on 5-bromo-4-chloro-3-indolyl-β-D-glucuronide contained in the growing medium (Table 10.1). Bacterial β-D-glucuronidases catalyze the hydrolysis of exogenous β-D-glucuronides occurring in diet (plant-derived constituents such as flavonoids and phytoestrogens), xenobiotics, and drugs as well as endogenous glucuronated compounds produced in the liver which can have potential health implications for the human host. Among the 19 positive clones identified, 17 were assigned to *Firmicutes*, including 15 associated with putative symporters in their genetic environment. This study also allowed the identification of an original β-D-glucuronidase possessing an extra C-terminal domain and distant sequence homologies with known β-glucuronidases.

In addition to these natural dietary fibers coming from vegetal foods, more and more humans in Asia, North America, and Europe consume functional food supplements in order to enrich their microbiota in microbes considered as beneficial for intestinal comfort and tentatively for prevention of metabolic and infectious diseases. Among these, prebiotics, defined as nondigestible oligosaccharides and polysaccharides that were initially thought to specifically support the growth and/or activity of health-promoting bacteria (in particular Bifidobacteria) in the gastrointestinal tract (Roberfroid 2007), do occupy a growing place on the world market of food ingredients, even if their health benefits have been sometimes controversial. However, until recently, only a few cultivated strains, representing a minor fraction of the gut microbial ecosystem, were identified as prebiotic metabolizers. In 2013, Cecchini et al. used activity-based metagenomic to explore the metabolization potential of prebiotics commercialized on the food ingredient market (inulin, fructo-oligosaccharides, xylo-oligosaccharides, transgalacto-oligosaccharides, and lactulose) by the overall microbiota (Table 10.1) (Cecchini et al. 2013). They demonstrated that many uncultivated colon- and ileum-colonizing bacteria are well equipped for prebiotic metabolization, together with *Bifidobacterium*, *Bacteroides*, *Eubacterium*, *Faecalibacterium*, and *Streptococcus* species. By mapping the identified sequences onto the metagenomic datasets released by the Human Microbiome Project and MetaHIT initiatives, they showed that most of the prebiotic metabolization pathways that were discovered in this study, including those assigned to still unknown bacteria, are highly prevalent in the gut microbiome and are probably often exchanged between gut bacteria via HGTs. The effects on human health of the growth stimulation of these unknown microbes by long-term consumption of specific prebiotics, which are easily accessible on the e-market without any medical prescription, remain to be studied.

The ability of the gut bacteria to adapt to diet-induced stresses such as osmotic stress has also been investigated by activity-based metagenomics. Culligan et al. (2012) focused their studies on salt tolerance from the human gut microbiota leading to the discovery of novel mechanisms of osmotolerance with potential use in biotechnology (Culligan et al. 2012). The authors identified 53 clones out of 23,040 able to grow on high salt concentration (higher than 6% NaCl) (Table 10.1) assigned to *Bacteroidetes*, *Actinobacteria*, *Proteobacteria*, *Verrucomicrobia*, and *Firmicutes*. Surprisingly, the members of the *Verrucomicrobia* phylum represented up to 8.2%, and the *Firmicutes* are underrepresented with 4.1%. Four of the positive clones possess six genes (*galE*, *murB*, *mazG*, *stlA*, *sdtR*, and *brpA*) which conferred a salt tolerance phenotype when expressed in *E. coli* (Culligan et al. 2012, 2013, 2014a, b). Two out of the six genes, *galE* and *sdtR*, showed high sequence homology with genes already linked with salt tolerance encoding a UDP-glucose 4-epimerase and a sigma-dependent transcriptional regulator, respectively (Culligan et al. 2012, 2014a). The authors also discovered a gut-specific rare gene, *sltA*, issued from a prophage, which might confer a particular advantage for low abundance species under stressful conditions (Culligan et al. 2013). The genes *murB*, *mazG*, and *brpA* were annotated as an UDP-*N*-acetylenolpyruvoylglucosamine reductase, a nucleoside triphosphate pyrophosphohydrolase, and a brp/blh-family β-carotene monooxygenase, respectively, suggesting potential new function of these genes (Culligan et al. 2012, 2014b).

Finally, in 2008, Jones et al. screened about 90,000 fosmid clones derived from a human fecal sample for bile salt hydrolase activity (Table 10.1) (Jones et al. 2008). Bile salts are not dietary constituents but directly derive from lipid metabolization. Among the 142 positive clones, they identified functional bile salt hydrolases in all the major bacterial divisions and archaeal species in the gut, with a high sequence identity between the different groups, indicating potential acquisition by HGT. This functional redundancy could be an advantage for functional stability through the gut (Jones 2010). However, depending on the phylogenetic groups, a specific trend in the substrate range has been observed. While clones assigned to *Firmicutes* and *Actinobacteria* were able to degrade all tested conjugated bile acids and human bile, those from *Bacteroidetes* were more specific to tauro-conjugated bile acids (Jones et al. 2008).

10.3 Host-Microbiota Cross Talk

The monolayer of the intestinal epithelial cells is lined by a mucosal barrier of host-derived glycans forming the cell surface glycocalyx and the extracellular secreted mucus. To gain information about the cross talk between the epithelial cells and gut microbiota, innovative strategies have been developed, allowing the high-throughput screening of modulation of cell proliferation or regulation of signaling pathways.

One of the first approaches developed consisted of quantifying the modulation of cell growth by applying full lysates of metagenomic clones directly onto the human cell lines and then measuring growth using crystal violet staining or luminescence-based intracellular ATP quantification (Gloux et al. 2007). In this study, the authors screened over 20,000 metagenomic clones derived from healthy subjects or Crohn's disease patients for modulation of growth of two human cell lines, HT-29 (colonic tumor cells) and CV-1 (kidney fibroblast cells) (Table 10.1). The authors identified both growth inhibitory inserts and stimulatory ones. For 5 out of the 59 (2 inhibitory and 3 stimulatory inserts), they used transposon mutagenesis and subcloning to identify the loci responsible for the modulation activity. Six out of the seven identified genes encoded for putative ATP-binding cassette (ABC) transporter systems, a RecD gene homologue, a glutamate synthase subunit, a V-type ATPase subunit, and a specific 16S rRNA gene, which are involved in processes impacting bacterial growth. For example, the ABC transporters, which are ubiquitous in nature, constitute the largest and most highly conserved superfamily and play crucial physiological functions (for review, see Davidson et al. 2008; Wilkens 2015). The last gene out of the seven was the only one encoding a hypothetical protein with unknown function. Moreover, the 59 modulatory clones were taxonomically assigned to the main gut phyla *Bacteroidetes*, *Firmicutes*, *Proteobacteria*, and *Actinobacteria*, as well as to uncultured bacteria (76%), suggesting a widespread potential in gut bacteria to modulate eukaryotic cell growth.

In 2010, Lakhdari et al. developed a high-throughput functional screening method to investigate new determinants of the immune and inflammatory responses in the gut, especially the NF-κB modulation. The authors constructed a NF-κB reporter cell line in HT-29 and Caco-2 cell lines to test the lysates of the metagenomic clones issued from Crohn's disease patients who have a defective intestinal epithelial barrier function (Table 10.1) (Lakhdari et al. 2010). While 5.6% of the clone lysates down-activated the reporter system, 0.8% had an enhancing effect. Interestingly, they observed a cell line-dependent effect, since the modulatory clones had no effect on the other constructed Caco-2 reporter cell line. The in-depth study of one stimulatory clone by transposon insertion allowed the identification, again, of an ABC transport system belonging to the LolD family of lipoprotein transporter and a putative lipoprotein with unknown function, which could work together to modulate NF-κB transcription. This study paved the way for automated screening and potential extension of this approach to other signaling pathways. In 2014, de Wouters et al. indeed optimized this technology, in order to improve reproducibility and hit rate accuracy (de Wouters et al. 2014). They completed their optimization strategy by incorporating a suitable data treatment which increased the robustness and sensitivity of the analysis. These improvements will be adaptable for the use of a broad range of reporter assays and for the identification of new bioactive molecules to better understand interactions between host and microbiota. Moreover, the expression of metagenomic DNA has been tested in *Bacillus subtilis* using an *E. coli-B. subtilis* shuttle expression vector in order to perform compatible cell-based screening for immunity modulation, but this technique has to be further developed for future high-throughput utilization (Dobrijevic et al. 2013). Recently, Cohen et al. (2015) developed a high-throughput microscopy screen using NF-κB reporter cell line coupled with GFP in HEK293 cell line allowing to transfer directly the culture of the metagenomic clone on the human cells to observe the production of NF-κB activators (Table 10.1) (Cohen et al. 2015). They created metagenomic libraries using DNA isolated from the stool of a healthy patient, a patient with Crohn's disease, and a patient with ulcerative colitis to enlarge the diversity of bacterial species. A total of 143 clones (over 75,003 screened) with the ability to reproducibly activate GFP expression were obtained of which 26 unique commensal bacteria effector genes were identified by transposition experiments. They in-depth characterized an active metabolite produced by one positive clone from each cohort, named *N*-acetyl-3-hydroxypalmitoyl-glycine or trivially commendamide. They demonstrated that the metabolite specifically activates the G protein-coupled receptor GPCR132/G2A mediating immune functions.

Besides, physical and trophic interactions with mucin are undoubtedly an important trait in understanding gut ecology. Indeed, by producing mucus-degrading enzymes and mucus-binding extracellular proteins, the mucosa-associated microbiota possesses a colonization advantage to compete by thriving in this glycan-rich environment. The mucosa-associated microbiota also has crucial roles in nutrient exchange, resistance against invading pathogens, communication with the host, and development of the immune system (for review, see Ouwerkerk et al. 2013). However, despite its importance in inflammation and epithelium health, the functional specificities of the mucosa-associated microbial community remain unknown. Indeed, very few mucin-degrading strains were isolated (Derrien et al. 2004; Crost et al. 2013), and very few papers report on the in vitro characterization of bacterial adhesion to mucin (Van den Abbeele et al. 2009; Dague et al. 2010). Recently, sampling of the human mucosa-associated bacteria for subsequent metagenomic studies has been optimized by using laser capture microdissection, in order to avoid significant contamination by host DNA and the too low amounts of extracted microbial DNA, compared to what was obtained by scraping the bacteria adherent to biopsy samples (Wang et al. 2010). Nevertheless, despite this technical improvement, high-throughput activity-based metagenomics was never used up to now to mine the human gut mucosa for new mucin-degrading enzymes and mucin-adhesive protein targets. Microplate assays have been developed in recent years for this purpose, based on quantification of mucus adherent bacterial cells by crystal violet staining. This approach has been applied for mouse functional metagenomic (Yoon et al. 2013). In this study, the authors screened 5472 clones derived from the large-bowel microbiota of an inbred BALB/c mouse strain for biofilm producers. They discovered two operons consisting of three to four genes each, encoding proteins able to enhance in vitro adherence of *E. coli* host cells to surfaces and in vivo bacterial colonization in the mouse intestine. The first operon possesses one gene encoding, a lysozyme-like protein, while the other encodes proteins belonging to well-characterized families including a competence protein family, a shikimate dehydrogenase superfamily domain, and an M13 peptidase domain. The genes encoding lysozyme and competence protein showed a relative abundance in the mouse gut microbiota metagenome of 24% and 31%, respectively.

Finally, if some gut bacteria certainly bind mucus, they are also able, with others, to metabolize host glycans for carbohydrate foraging. As described in the activity-based metagenomic study previously cited regarding food microbiota interactions (Tasse et al. 2010), a very original enzyme belonging to a new CAZy family was discovered. This mannoside phosphorylase is indeed able to cleave both dietary mannans (a constituent of plant hemicelluloses, one of the main sources of dietary fibers) and the host mannose-containing *N*-glycans (Ladevèze et al. 2013). This constitutes the first example of carbohydrate modification activity discovered by meta-omics, regardless of the technological approach or the targeted ecosystem. In addition, an integrative study, based on analysis of these biochemical properties and of genomic and metagenomic data at the scale of the intestinal ecosystem, highlighted that this enzyme and its homologues are encoded by highly prevalent genes in the human gut microbiome, especially that of patients suffering from inflammatory bowel diseases. Many gut bacteria are thus well equipped for metabolization of both dietary fibers and host glycans, which would confer a competitive advantage in this ecosystem and allow them to participate in the modification of the intestinal epithelium. The solving of the tridimensional structure of this enzyme allowed identification of the molecular determinants of its specificity toward host glycans (Ladevèze et al. 2015). Finally, this enzyme was also shown to be able to synthesize, by reverse phosphorolysis, oligosaccharides commercialized at a price exceeding $10,000 per milligram. This study, which is one of the most accomplished so far from metagenomic data to protein structure, highlights the promising future of functional metagenomic studies and of data integration to investigate gut microbiota functions, from molecular to ecosystem scales, and to promising high potential targets for health and biotechnological interests.

10.4 Microbial Interactions

To face the permanent fight for survival they face in competition with other microbes of their dense habitat, certain bacteria produce metabolites, namely, antibiotics, which can kill other microbes or inhibit their growth. Environmental stresses trigger a complex network of regulatory signals modulating the antibiotic synthesis pathways involving a broad range of compounds (Foerstner et al. 2008). On the other hand, due to the constant exposure of humans to antibiotics through medicines or indirectly via agriculture and daily products or competition between microbes, bacteria of the human gut microbiota have acquired resistance determinants forming the "resistome" (D'Costa et al. 2006). The emergence of multiresistant bacteria is a severe threat to human health (Högberg et al. 2010). A better understanding of the diversity, the mode of acquisition, and the identification of emergent resistances will be crucial for prediction of future specific anti-infectious treatments or to mitigate the spread of resistance.

Previously PCR-based studies allowed the screening of known resistance genes (Gueimonde et al. 2006) but did not allow the discovery of new antibiotic resistance genes (ARGs). For that purpose, activity-based metagenomics has proved to be the best option. By identifying genes through their functions and not by sequence homology, novel ARGs have been identified. Sommer et al. (2009) characterized the resistance reservoir in saliva and fecal samples from two unrelated individuals and screened around 9.3 Gbp of metagenomic DNA for 13 different antibiotic resistance phenotypes (Table 10.1) (Sommer et al. 2009). Among the 95 metagenomic inserts containing functional ARGs, they obtained 22% of genes with high homology (>90% amino acid identity) to known genes. By contrast, the ARGs identified from saliva isolates were closely related to the resistance genes found in pathogen species with almost half of these genes being 100% identical. In addition, this study allowed the identification of ten new beta-lactamase sequence families and revealed a considerable diversity of ARGs in microbes. Similarly, Cheng et al. (2012) screened about 415,000 fosmid clones derived from a human fecal sample (representing 12.5 Gbp large-insert metagenomic DNA) for seven different antibiotic resistance phenotypes (Table 10.1) (Cheng et al. 2012). They obtained eight new ARGs of which one was resistant to amoxicillin, six to D-cycloserine, and one to kanamycin. By subsequently subcloning the N- and C-terminal domains of the new ARG active against kanamycin, the authors highlighted the existence of an unknown resistance determinant as only the N-terminal domain conferred kanamycin resistance, while the C-terminal domain function remained unclear. In 2013, Moore et al. explored pediatric gut-associated resistomes using fecal samples from 22 infants and children for functional screens of 18 different antibiotic resistances (Table 10.1) (Moore et al. 2013). In addition to the identification of three novel resistance genes, they reported for the first time a new resistance mechanism against folate-synthesis inhibitors conferred by a predicted Nudix hydrolase which catalyzes an essential step in the folate-synthesis pathway.

Recently, a correlation has been observed between the country level of resistance potential and the country level of antibiotic exposure (Hu et al. 2013; Forslund et al. 2013). Moreover by optimizing the analysis of the antibiotic resistance profile, Gibson et al. (2015) established the environmental impact on the human-associated microbial resistome with specific antibiotic resistance determinants (Gibson et al. 2015). Furthermore, Moore et al. (2013) observed that the individual resistomes of healthy infants and children were quite similar, suggesting a diversity established earlier in lifetime than previously described (Moore et al. 2013). Thus, the use of antibiotics cannot explain by itself the resistance potential of the gut microbiota, since, for example, a diverse and robust resistome has been identified in the gut microbiome of a hunter-gatherer community of Tanzania (the Hazda) with antibiotic exposure mainly due to environment (Rampelli et al. 2015). Association of ARGs with mobile elements may explain their spread between different environments. Forsberg et al. (2012) demonstrated that exchange of multiple classes of ARGs has occurred between soil bacteria and clinical pathogens (Forsberg et al. 2012). In most of the studies, inter-bacterial transfer of ARGs is believed to be one of the main mechanisms of acquisition with a lot of mobile elements present in the vicinity of ARGs. For example, Sommer et al. (2009) observed that 14% of the ARGs identified were flanked by mobile elements. Moreover, Kazimierczak et al. (2008) performed an activity-based metagenomic screening of the fecal sample from a donor who had received repeated doses of tetracycline for many years and for which more than 90% of fecal bacteria remained resistant 3 years later (Kazimierczak et al. 2008). They observed a high incidence of tetracycline resistance genes with up to three genes in each positive clone with 2% of the metagenomic clones conferring resistance to 10 μg/mL of tetracycline (Table 10.1). Analysis of flanking regions for a new tetracycline resistance gene revealed mobile elements which could explain the spread within gut bacteria. All in all, the diversity of the gut-associated resistomes might have been underestimated, and future studies will be needed to improve the methods of investigation.

10.5 Conclusion and New Challenges for Activity-Based Meta-omics

In the past few years, the human gut microbiome has been the focus of intense efforts of functional characterization through meta-omic approaches. Overall these data, coupled to those delivered by genomic studies targeting the most prevalent human gut symbionts and pathogens, make it the best characterized microbial ecosystem to date. It is thus a target of choice to test the latest developments in systemic analyses and for integrating the data issued from various technological approaches, including activity-based meta-omics. Indeed, if this technique allows investigation of this ecosystem through a highly focused scope, covering only few Gbp of sequences, and implies an important bias regarding heterologous gene expression, this is the only one which allows to assign biochemically proven functions to individual proteins or to entire metabolic pathways, at a throughput of hundreds of thousands of assays per week. However, this approach has been used only in rare examples to improve the understanding of ecosystem functions. Indeed, numerous patches of the gut microbiome have still not been specifically targeted, including:

- Catabolic pathways of native and synthetic human milk oligosaccharides, while these compounds play a key role in the development of a healthy microbiota in the first year of life
- Enzymes required to breakdown poly-aromatic compounds, especially lignin which constitutes a significant part of dietary fibers, and other polyphenolics,

which are directly associated with the beneficial effects of numerous fruits and vegetables

- Enzymes implicated in the breakdown of animal protein motifs, especially those established as immunogenic or enzymes transforming amino acids into cell mediators
- Pathways involved in the metabolization and in increasing bioavailability of bioactive compounds, including drugs and xenobiotics
- Anabolic pathways, in particular those relative to the biosynthesis of (a) antimicrobial compounds, for antibiotic discovery, and (b) microbial cell envelopes, which mediate microbial interactions and bacterial recognition by host cells
- Biosynthetic pathways of vitamins, which play a key role in human health, and their bioavailability
- Microbial signaling and interactions, since, besides antibiotics, quorum sensing can play an important part in microbial function at a niche level

In order to investigate these functions by using activity-based metagenomics, some technological locks have to be blown. Ultrafast screening using microfluidic technologies has started to enter the metagenomic area (for review, see Ufarté et al. 2015). They will without any doubt allow (a) exploration of a much larger space of metagenomic sequences without requiring clone isolation and arrangement in microplate format and (b) reduction by several orders of magnitude the amount of substrate required for screening, rendering possible the study of interactions between gut bacteria and highly expensive compounds like human glycans. On the particular question of carbohydrate foraging, glycoarrays (Gildersleeve et al. 2012) would constitute highly valuable screening tools to expand the diversity of targets and thus mimic what happens in vivo and for improvement of screening sensitivity and assay miniaturization. Finally, activity-based metatranscriptomics, which has been recently developed to mine eukaryotic ecosystems by circumventing the existence of introns (Findley et al. 2011), would be highly interesting to apply to the human gut microbiome, in order to reduce the sequence diversity to the sole pathways which are expressed in particular physiological contexts.

To better understand the role played by the discovered proteins in the host-microbiota-food/drug interactions, functional screening data must be completed by analysis of gene prevalence and abundance, to focus the efforts dedicated to molecular characterization onto proteins that play a key role in the ecosystem. The sequencing of numerous gut bacterial reference genomes also constitutes a specificity of this microbiota and a definite asset to inspire one from genomic context analysis to guide the fine functional characterization of the targets. Thanks to the sequencing depth of recent studies and progress in read assembly to obtain ever longer contiguous sequences, it is likely that sequence-based metagenomics will overhaul the genomic datasets to facilitate the identification of the most prevalent multigenic systems and, thus, of all the protein partners implicated in the targeted pathways.

Comparative meta-omics targeting healthy or patient microbiomes or even species consortia or synthetic ecosystems should also lead to the identification of an increasing number of biomarker proteins and entire metabolic pathways. The in silico prediction of their function, based on genomic context analysis and structural models, will constitute a key requisite to guide the development of new activity screens and to obtain biochemical proofs of function. Ultimately, characterization of the in vivo expression of these target proteins and solving of their tridimensional structures should allow one to develop specific elicitors or inhibitors in order to control the functioning of the gut microbiota in health or in pathological contexts.

References

Abrahamsson TR, Jakobsson HE, Andersson AF et al (2014) Low gut microbiota diversity in early infancy precedes asthma at school age. Clin Exp Allergy 44:842–850. doi:10.1111/cea.12253

Anderson KL, Salyers AA (1989a) Genetic evidence that outer membrane binding of starch is required for starch utilization by Bacteroides thetaiotaomicron. J Bacteriol 171:3199–3204

Anderson KL, Salyers AA (1989b) Biochemical evidence that starch breakdown by Bacteroides thetaiotaomicron involves outer membrane starch-binding sites and periplasmic starch-degrading enzymes. J Bacteriol 171:3192–3198

Cecchini DA, Laville E, Laguerre S et al (2013) Functional metagenomics reveals novel pathways of prebiotic breakdown by human gut bacteria. PLoS One 8:e72766. doi:10.1371/journal.pone.0072766

Cheng G, Hu Y, Yin Y et al (2012) Functional screening of antibiotic resistance genes from human gut microbiota reveals a novel gene fusion. FEMS Microbiol Lett 336:11–16. doi:10.1111/j.1574-6968.2012.02647.x

Cohen LJ, Kang H-S, Chu J et al (2015) Functional metagenomic discovery of bacterial effectors in the human microbiome and isolation of commendamide, a GPCR G2A/132 agonist. Proc Natl Acad Sci 112:E4825–E4834. doi:10.1073/pnas.1508737112

Crost EH, Tailford LE, Le Gall G et al (2013) Utilisation of mucin glycans by the human gut symbiont Ruminococcus gnavus is strain-dependent. PLoS One 8:e76341. doi:10.1371/journal.pone.0076341

Culligan EP, Marchesi JR, Hill C, Sleator RD (2014a) Combined metagenomic and phenomic approaches identify a novel salt tolerance gene from the human gut microbiome. Front Microbiol 5:189. doi:10.3389/fmicb.2014.00189

Culligan EP, Sleator RD, Marchesi JR, Hill C (2012) Functional metagenomics reveals novel salt tolerance loci from the human gut microbiome. ISME J 6:1916–1925. doi:10.1038/ismej.2012.38

Culligan EP, Sleator RD, Marchesi JR, Hill C (2014b) Metagenomic identification of a novel salt tolerance gene from the human gut microbiome which encodes a membrane protein with homology to a brp/blh-family β-carotene 15,15′-monooxygenase. PLoS One 9:e103318. doi:10.1371/journal.pone.0103318

Culligan EP, Sleator RD, Marchesi JR, Hill C (2013) Functional environmental screening of a metagenomic library identifies stlA; a unique salt tolerance locus from the human gut microbiome. PLoS One 8:e82985. doi:10.1371/journal.pone.0082985

Cuskin F, Lowe E, Temple M, Zhu Y (2015) Human gut Bacteroidetes can utilize yeast mannan through a selfish mechanism. Nature 517(7533):165–169. doi:10.1038/nature13995

D'Costa VM, McGrann KM, Hughes DW, Wright GD (2006) Sampling the antibiotic resistome. Science 311:374–377. doi:10.1126/science.1120800

Dague E, Le DTL, Zanna S et al (2010) Probing in vitro interactions between Lactococcus lactis and mucins using AFM. Langmuir 26:11010–11017. doi:10.1021/la101862n

David LA, Maurice CF, Carmody RN et al (2014) Diet rapidly and reproducibly alters the human gut microbiome. Nature 505:559–563. doi:10.1038/nature12820

Davidson AL, Dassa E, Orelle C, Chen J (2008) Structure, function, and evolution of bacterial ATP-binding cassette systems. Microbiol Mol Biol Rev 72:317–364. doi:10.1128/MMBR.00031-07
de Wouters T, Ledue F, Nepelska M et al (2014) A robust and adaptable high throughput screening method to study host-microbiota interactions in the human intestine. PLoS One 9:e105598. doi:10.1371/journal.pone.0105598
Derrien M, Vaughan EE, Plugge CM, de Vos WM (2004) Akkermansia muciniphila gen. nov., sp. nov., a human intestinal mucin-degrading bacterium. Int J Syst Evol Microbiol 54:1469–1476. doi:10.1099/ijs.0.02873-0
Dobrijevic D, Di Liberto G, Tanaka K et al (2013) High-throughput system for the presentation of secreted and surface-exposed proteins from Gram-positive bacteria in functional metagenomics studies. PLoS One 8:e65956. doi:10.1371/journal.pone.0065956
El Kaoutari A, Armougom F, Gordon JI et al (2013) The abundance and variety of carbohydrate-active enzymes in the human gut microbiota. Nat Rev Microbiol 11:497–504. doi:10.1038/nrmicro3050
Ferrer M, Beloqui A, Vieites JM et al (2009) Interplay of metagenomics and in vitro compartmentalization. Microb Biotechnol 2:31–39. doi:10.1111/j.1751-7915.2008.00057.x
Findley SD, Mormile MR, Sommer-Hurley A et al (2011) Activity-based metagenomic screening and biochemical characterization of bovine ruminal protozoan glycoside hydrolases. Appl Environ Microbiol 77:8106–8113. doi:10.1128/AEM.05925-11
Foerstner KU, Doerks T, Creevey CJ et al (2008) A computational screen for type I polyketide synthases in metagenomics shotgun data. PLoS One 3:e3515. doi:10.1371/journal.pone.0003515
Forsberg KJ, Reyes A, Wang B et al (2012) The shared antibiotic resistome of soil bacteria and human pathogens. Science 337:1107–1111. doi:10.1126/science.1220761
Forslund K, Sunagawa S, Kultima JR et al (2013) Country-specific antibiotic use practices impact the human gut resistome. Genome Res 23:1163–1169. doi:10.1101/gr.155465.113
Frank D, Amand A (2007) Molecular-phylogenetic characterization of microbial community imbalances in human inflammatory bowel diseases. Proc Natl Acad Sci 104:13780–13785. doi:10.1073/pnas.0706625104
Gibson MK, Forsberg KJ, Dantas G (2015) Improved annotation of antibiotic resistance determinants reveals microbial resistomes cluster by ecology. ISME J 9:207–216. doi:10.1038/ismej.2014.106
Gildersleeve JC, Wang B, Achilefu S et al (2012) Glycan array analysis of the antigen repertoire targeted by tumor-binding antibodies. Bioorg Med Chem Lett 22:6839–6843. doi:10.1016/j.bmcl.2012.09.055
Gill SR, Pop M, Deboy RT et al (2006) Metagenomic analysis of the human distal gut microbiome. Science 312:1355–1359. doi:10.1126/science.1124234
Gloux K, Berteau O, El Oumami H et al (2011) A metagenomic β-glucuronidase uncovers a core adaptive function of the human intestinal microbiome. Proc Natl Acad Sci U S A 108:4539–4546. doi:10.1073/pnas.1000066107
Gloux K, Leclerc M, Iliozer H et al (2007) Development of high-throughput phenotyping of metagenomic clones from the human gut microbiome for modulation of eukaryotic cell growth. Appl Environ Microbiol 73:3734–3737. doi:10.1128/AEM.02204-06
Gueimonde M, Salminen S, Isolauri E (2006) Presence of specific antibiotic (tet) resistance genes in infant faecal microbiota. FEMS Immunol Med Microbiol 48:21–25. doi:10.1111/j.1574-695X.2006.00112.x
Handelsman J (2004) Metagenomics: application of genomics to uncultured microorganisms. Microbiol Mol Biol Rev 68:669–685. doi:10.1128/MMBR.68.4.669-685.2004
Hehemann J-H, Correc G, Barbeyron T et al (2010) Transfer of carbohydrate-active enzymes from marine bacteria to Japanese gut microbiota. Nature 464:908–912. doi:10.1038/nature08937
Högberg LD, Heddini A, Cars O (2010) The global need for effective antibiotics: challenges and recent advances. Trends Pharmacol Sci 31:509–515. doi:10.1016/j.tips.2010.08.002
Hu Y, Yang X, Qin J et al (2013) Metagenome-wide analysis of antibiotic resistance genes in a large cohort of human gut microbiota. Nat Commun 4:2151. doi:10.1038/ncomms3151

Jones BV (2010) The human gut mobile metagenome: a metazoan perspective. Gut Microbes 1:415–431. doi:10.4161/gmic.1.6.14087

Jones BV, Begley M, Hill C et al (2008) Functional and comparative metagenomic analysis of bile salt hydrolase activity in the human gut microbiome. Proc Natl Acad Sci U S A 105(36):13580–13585. doi:10.1073/pnas.0804437105

Karlsson FH, Tremaroli V, Nookaew I et al (2013) Gut metagenome in European women with normal, impaired and diabetic glucose control. Nature 498:99–103. doi:10.1038/nature12198

Kazimierczak KA, Rincon MT, Patterson AJ et al (2008) A new tetracycline efflux gene, tet(40), is located in tandem with tet(O/32/O) in a human gut firmicute bacterium and in metagenomic library clones. Antimicrob Agents Chemother 52:4001–4009. doi:10.1128/AAC.00308-08

Ladevèze S, Cioci G, Roblin P et al (2015) Structural bases for N-glycan processing by mannoside phosphorylase. Acta Crystallogr D Biol Crystallogr 71:1335–1346. doi:10.1107/S1399004715006604

Ladevèze S, Tarquis L, Cecchini DA et al (2013) Role of glycoside phosphorylases in mannose foraging by human gut bacteria. J Biol Chem 288:32370–32383. doi:10.1074/jbc.M113.483628

Lakhdari O, Cultrone A, Tap J et al (2010) Functional metagenomics: a high throughput screening method to decipher microbiota-driven NF-κB modulation in the human gut. PLoS One 5:1–10. doi:10.1371/journal.pone.0013092

Larsbrink J, Rogers TE, Hemsworth GR et al (2014) A discrete genetic locus confers xyloglucan metabolism in select human gut Bacteroidetes. Nature 506:498–502. doi:10.1038/nature12907

Le Chatelier E, Nielsen T, Qin J et al (2013) Richness of human gut microbiome correlates with metabolic markers. Nature 500:541–546. doi:10.1038/nature12506

Li J, Jia H, Cai X et al (2014) An integrated catalog of reference genes in the human gut microbiome. Nat Biotechnol 32:834–841. doi:10.1038/nbt.2942

Moore AM, Patel S, Forsberg KJ et al (2013) Pediatric fecal microbiota harbor diverse and novel antibiotic resistance genes. PLoS One 8:e78822. doi:10.1371/journal.pone.0078822

Ouwerkerk JP, de Vos WM, Belzer C (2013) Glycobiome: bacteria and mucus at the epithelial interface. Best Pract Res Clin Gastroenterol 27:25–38. doi:10.1016/j.bpg.2013.03.001

Park AJ, Collins J, Blennerhassett PA et al (2013) Altered colonic function and microbiota profile in a mouse model of chronic depression. Neurogastroenterol Motil 25:733–e575. doi:10.1111/nmo.12153

Qin J, Li R, Raes J et al (2010) A human gut microbial gene catalogue established by metagenomic sequencing. Nature 464:59–65. doi:10.1038/nature08821

Rampelli S, Schnorr SL, Consolandi C et al (2015) Metagenome sequencing of the Hadza hunter-gatherer gut microbiota. Curr Biol 25(13):1682–1693. doi:10.1016/j.cub.2015.04.055

Roberfroid M (2007) Prebiotics: the concept revisited. J Nutr 137:830S–837S

Sommer MOA, Dantas G, Church GM (2009) Functional characterization of the antibiotic resistance reservoir in the human microflora. Science 325:1128–1131. doi:10.1126/science.1176950

Steele HL, Jaeger K-E, Daniel R, Streit WR (2009) Advances in recovery of novel biocatalysts from metagenomes. J Mol Microbiol Biotechnol 16:25–37. doi:10.1159/000142892

Tasse L, Bercovici J, Pizzut-Serin S et al (2010) Functional metagenomics to mine the human gut microbiome for dietary fiber catabolic enzymes. Genome Res 20:1605–1612. doi:10.1101/gr.108332.110

Thomas F, Hehemann J-H, Rebuffet E et al (2011) Environmental and gut bacteroidetes: the food connection. Front Microbiol 2:93. doi:10.3389/fmicb.2011.00093

Turnbaugh PJ, Bäckhed F, Fulton L, Gordon JI (2008) Diet-induced obesity is linked to marked but reversible alterations in the mouse distal gut microbiome. Cell Host Microbe 3:213–223. doi:10.1016/j.chom.2008.02.015

Turnbaugh PJ, Quince C, Faith JJ et al (2010) Organismal, genetic, and transcriptional variation in the deeply sequenced gut microbiomes of identical twins. Proc Natl Acad Sci U S A 107:7503–7508. doi:10.1073/pnas.1002355107

Turnbaugh PJ, Ridaura VK, Faith JJ et al (2009) The effect of diet on the human gut microbiome: a metagenomic analysis in humanized gnotobiotic mice. Sci Transl Med 1:6ra14. doi:10.1126/scitranslmed.3000322

Ufarté L, Potocki-Veronese G, Laville É (2015) Discovery of new protein families and functions: new challenges in functional metagenomics for biotechnologies and microbial ecology. Front Microbiol 6:563. doi:10.3389/fmicb.2015.00563

Van den Abbeele P, Grootaert C, Possemiers S et al (2009) In vitro model to study the modulation of the mucin-adhered bacterial community. Appl Microbiol Biotechnol 83:349–359. doi:10.1007/s00253-009-1947-2

Wang Y, Antonopoulos DA, Zhu X et al (2010) Laser capture microdissection and metagenomic analysis of intact mucosa-associated microbial communities of human colon. Appl Microbiol Biotechnol 88:1333–1342. doi:10.1007/s00253-010-2921-8

Wilkens S (2015) Structure and mechanism of ABC transporters. F1000Prime Rep 7:14. doi:10.12703/P7-14

Wu GD, Chen J, Hoffmann C et al (2011) Linking long-term dietary patterns with gut microbial enterotypes. Science 334:105–108. doi:10.1126/science.1208344

Yoon MY, Lee KM, Yoon Y et al (2013) Functional screening of a metagenomic library reveals operons responsible for enhanced intestinal colonization by gut commensal microbes. Appl Environ Microbiol 79(12):3829–3838. doi:10.1128/AEM.00581-13

Chapter 11
Metagenomics of Plant Microbiomes

G. Brader, E. Corretto, and A. Sessitsch

Abstract The collective genomes of the holobiont plant comprising diverse microbiota encode a number of functions required for the host as well as for supporting the interaction between the plant and its associated microbiome. This chapter reviews various plant habitats for microorganisms, microbiome functions, and functional as well as sequence-based metagenomics screening approaches, which can be used to elucidate holobiont functioning.

11.1 Introduction

Animals and plants are inhabited by a large number of microorganisms including archaea, bacteria, fungi, oomycetes, and viruses. The host organism including the associated symbiotic microorganisms can be seen as a "holobiont" entity that is exposed to evolutionary processes. As such, all animals and plants are not heralded as single organisms but in combination with their microbial associates. The collective genomes of the holobiont are then referred to as "hologenome," which encodes the functions necessary for both the host and its associated microbiome (Bordenstein and Theis 2015; Rosenberg et al. 2009). The concept of the hologenome implies certain functionalities that are typical and expected in the host-microbial association that can be exploited in functional and sequence-based metagenome screening approaches. This chapter will focus on plant-microbial interactions within the context of using metagenomics approaches to explore their hologenome.

G. Brader (✉) • E. Corretto • A. Sessitsch
Center for Health and Bioresources, AIT Austrian Institute of Technology GmbH, Bioresources Unit, Konrad-Lorenz-Straße 24, 3430 Tulln, Austria
e-mail: guenter.brader@ait.ac.at

T.C. Charles et al. (eds.), *Functional Metagenomics: Tools and Applications*,
DOI 10.1007/978-3-319-61510-3_11

11.2 Differentiation of the Plant Microbiome

11.2.1 Symbiotic, Commensal, and Pathogenic Fractions of the Plant Microbiome

In addition to symbiotic and beneficial microorganisms living in association with plants, plant-associated microorganisms that are commensal and pathogenic can be expected in and on the plant environment. Metadata analysis of plant-associated genomes indeed point to a differentiation in functions dependent on a nonpathogenic (e.g., endophytic) lifestyle and pathogenic associations (Hardoim et al. 2015). Depending on the host plant and other external factors, endophytic microorganisms can interact in all ranges from beneficial to pathogenic with their host plants (Mendes et al. 2013; Ryan et al. 2009).

11.2.2 Plant Habitats

Apart from the type of interactions, terrestrial plants provide three clearly distinct habitats with differentiation of the microbiome: the rhizoplane for the organisms living on root surfaces, the phyllosphere for organisms inhabiting the aerial parts of plants (surfaces of aerial roots, stems, twigs, leaves, and reproductive organs), as well as the endosphere for microorganisms living inside plant tissues as endophytes. Depending on the respective niches, different sets of traits and competences are required for successful colonialization of the distinct habitats (Compant et al. 2010; Vorholt 2012). Several plant-associated microorganisms are strictly bound to their hosts such as a number of bacterial species, mycorrhizal fungi, and fungi of the members of the family Clavicipitaceae (Ascomycota) with the genera *Epichloë* and *Neotyphodium* (Parniske 2008; Schardl et al. 2004). Many of these obligate endophytic organisms are transmitted vertically from mother to daughter plants as is the case with *Neotyphodium* spp., *Epichloë* spp., or Betaproteobacteria belonging to *Candidatus* Burkholderia spp. (Carlier and Eberl 2012). Some bacteria such as *Candidatus Liberibacter europaeus* may also be horizontally transmitted either by insect vectors (Camerota et al. 2012) or by wind infections through spores, as has been reported for *Epichloë* spp. (Bush et al. 1997). A higher number of plant-associated microorganisms are not obligate endophytes but sporadically enter plant tissues as "opportunistic" or "facultative" endophytes and have also competence to survive on aerial plant surfaces or in the rhizosphere (Hardoim et al. 2015). Bacterial genera such as *Pseudomonas* and *Azospirillum* or strains of the fungal genus *Trichoderma* are well known for their rhizosphere competence and facultative colonization of the plant endosphere (Lugtenberg et al. 2001; Fibach-Baldi et al. 2012; Singh et al. 2014). The different niches and types of interactions with the plant are expected to elicit different functions and characteristics encoded by the microbial genomes. Indeed, comparison of the gene sets in different plant niches show differentiation of functional categories between leaf and root bacterial isolates in *Arabidopsis* and grapevine (Bai et al. 2015). Also rhizoplane, endosphere, and phyllosphere show distinct microbiota, and the epiphytic flora of flowers and leaves may be clearly differentiated on the same plant (Junker et al. 2011). There are temporal and spatial differences in microbiota in the roots that can be observed, as roots, root segments, and the rhizosphere do not provide a homogenous environment, and differences can in part be explained by different root exudation patterns (Compant et al. 2010). Endophytic colonization of plants is also not homogenous, and aerial plant parts and especially reproductive tissues are much less colonized by endophytes than the interior of roots.

11.2.3 Density and Diversity of the Plant Microbiome

Additional factors which can increase the total microbiome diversity of a given plant species are regional differences and differences observed in plants growing on different soils. Edaphic factors serve as crucial factors for determining the plant-associated microbiome both in the dicotyledons *Arabidopsis* (Bulgarelli et al. 2012; Lundberg et al. 2012), grapevine (Zarraonaindia et al. 2015) and potato (Weinert et al. 2011), and in the monocotyledons barley (Bulgarelli et al. 2015) and maize (Peiffer et al. 2013). Nevertheless, the rhizosphere and the rhizoplane have a specifically differentiated microbiome compared to the surrounding soils and are dominated mainly by bacteria of the phyla Actinobacteria, Bacteroidetes, Firmicutes, and Proteobacteria (Buee et al. 2009; Bulgarelli et al. 2013, 2015; Zarraonaindia et al. 2015). These phyla are also mainly present in the phyllosphere, where the genera *Arthrobacter, Bacillus, Massilia, Methylobacterium, Pantoea, Pseudomonas*, and *Sphingomonas* dominate many phyllosphere communities (Rastogi et al. 2013).

Generally, plant-associated microbial communities are less diverse than the surrounding soil, and differentiation also takes place between different plant organs. The genotype of the plant apparently has a smaller influence on microbiome composition compared to the influence of the source environments. Nevertheless, differences are clearly more pronounced between different plant species and more distantly related plants as compared to genotypes of the same species (Bulgarelli et al. 2015). The shaping effect of the host plant on the rhizosphere microbial community is even more manifested in the transcriptome, where a clear differentiation between peas, oat, and wheat can be observed and even the relative amounts of eukaryotic, fungal, and bacterial transcripts were found to be determined by the host plants (Turner et al. 2013). In addition to the plant genotype, also the vegetation stage is a major driver of plant-associated microbiota (Rasche et al. 2006a, b). Apart from vascular land plants, endophytic and epiphytic microorganisms have been also described from mosses (Bragina et al. 2015) and algae (Zuccaro et al. 2008), and potentially most, if not all, plants live with associated microorganisms (Hardoim et al. 2015).

Based on 16S rRNA gene sequences, rhizosphere microbial communities reach diversities of more than 10^4 species per gram, and microbial density reaches 10^8 to 10^9 bacteria per gram (Berendsen et al. 2012). Unlike bulk soils, a larger fraction of rhizosphere microbes were reported to be metabolically active (Lebeis 2015). Also, pyrosequencing of nonribosomal peptide adenylation and polyketide ketosynthase domains involved in the production of secondary metabolites points to a similar high diversity in the rhizosphere (Charlop-Powers et al. 2015). Different soils and different regional proveniences increase this diversity. The majority (95–99%) of soil microorganisms are considered as not cultivable by using conventional cultivation methods (Raaijmakers and Mazzola 2012). However, recent systematic attempts to recover bacterial strains from the rhizoplane and endosphere resulted in more than 50% recovered taxa of the 100 most abundant OTUs found by the pyrosequencing of 16S rRNA genes (Bai et al. 2015). This does not necessarily point to a complete cultivable microbiome, as clearly not all species and strains representing an OTU have been retrieved, but this does indicate that a substantially higher fraction of cultivable strains are obtained from plant-associated microbiota using systematic cultivation approaches. Nevertheless, for specific plant-associated taxa, such as members of *Candidatus* Burkholderia, *Candidatus* Liberibacter, and *Candidatus* Phytoplasma, cultivation conditions are yet unknown, but this may change in the future since intensive cultivation attempts are ongoing and preliminary success has been reported (Bertaccini et al. 2014). Possibly, a focus on cultivation conditions better reflecting the natural habitats and improving our understanding of growth-promoting factors of different bacterial groups will lead to a more successful cultivation of *hitherto* uncultured bacteria (Puspita et al. 2012). Nevertheless, functional metagenomics using heterologous hosts for screening environmental DNA allows a more comprehensive investigation and exploitation of specific functions exhibited by microbiome members.

11.3 Functions of the Microbiome

11.3.1 Adaptions of Microorganisms to Survive in the Plant Environment

Plant-associated microorganisms require a specific set of functions for establishment and survival in the plant environment. Besides the functions needed in all plant niches such as inter- and intraspecies communication and interference with plant hormone signaling, these functions may be also differentiated for microbiota associated with the rhizoplane, endosphere, phyllosphere, or even with different plant organs depending on the specific needs in the respective environment. For example, the epiphytic colonialization of leaves allows the usage of an important carbon source in the form of methanol, which is formed by pectin methyl esterases as a side product of cell wall metabolism of leaves. This carbon source is utilized in the

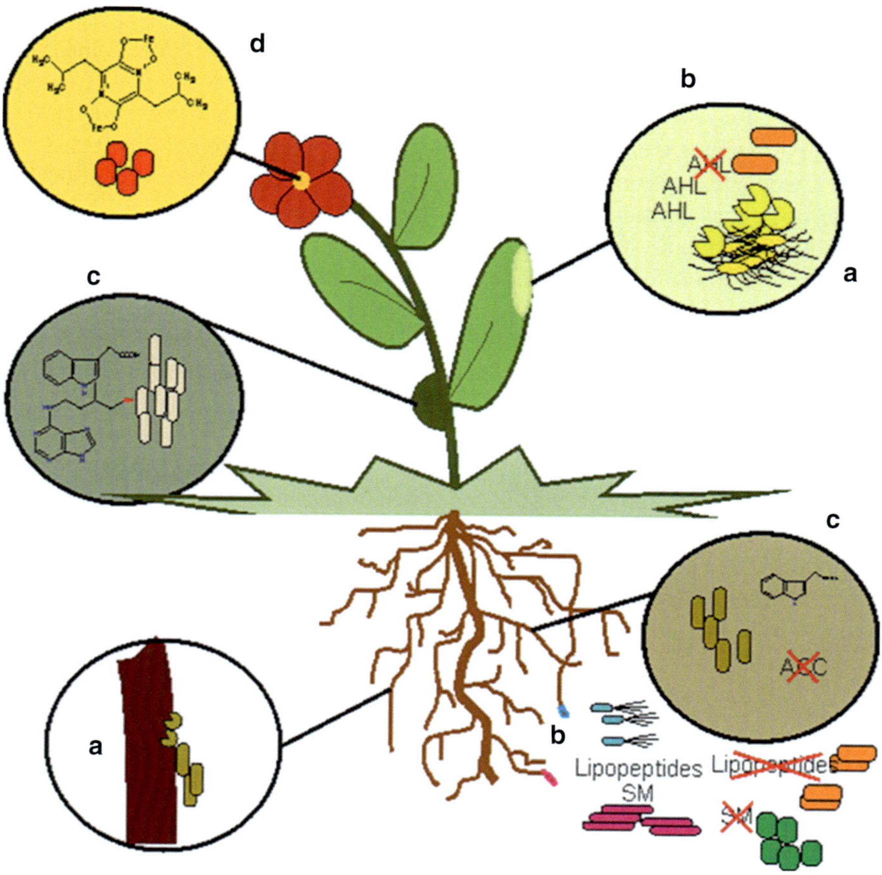

Fig. 11.1 Potential targets for metagenomic screens. Examples for potential screening targets in the plant microbiome deduced from defined traits necessary for survival in and on the plant. (**a**) Enzymes necessary for nutrient uptake, e.g., hydrolases and lyases required for digestion of polysaccharides, which are produced in pathogenic bacteria (*yellow rods*) and endophytes (*brown rods*). (**b**) Traits for establishment in the plant environment include signaling compounds (e.g., AHLs in pathogens), metabolites required for biofilm formation (e.g., lipopeptides *pink* and *blue rods*), quenching signal molecules (*orange rods*), and detoxifying secondary metabolites (*green cocci*). (**c**) Pathogenic microorganisms interfere with plant hormone signaling by production of phytohormones such as auxins and cytokinins (*pink rods*). Plant-growth-promoting bacteria (*brown rods*) can produce auxin and degrade the ethylene precursor ACC. (**d**) Iron scavenging by siderophores under limiting conditions (e.g., pulcherrimin produced by yeasts on fruits)

phyllosphere by the methylotrophic bacterium *Methylobacterium extorquens* or by the methylotrophic yeast *Candida boidinii* (Vorholt 2012).

There are specific microbial traits that are expected in plant-associated environments (Fig. 11.1). First, microorganisms use the specific nutrients provided in the environment of the host and metabolic capacities to exploit plant polysaccharides, and other plant-derived metabolites are to be expected. In addition, there might be competition for certain nutrients between microorganisms and metabolic capacities to thrive secondarily on other microorganisms or their products. Second, microorganisms possess traits that are necessary to compete and survive in their environment. This includes production of signaling compounds and polysaccharides necessary to establish multicellular-like behavior as required for biofilm formation. Particularly in highly competitive environments with a rich nutrient supply such as the rhizoplane, production of antibiotic compounds plays a role for establishment in the root environment. Moreover, degradation of toxic factors and antibiotic resistance are important factors. Third, microbes also interact with the plant host by producing plant hormones or interfering with plant hormone signaling to adapt to the plant environment for their own needs. And fourth, microorganisms have developed traits to survive stresses encountered in the plant environment, such as shortage of nutrients, particularly iron, high UV radiation in exposed parts of the phyllosphere, and reactive oxygen species (ROS), particularly in the endosphere. These traits do not necessarily belong to only a single of the above categories. For example, hydrolytic enzymes may play a role in nutrient release in pathogenic and saprophytic lifestyles but might also allow niche occupation of commensal endophytes allowing entry or relocation in the plant (Mitter et al. 2013a) and might also play a role in competence with other organisms by degradation of cell walls of competitors (Mitter et al. 2013b).

11.3.2 Adaptations to Nutrient Supply

Apart from phototrophic microorganisms such as cyanobacteria living as epiphytes on vascular plants, mosses, and algae (Singh 2014), microorganisms living in plant associations usually depend on the primary delivery of organic compounds by the host plant. The largest part of energy source from plants are polysaccharides exploited by microorganisms such as plant pathogens with necrotrophic lifestyle, e.g., *Pectobacterium carotovorum*, containing a wide array of plant cell wall-degrading enzymes (Toth et al. 2003). Also nonpathogenic plant-associated microorganisms may have the enzymatic capacity to degrade polysaccharides. In the rhizoplane and rhizosphere microbiome of cucumbers and wheat, genes involved in the utilization of cellulose, starch, mannan, arabinan, and pectin are enriched compared to the surrounding soil (Ofek-Lalzar et al. 2014). Metagenome analysis of the rice endosphere showed high abundance of cellulases, xylanases, and pectinases pointing to the importance of these enzymatic capacities for endophytic colonialization (Sessitsch et al. 2012). Plant roots provide a rich environment for bacterial growth as root caps secrete polysaccharide mucilage, and rhizodermis cells have been reported to secrete a wide range of inorganic ions and organic compounds such as sugars, vitamins, amino acids, purines, and nucleosides (Bulgarelli et al. 2013; Compant et al. 2010). Important traits expected to be found in the rhizoplane also include bacterial movement and chemotaxis reflecting the attraction to the nutrient sources. In contrast to the root environment, nutrients are scarce on leaf surfaces with diverse carbohydrates, amino acids, organic acids, and sugar alcohols as energy source (Vorholt 2012). The challenges for epiphytes and endophytes to gain nutrients from their environments might be reflected in an enrichment of the category "carbohydrate metabolism" in leaf-associated bacteria of *Arabidopsis* compared to their rhizosphere counterparts. The more complex carbohydrate supply in aerial parts of the plants is compensated by a more diverse set of genes involved in utilization of diverse or rare carbon sources (Bai et al. 2015).

Besides the classical nutrients, plants produce an enormous diversity of secondary metabolites with more than 50,000 elucidated structures (Osbourn et al. 2003). Associated microorganisms have also adapted to utilize such complex carbon sources. These enzymatic capacities might be even exploited by herbivorous insects, which benefit from toxin-degrading capacities of microorganisms (Mason et al. 2014). Interestingly, a number of traits for the utilization of plant-derived compounds are prevalent in plant-associated organisms including the degradation of phenolic compounds and polysaccharides that can also be found in the gut microbiome (Ramirez-Puebla et al. 2013). Differentiation in metabolic capacities can be expected due to the availability of different metabolites in plants and animals. However, specific degradation capacities may be transferred from plant to gut microbiota as bacteria degrading a toxic metabolite of *Mimosa pudica*, mimosine, were found on the plant and in cows feeding on *M. pudica* (Ramirez-Puebla et al. 2013).

11.3.3 Microbial Traits Required for Establishment in and on Plants

Bacterial signaling and quorum sensing have been described to play key roles in colonization of the phyllosphere, endosphere, and rhizoplane. The best characterized quorum-sensing system found in many Proteobacteria taxa depends on the production of *N*-acyl homoserine lactones (AHL). For phyllosphere colonization of nonpathogenic *Pseudomonas* spp., AHLs are needed for the establishment of the microorganisms on the leaf surface for coping with water stress, for biofilm formation, and for regulation of the production of extracellular polymeric substances involved in these processes (Vorholt 2012). Similarly, AHL systems have been also reported to be involved in the synthesis of extracellular polysaccharides as well as in the regulation of virulence factors of pathogenic enterobacteria such as *Dickeya*, *Pantoea*, and *Pectobacterium* (Toth et al. 2003). In addition to the well-described AHLs, a number of other quorum-sensing systems and signals have been described and depend on various peptides, fatty acid derivatives, and diverse secondary metabolites. The presence of multiple quorum-sensing systems in the endosphere of rice roots has been shown and included AHL signaling and a diffusible signal factor system in the rice root metagenome (Sessitsch et al. 2012). Similarly, individual plant-associated microorganisms such as *Burkholderia phytofirmans* may encode multiple quorum-sensing systems, e.g., based on AHLs and 3-hydroxypalmitic acid methyl esters (Mitter et al. 2013a). Root-associated *Pseudomonas* spp. and *Bacillus* spp. use lipopeptides for the establishment of root biofilms (Raaijmakers et al. 2010). Interference with quorum sensing has long been described for AHL signaling (Fuqua et al. 2001), but far less is known concerning signal manipulation of the other diverse signaling components. As a highly competitive environment, the rhizoplane is rich in microorganisms with the capacity to produce antibiotic substances (Compant et al. 2010; Mitter et al. 2013b). The capacity of degradation or tolerance to both plant- and microbial-derived secondary metabolites is an important trait for niche occupation. Studies with avenacin mutants of oat showed little effects of these compounds on bacterial microbiome composition but induced a strong effect on eukaryotic microorganisms (Turner et al. 2013).

11.3.4 Manipulation of Plant Metabolism and Plant Hormone Signaling by Plant-Associated Bacteria

Plant-associated microorganisms may affect their host by interfering with their hormonal signaling. Pathogenic *Agrobacterium tumefaciens* is responsible for shaping auxin and cytokinin production in plants and for establishing tumor formation as a niche for bacterial growth (Escobar and Dandekar 2003). Similarly, nodule formation after infection with symbiotic rhizobia is regulated by bacteria-mediated influence on cytokinin and auxin signaling (Suzaki et al. 2013). Pathogens are also known to interfere with defense signaling and even to suppress salicylate signaling by producing the jasmonate analogue coronatine (Lee et al. 2013). Bacteria living in all plant-associated niches are known to produce phytohormones or interfere with phytohormone signaling. Importantly, the ratio between hormones such as cytokinins and auxins has a strong influence on plant development (Moubayidin et al. 2009), and plant-associated rhizobacteria with their capacities to produce phytohormones might influence this ratio.

Exogenous auxins have a strong effect on plant development, and different concentrations of auxins shape primary and secondary root development and elongation. Auxins are produced by a number of rhizoplane-associated microorganisms, and several biosynthetic pathways from different metabolic intermediates have been described (Vacheron et al. 2013). In the microbial metagenome of the endosphere of rice, genes responsible for three different auxin biosynthesis pathways have also been detected (Sessitsch et al. 2012). Microorganisms might also indirectly influence plant auxin signaling, either by producing auxin mimicking compounds, auxin antagonists such as cytokinins, or by producing protein effectors, which interfere with auxin signaling and result, e.g., in witches broom formation in *Phytoplasma*-infected plants (Hoshi et al. 2009). Other microbial compounds interfering with auxin signaling are nitric oxide produced by *Azospirillum brasilense* strains and 2,4-diacetylphloroglucinol produced by fluorescent pseudomonads (Vacheron et al. 2013). Cytokinins which act as auxin antagonists are involved in cell division and root and shoot development and have been described from a wide array of Gram-positive and Gram-negative bacteria in addition to rhizobia and *Agrobacterium*, albeit in many cases, cytokinin genes have been only described in silico. For the root-colonizing fungus *Piriformospora indica*, the importance of cytokinin production on growth-promoting effects has been demonstrated (Vadassery et al. 2008).

Gibberellins are involved in plant growth and development and interact with ethylene and abscisic acid signaling in the development of reproductive organs and response to abiotic stress (Swain and Singh 2005). The ability to produce gibberellins has been described for various endophytic and root-associated microorganisms (Vacheron et al. 2013; Hardoim et al. 2015). Ethylene is a hormone involved in germination, root development, modulating senescence, and fruit ripening and is important for defense. Many endophytes are able to lower ethylene levels of plants by degrading the ethylene precursor 1-aminocyclopropane-1-carboxylate (ACC) and preventing ethylene signaling. These functions seem important for endophytic colonialization and lowering plant stress (Hardoim et al. 2015).

Arbuscular mycorrhizal fungi (AMF) interacting with plant roots are influenced by strigolactones of plants (Besserer et al. 2006), and so far very little is known how other plant-associated microorganisms might modulate strigolactone signaling. Other recognized phytohormones are abscisic acid, brassinosteroids, and jasmonates. For these phytohormones the influence of plant-associated microorganisms are less well described. Nevertheless, plant-associated microorganisms produce a wide range of volatiles and diverse secondary metabolites including antibiotics, lipopeptides, and siderophores with an effect on induced systemic resistance in plants (Berendsen et al. 2012; Hardoim et al. 2015) and numerous effects of microbiome-derived signals on the fine-tuning, and signal interactions in phytohormone signaling still need to be investigated.

11.3.5 Adaptations to Stress Factors

Bacteria and fungi produce structurally and biosynthetically different types of siderophores for sequestering iron (Barry and Challis 2009). Iron as an essential nutrient can be a limiting factor for bacterial colonialization, albeit plant-specific differences in iron availability have been suggested (Vorholt 2012). Also phosphorus mobilization in the rhizosphere (Bulgarelli et al. 2013) and nitrogen limitation in the endosphere play a role as suggested by the high density of diazotrophic endophytes and by the number of genes involved in nitrogen cycling in the rice root endosphere (Sessitsch et al. 2012).

Epiphytic and endophytic microorganism must cope with oxidative stress and ROS signaling in the plant environment. The bacterial endophyte *Burkholderia phytofirmans* contains a number of detoxification factors involved in oxidative stress (Mitter et al. 2013b), and also metagenome analysis shows metabolic adaptations to ROS (Sessitsch et al. 2012). Degradation capacities of xenobiotics can be found in the rice endophyte metagenome, possibly reflecting the ability to cope with plant secondary metabolism (Sessitsch et al. 2012). By comparing the genomic capacities of rhizosphere and endophytic capacities of *Arabidopsis*, it has been shown that the functional category "xenobiotics biodegradation and catabolism" is enriched in the rhizosphere. Not a single taxon is responsible for these significant differences but seems a descriptor for the ecological niche (Bai et al. 2015). It seems that here, rather the wide arsenal of secondary metabolites produced in the species-rich environment are responsible for this enrichment.

11.4 Metagenomic Methodologies

11.4.1 Technical Challenges

To investigate and exploit the functionalities of microbiomes associated with plants as a whole, the challenges of studying these functions without the need for cultivation must be overcome. For this purpose, both sequence-based DNA mining methods as well as functional screening in heterologous hosts have been developed (Engel et al. 2013). There are many technical challenges that have to be overcome for successful functional metagenomic exploration of plant-associated metagenomes, such as library preparation, the preparation of vectors suitable for different heterologous hosts, expression challenges for functional screening (e.g., different codon usage), and the requirement of different cofactors or precursors in different hosts (Engel et al. 2013; Neufeld et al. 2011). Furthermore, specific technological challenges in investigating and exploiting the metagenomes of plant-associated microorganisms are inherent due to the presence of the plant-host background. Endophytic bacteria and highly adapted microorganisms might be difficult to separate in an unbiased way from the plant host. Although the number and genomic content of microorganisms associated with plants usually exceeds the genomic content of its host, the concentration of DNA of these microorganisms, at least in some plant tissues, is much less than the eukaryotic host DNA content. To understand the functionality of the plant microbiome, it has to be taken into consideration that with one metagenome sampling and analysis, only a snapshot of the microbiome can be assessed, and temporal and spatial variations are not considered. As seen with transcriptome studies, a more complete assessment of the functional capacities of a given plant microbiome within this dynamic microbiome would be highly work intensive and costly. Nevertheless, this approach can provide numerous new genes of *hitherto* poorly characterized or uncultivated microorganisms, which can subsequently be examined over many time points.

11.4.2 Isolation of Metagenomic DNA

Depending on the plant niche, different techniques for separation of associated microorganisms from their host are necessary and have been described. Rhizosphere and rhizoplane microorganisms have been mechanically separated and washed from root surfaces prior to microbial DNA and RNA isolation (Chaparro et al. 2014). The subsequent challenge lies in the purification of high-quality DNA from the soil-containing samples. Especially DNA preparation for large insert fosmid or cosmid libraries might be hampered by residual phenol and humic acids in the DNA preparations. Different protocols for obtaining good-quality DNA in sufficient amounts have been described, and by comparing six isolation methods, for example, Engel et al. (2012) described the PowerSoil DNA Isolation Kit (Mo Bio) and nonlinear electrophoresis as methods with the highest recovery and purity. Phyllosphere microbial communities have been also collected by washing from leaf surfaces using detergents and DNA trapping methods for enriching DNA in sufficient quantities (Yang et al. 2001). DNA isolation from endophytic microorganism has been obtained after substantial enrichment of endophytic bacterial cells enriched substantially before DNA isolation for soils using mechanical isolation by shaking and glass beads followed by removing eukaryotic cells and debris by filtration over 5 μM pore size filters (Sessitsch et al. 2012). Using shotgun library sequencing, this yielded almost 40% bacterial reads reflecting a large enrichment of bacterial DNA. Microcosm studies also provide the opportunity to streamline the discovery of novel activities of plant-associated organisms by enriching specific microorganisms of interest. Enrichments for (specific) active bacterial fractions can also include DNA stable isotope probing with substrates for which metabolic activities are required. Combined with multiple-displacement amplification, this can subsequently increase the hit rate in metagenomic screens (Verastegui et al. 2014). The development of massive sequencing and whole-genome amplification techniques allows the obtaining of the genome information of single cells. Methods such as micromanipulation, laser capture microdissection, Raman tweezers, fluorescence-activated cell sorting (FACS), and microfluidics have been performed to isolate single cells, of which DNA is subsequently amplified by multiple displacement and subjected to whole-genome sequencing (Blainey 2013; Yilmaz and Singh 2012). For functional screening purposes, the genome amplification techniques to obtain DNA of sufficient length, amount, and quality still remains a bottleneck. However, microcosm experiments, stable isotope probing, specific isolation of plant-associated microbial cells, and genome amplification might all introduce biases in subsequent analyses that need to be kept in mind for the interpretation of the functions of the microbiome in the plant environment. Nevertheless, these techniques provide valuable tools to enhance the hit rates in functional metagenome screens.

11.4.3 Functional Screening

The discussed microbiome functions associated with plants are accessible via functional screening to different extents. Screening of DNA of plant-associated microorganisms expressed in heterologous laboratory hosts allows the parallel screen for a given function in large numbers of microorganisms. However, complex multilocus traits such as adaptation to UV radiation are not easily accessible using functional screens. Also, complex secondary metabolites such as nonribosomal polyketides spanning up to 50–120 kb and possibly requiring cofactors are difficult to access. Nevertheless, a number of traits discussed above can be screened for functionalities from complex microbial communities. Especially catabolic genes involved in degradation of all kinds of plant polysaccharides, in degradation of secondary metabolites and toxins, in modulation of signaling compounds, and also simple anabolic pathways using readily available substrates are rewarding targets. Apart from well-known hydrolytic enzymes targeting common polysaccharides, the plant-specific environment with its specific cell wall composition (Vorwerk et al. 2004), especially of non-vascular plants, can be an interesting environment for bioprospecting the degradation of unusual polymers. The diverse repertoire of plant-derived, fungal and bacterial secondary metabolites and toxins found in certain plant environments, for which degradation is readily exploited with cultivation-dependent techniques (McCormick 2013), represent targets for functional metagenome screens with a high potential for discovering novel natural products. Also, interference with bacterial communication beyond the well-described AHL signaling for which several interference mechanisms have been described awaits exploration. Recent developments in screening gene functions in different heterologous hosts have resulted in the development of vectors with the capacity to multiply in diverse bacterial hosts (Cheng et al. 2014). The composition of plant-associated microbiomes with mainly Actinobacteria, Bacteroidetes, Firmicutes, and diverse Proteobacteria as largely dominant bacterial taxa (Bulgarelli et al. 2013; Hardoim et al. 2015) allows the theoretical functional screening of the largest proportion (>80% of the diversity; Bai et al. 2015; Bulgarelli et al. 2015) of the plant-associated microbiome.

11.4.4 Sequence-Based Screening

The fast development of next generation techniques in recent years (Caporaso et al. 2012) has fundamentally changed the ways in which microbiomes can be analyzed. Small genomes such as those of phytoplasmas have been already sequenced and assembled without prior isolation or enrichment from its plant environment (Mitrovic et al. 2014). Nevertheless, for the sequencing and assembly of whole plant-associated microbiomes data handling and bioinformatics are still bottlenecks. The necessary computational infrastructures are combined with substantial costs but also require continuous improvement of efficiency, accuracy, and reproducibility. Recent efforts in the open-source community for improving and developing metagenome sequence analysis have been summarized by Bulgarelli et al. (2013) and include CloVR (Cloud Virtual Resource), MG-RAST (Metagenomics Rapid Annotation Using Subsystem Technology), and QIIME (Quantitative Insights into Microbial Ecology). The substantial bottleneck for sequence-based screens is the large fraction of hypothetical proteins and proteins without known function, especially in less well-investigated taxa. Recent efforts to improve metagenome annotations also include efforts for predicting operon structures in metagenome sequences (Vey and Charles 2014). An additional challenge is the necessity to generate a large amount of sequence data (in the terabyte range) in order to assess expressed genes from the plant-associated microbial metatranscriptome, due to large amounts of contaminating host RNAs. Nevertheless, the transcriptomes of highly concentrated plant pathogenic organisms with small genome sizes such as phytoplasmas have been already investigated by mRNA sequencing without prior enrichment of bacterial cells (Abbà et al. 2014; Siewert et al. 2014).

11.5 What Have We Learnt from Function- and Sequence-Based Metagenomics of Plant Microbiomes?

11.5.1 Process Understanding in the Plant Environment

The rhizosphere is a hot spot of microbial diversity and activity. Plant roots provide favorable conditions for microbial growth due to the provision of nutrients. Diverse microbial communities colonize the root zone, and it is well established that root exudates, to a great extent, shape microbial communities. Many of these microorganisms interact with the plant, whereas others are involved in nutrient cycling and in degradation processes. Some of these functions have been analyzed in isolated microorganisms, but at a community level, only few functions have been addressed. For some processes such as for N cycling, gene information of the most important enzymes (e.g., involved in nitrification, denitrification, or N fixation) is available from many different types of microorganisms. In fact, some of these genes are even highly conserved, and genes encoding key enzymes can serve as phylogenetic and functional markers, i.e., their analysis provide information on the taxon as well as on the function. However, apart from these exceptions, most of the (e.g., nutrient cycling or degrading of soil organic matter) pathways occurring in the plant environment have not been investigated yet, and limited information on the mechanisms or involved genes is available. Metagenomic analysis can provide information on as-yet-unidentified genes that are important in a particular environment and may pinpoint to the mechanisms involved. One example is the discovery of ammonium oxidation by archaea (Leininger et al. 2006), which has been for a long time believed to be mediated solely by bacteria.

Studies aiming at the elucidation of microbial processes in the plant environment have so far mostly addressed the rhizosphere compartment using a sequence-based metagenomic approach. Bai et al. (2014) studied microbial diversity and functions in the reed rhizosphere of a constructed wetland. Wetlands are constructed to improve the quality of wastewater as they have the potential to remediate various toxic pollutants from a waterbody. These processes are mediated mostly by rhizosphere microorganisms (Stottmeister et al. 2003). Although functional gene-based analyses have been performed (Calheiros et al. 2009; Imfeld et al. 2010), Bai et al. (2014) aimed to obtain a better understanding on how microbial communities transform nutrients and degrade pollutants in a constructed wetland by using a metagenomic approach.

Illumina sequencing of the reed rhizosphere as well as of influent water was performed, and microbial community functioning was assessed. The authors focused on the metabolism of nitrogen, xenobiotics and metals, and the complete set of nitrification and denitrification genes was encountered. However, no archaeal ammonium oxidizers and only a limited diversity of nitrifiers (mostly *Nitrosomonas eutropha*) were found. Degradation pathways of polycyclic aromatic compounds included the benzoate and atrazine degradation pathways. Metagenomic analysis indicated oxidation and subsequent dechlorination of the prevailing chlorinated compounds. Data also suggested biological oxidation of Mn^{2+} and chemical oxidation of Fe^{2+}.

Another investigation by Cobo-Díaz et al. (2015) addressed microbial processes involved in the recovery of soils after a wildfire event and those helping plants to reestablish. The authors focused on N cycling, as available N and water are the most limiting factors in restoration (Allen et al. 2002). Rhizosphere soils of young roots of holm oaks of burned and undisturbed soil N metabolism were analyzed by shotgun sequencing, which revealed that both types of rhizospheres contain similar N cycling pathways with a larger number of sequences related to N incorporation and a smaller number of sequences related to N output (Cobo-Díaz et al. 2015). A higher number of sequences related to N input (plant-derived allantoin degradation, N fixation) and a lower number of genes involved in nitrification and denitrification were found in burned rhizosphere soils. Souza et al. (2013) addressed the impact of different soil management practices, conventional tillage and no tillage in a crop succession and rotation of soybean and wheat, on the functional potential of soil microbial communities. Conventional tillage favored bacteria with the capacity to degrade organic matter and/or xenobiotics and involved in N cycling, whereas no tillage rather enriched bacteria with a potential to fix atmospheric N.

A functional metagenomic approach was applied by Chhabra et al. (2013) to study mineral phosphate solubilization genes in the rhizosphere of barley, which had not received phosphate fertilizer for 15 years. The rhizosphere metagenome was cloned into fosmids resulting in 18,000 clones (app. 666 Mb of DNA) and screened for mineral phosphate solubilization. Six fosmids were identified showing mineral phosphate solubilization, which were then sequenced, and several genes/operons for P uptake, mineralization, and solubilization could be identified.

11.5.2 Understanding Plant-Microb(iom)e Interactions

Metagenomics holds the potential to reveal novel information on how plants and as-yet-uncultivated microorganisms can interact. The encoded functions in plant-associated microorganisms may be identified in a culture-independent approach. These may either include full gene operons of principally known gene (families) or new genes encoding for a particular function. Sequence-based metagenomics may also lead to the assembly of whole genomes of uncultured microorganisms. Only a few studies have applied functional metagenomics to better understand plant-microbe interactions, probably due to the only few (simple) screening targets such as

P-solubilization (Chhabra et al. 2013) or ACC deaminase (Nikolic et al. 2011). In both studies gene operons of rhizosphere bacteria involved in P uptake and mineralization and of an endophyte encoding ACC deaminase, respectively, were identified by cloning, high-throughput screening, and subsequent sequencing.

A few sequence-based metagenomic studies have addressed the functions of root epi- and endophytes. Tsurumaru et al. (2015) studied plant-growth-promoting traits encoded in the metagenome of the microbiome associated with the taproot of sugar beet. A high number of genes coding for ß-1,3-glucanases, ACC deaminase, and genes involved in phosphate solubilization, siderophore production, and methanol utilization were detected, whereas genes encoding IAA biosynthesis or those involved in N_2-fixation were less abundant or not found at all. Few studies addressed the metagenomes of lichen- and moss-associated bacteria (Erlacher et al. 2015; Grube et al. 2015; Bragina et al. 2014). More than 800 bacterial species were found to contribute multiple aspects of the lichen-algae symbiosis. The bacterial metagenome of the lung lichen *Lobaria pulmonaria* revealed the potential to be involved in hormone production, nutrient supply, pathogen defense, resistance against abiotic factors, detoxification, and degradation of older parts of the lichen thallus (Grube et al. 2015). *Sphagnum* mosses are known to be associated with a very high microbial diversity (Bragina et al. 2012, 2013). A recent metagenome sequence analysis revealed that the *Sphagnum*-associated microbiome harbors features, which are different from those found in microbiomes of higher plants or peat soils (Bragina et al. 2014). A high proportion of genes involved in N cycling and recycling of organic material was encountered in the *Sphagnum* microbiome (Bragina et al. 2014). Similarly, the sequence of the rice root metagenome revealed that endophytes host genes encoding nitrogen fixation, denitrification, and nitrification (Sessitsch et al. 2012). Furthermore, genes responsible for flagella formation, plant polymer degradation, iron acquisition and storage, and detoxification of reactive oxygen species were found.

11.5.3 Metagenomics and Plant Health

Information on the presence of potential phytopathogens and their genomes is restricted to sequence-based metagenomics. Nevertheless, novel phytopathogens, e.g., a new luteovirus in nectarine trees (Bag et al. 2015), have been detected demonstrating the potential of this approach. This luteovirus was detected in 5-year-old trees in the USA and probably derived from imported budwood, which passed the quarantine procedure. Extensive pitting of the wood cylinder was found and the new virus was discovered based on sequencing of total cDNA derived from the wood. The presence of this virus in diseased trees was confirmed. A deep sequencing approach to discriminate plant parasitic nematodes from complex soil microbiota was also proposed by Porazinska et al. (2014). Sequence-based metagenomics has been proposed as a universal diagnostic tool in plant virology (Adams et al. 2009) and as a new avenue to predict emerging plant diseases (Roossinck et al. 2015).

Whole-genome sequences have been obtained from phytopathogens (Duan et al. 2009; Lin et al. 2011; Kakizawa et al. 2014) without cultivation, demonstrating that sequence-based metagenomics has the potential to advance our understanding of pathogens such as phytoplasmas or the bacterium causing citrus greening disease, Huanglongbing.

11.5.4 Bioprospection

In the last decade, functional metagenomics has been applied to discover novel enzymes and bioactive compounds potentially valuable for the industry. Hydrolytic enzymes were the major target of screening efforts as these enzymes find application in a wide array of industrial processes. They have become equally important for the degradation of recalcitrant plant material to be used in biomaterial and biofuel production valorizing renewable resources (Gilbert et al. 2012; Guerriero et al. 2015). Seaweeds and their associated microbiota are exposed to quite extreme conditions such as high salinity, high pressure, and low nutrient availability. They were therefore chosen to construct metagenomic plasmid-based libraries in *E. coli* containing app. 4.5 kb inserts (Martin et al. 2014). From app. 40,000 recombinant clones, 11 clones showed lipase activity, one clone encoding a cellulase, and one clone encoding a beta-glucosidase activity. The latter was found to have less than 50% identity to sequences of known cellulases and showed activity at low temperatures and in concentrated salt solutions (Martin et al. 2014). Similarly, a metagenome library comprising 29,600 clones with an average insert size of 3.5 kb derived from sugarcane soil was constructed in *E. coli* and screened for cellulolytic activities (Alvarez et al. 2013a, b). A novel endoxylanase family was identified being highly active against xylan from beechwood and with optimal enzyme activity at pH 6.0 and 45 °C (Alvarez et al. 2013a). Furthermore, a novel cellulase with a unique glycoside hydrolase domain belonging to the GH5 family and with unusual catalytic properties was obtained from the same library (Alvarez et al. 2013b). The recombinant enzyme showed remarkable activity at alkaline conditions that is attractive for industrial applications in which conventional acidic cellulases are not suitable. Metagenomic fosmid *E. coli* libraries were prepared from red pepper and strawberry rhizosphere samples (Lee et al. 2010). From 142,900 clones, 35 clones with lipolytic activity were detected. One clone belonged to a novel family of lipolytic enzymes and was probably derived from a *Phenylobacterium zucineum* relative. Furthermore, fluids from pitcher plants (*Nepenthes hybrida*) were used to construct metagenomic libraries in *E. coli* (Morohoshi et al. 2011). From 55,500 plasmid clones, two clones showed lipolytic activities in acid conditions, and both enzymes were found to belong to a novel family or subfamily of lipases. A different approach was used to address potential novel chitinases (Cretiou et al. 2012). Here, different habitats including the rhizospheres of Arctic plants were screened for overall chitinolytic activity as well as for the abundance and diversity of chitinase-encoding genes (*chiA*). The rhizosphere soil of the Arctic plant *Oxyria digyna* was identified as one of the promising habitats for further bioexploration.

An additional frequent screening target of metagenomic libraries are genes encoding antibiotic production genes. Such activities are highly important for medical as well as for agricultural applications and rather simple to screen for. However, genes coding for the production of secondary metabolites encompass large gene clusters requiring suitable libraries with large insert sizes. Lee et al. (2004) constructed such libraries with insert sizes of 35 kb from pine tree rhizosphere and forest top soil. The rhizosphere soil library comprising 33,700 clones did not reveal any activity against *Saccharomyces cerevisiae* or *Agrobacterium tumefaciens*, whereas one clone was found in the topsoil library encompassing a higher number of clones. One reason for the low number of positive hits might be that *E. coli* is not a suitable host for secondary metabolite gene clusters derived from unrelated microorganisms. A recent review (Singh et al. 2015) suggested metagenomics as a promising tool to explore seaweed-associated microbiota for novel antimicrobial compounds.

11.6 Conclusions

The rapid development of microbiome research in recent years has clearly revealed that most if not all multicellular eukaryotic organisms live in association with diverse microorganisms. These microorganisms interact with their host plants in various ways and proliferate as one entity termed a "holobiont," and evolutionary processes concern all holobiont members. To exploit the plant-associated microbiome, it is worthwhile to consider microbial functions required to survive in this environment. Although a large array of beneficial effects of microorganisms for plants have been described, the primary interest—from a microbial point of view—is to survive in its niche. For this, microorganisms possess a large array of metabolites and enzymatic activities, which can be considered for metagenome exploitation. So far, a limited number of metagenome studies have been performed for plant-associated microorganisms, mainly using sequence-based metagenomic approaches. The diverse niches represented in different plant species and plant environments with different metabolic profiles and growing in different habitats and climate zones provide an enormous potential for the discovery of novel natural products.

References

Abbà S, Galetto L, Carle P et al (2014) RNA-Seq profile of flavescence dorée phytoplasma in grapevine. BMC Genomics 15:1088

Adams IP, Glover RH, Monger WA et al (2009) Next-generation sequencing and metagenomic analysis: a universal diagnostic tool in plant virology. Mol Plant Pathol 104:5375–5345

Alvarez TM, Goldbeck R, Santos CR et al (2013a) Development and biotechnological application of a novel endoxylanase family GH10 identified from sugarcane soil metagenome. PLoS One 8:e70014

Alvarez TM, Paiva JH, Ruiz DM et al (2013b) Structure and function of a novel cellulase 5 from sugarcane soil metagenome. PLoS One 8:e83635
Bag S, Al Rwahnih M, Li A et al (2015) Detection of a new luteovirus in imported nectarine trees: a case study to propose adoption of metagenomics in post-entry quarantine. Phytopathology 105:840–846
Bai Y, Liang J, Liu R et al (2014) Metagenomic analysis reveals microbial diversity and function in the rhizosphere soil of a constructed wetland. Environ Technol 35:2521–2527
Bai Y, Müller DB, Srinivas G et al (2015) Functional overlap of the *Arabidopsis* leaf and root microbiota. Nature 528:364–369
Barry SM, Challis GL (2009) Recent advances in siderophore biosynthesis. Curr Opin Chem Biol 13:205–215
Berendsen RL, Pieterse CM, Bakker PA (2012) The rhizosphere microbiome and plant health. Trends Plant Sci 17:478–486
Bertaccini A, Duduk B, Paltrinieri S, Contaldo N (2014) Phytoplasmas and *Phytoplasma* diseases: a severe threat to agriculture. Am J Plant Sci 5:1763–1788
Besserer A, Puech-Pages V, Kiefer P et al (2006) Strigolactones stimulate arbuscular mycorrhizal fungi by activating mitochondria. PLoS Biol 4:e226
Blainey PC (2013) The future is now: single-cell genomics of bacteria and archaea. FEMS Microbiol Rev 37:407
Bordenstein SR, Theis KR (2015) Host biology in light of the microbiome: ten principles of holobionts and hologenomes. PLoS Biol 13:e1002226
Bragina A, Maier S, Berg C et al (2012) Similar diversity of alphaproteobacteria and nitrogenase gene amplicons on two related *Sphagnum* mosses. Front Microbiol 2:275
Bragina A, Berg C, Müller H (2013) Insights into functional bacterial diversity and its effect on Alpine bog ecosystem functioning. Sci Rep 3:1955
Bragina A, Oberauner-Wappis L, Zachow C et al (2014) The *Sphagnum* microbiome supports ecosystem functioning under extreme conditions. Mol Ecol 23:4498–4510
Bragina A, Berg C, Berg G (2015) The core microbiome bonds the Alpine bog vegetation to a transkingdom metacommunity. Mol Ecol 24:4795–4807
Buee M, De Boer W, Martin F, van Overbeek L, Jurkevitch E (2009) The rhizosphere zoo: an overview of plant-associated communities of microorganisms, including phages, bacteria, archaea, and fungi, and of some of their structuring factors. Plant Soil 321:189–212
Bulgarelli D, Rott M, Schlaeppi K et al (2012) Revealing structure and assembly cues for *Arabidopsis* root-inhabiting bacterial microbiota. Nature 488:91–95
Bulgarelli D, Schlaeppi K, Spaepen S, van Themaat EVL, Schulze-Lefert P (2013) Structure and functions of the bacterial microbiota of plants. Annu Rev Plant Biol 64:807–838
Bulgarelli D, Garrido-Oter R, Münch PC et al (2015) Structure and function of the bacterial root microbiota in wild and domesticated barley. Cell Host Microbe 17:392–403
Bush LP, Wilkinson HH, Schardl CL (1997) Bioprotective alkaloids of grass-fungal endophyte symbioses. Plant Physiol 114:1–7
Calheiros CSC, Duque AF, Moura A et al (2009) Substrate effect on bacterial communities from constructed wetlands planted with *Typha latifolia* treating industrial wastewater. Ecol Eng 35:744–753
Camerota C, Raddadi N, Pizzinat A et al (2012) Incidence of '*Candidatus* Liberibacter europaeus' and phytoplasmas in *Cacopsylla* species (Hemiptera: Psyllidae) and their host/shelter plants. Phytoparasitica 40:213–221
Caporaso JG, Lauber CL, Walters WA et al (2012) Ultra-high-throughput microbial community analysis on the Illumina HiSeq and MiSeq platforms. ISME J 6:1621–1624
Carlier AL, Eberl L (2012) The eroded genome of a *Psychotria* leaf symbiont: hypotheses about lifestyle and interactions with its plant host. Environ Microbiol 14:2757–2769
Chaparro JM, Badri DV, Vivanco JM (2014) Rhizosphere microbiome assemblage is affected by plant development. ISME J 8:790–803

Charlop-Powers Z, Owen J, Reddy B et al (2015) Global biogeographic sampling of bacterial secondary metabolism. elife 4:e05048

Cheng J, Pinnell L, Engel K, Neufeld JD, Charles TC (2014) Versatile broad-host-range cosmids for construction of high quality metagenomic libraries. J Microbiol Method 99:27–34

Chhabra S, Brazil D, Morrissey J et al (2013) Characterization of mineral phosphate solubilization traits from a barley rhizosphere soil functional metagenome. Microbiol Open 2:717–724

Cobo-Díaz JF, Fernández-González AJ, Villadas PJ et al (2015) Metagenomic assessment of the potential microbial nitrogen pathways in the rhizosphere of a Mediterranean forest after a wildfire. Microb Ecol 69:895–904

Compant S, Cement C, Sessitsch A (2010) Plant growth-promoting bacteria in the rhizo- and endosphere of plants: their role, colonization, mechanisms involved and prospects for utilization. Soil Biol Biochem 42:669–678

Cretiou MS, Kielak AM, Abu Al-Soud W et al (2012) Mining of unexplored habitats for novel chitinases – *chiA* as a helper gene proxy in metagenomics. Appl Microbiol Biotechnol 94(5):1347–1358

Duan YP, Zhou LJ, Hall DG et al (2009) Complete genome sequence of citrus Huanglongbing bacterium, '*Candidatus* Liberibacter asiaticus' obtained through metagenomics. Mol Plant-Microbe Interact 22:1011–1020

Engel K, Pinnell L, Cheng J, Charles TC, Neufeld JD (2012) Nonlinear electrophoresis for purification of soil DNA for metagenomics. J Microbiol Method 88:35–40

Engel K, Ashby D, Brady SF et al (2013) Meeting report: 1st international functional metagenomics workshop May 7-8, 2012, St Jacobs, Ontario, Canada. Stand Genom Sci 8(1):106–111

Erlacher A, Cernava T, Cardinale M et al (2015) *Rhizobiales* as functional and endosymbiontic members in the lichen symbiosis of *Lobaria pulmonaria* L. Front Microbiol 6:53

Escobar MA, Dandekar AM (2003) *Agrobacterium tumefaciens* as an agent of disease. Trends Plant Sci 8:380–386

Fibach-Baldi S, Burdman S, Okon Y (2012) Key physiological properties contributing to rhizosphere adaptation and plant growth promotion abilities of *Azospirillum brasilense*. FEMS Microbiol Lett 326:99–108

Fuqua C, Parsek MR, Greenberg EP (2001) Regulation of gene expression by cell-to-cell communication: acyl-homoserine lactone quorum sensing. Annu Rev Genet 35:439–468

Gilbert J, Li LL, Taghavi S et al (2012) Bioprospecting metagenomics for new glycoside hydrolases. Methods Mol Biol 908:141–151

Grube M, Cernava T, Soh J et al (2015) Exploring functional contexts of symbiotic sustain within lichen-associated bacteria by comparative genomics. ISME J 9:412–424

Guerriero G, Hausman JF, Strauss J et al (2015) Destructuring plant biomass: focus on fungal and extremophilic cell wall hydrolases. Plant Sci 234:180–193

Hardoim PR, van Overbeek LS, Berg G et al (2015) The hidden world within plants: ecological and evolutionary considerations for defining functioning of microbial endophytes. Microbiol Mol Biol Rev 79:293–320

Hoshi A, Oshima K, Kakizawa S et al (2009) A unique virulence factor for proliferation and dwarfism in plants identified from a phytopathogenic bacterium. Proc Natl Acad Sci U S A 106:6416–6421

Imfeld G, Aragones CE, Fetzer I et al (2010) Characterization of microbial communities in the aqueous phase of a constructed model wetland treating 1,2-dichloroethene-contaminated groundwater. FEMS Microbiol Ecol 72:74–88

Junker R, Loewel C, Gross R et al (2011) Composition of epiphytic bacterial communities differs on petals and leaves. Plant Biol 13:918

Kakizawa S, Makino A, Ishii Y et al (2014) Draft genome sequence of "*Candidatus* Phytoplasma asteris" strain OY-V, an unculturable plant-pathogenic bacterium. Genome Announc 18:2

Lebeis SL (2015) Greater than the sum of their parts: characterizing plant microbiomes at the community level. Curr Opin Plant Biol 24:82–86

Lee SW, Kim HK, Lim HK et al (2004) Searching antimicrobial activities from plant rhizosphere metagenomics library. Phytopathology 94:S59

Lee MH, Hong KS, Malhotra S et al (2010) A new esterase EstD2 isolated from plant rhizosphere soil metagenome. Appl Microbiol Biotechnol 88:1125–1134

Lee S, Ishiga Y, Clermont K, Mysore KS (2013) Coronatine inhibits stomatal closure and delays hypersensitive response cell death induced by non-host bacterial pathogens. PeerJ 1:e34

Leininger S, Urich T, Schloter M et al (2006) Archaea predominate among ammonia-oxidizing prokaryotes in soils. Nature 442:806–809

Lin H, Lou B, Glynn JM et al (2011) The complete genome sequence of '*Candidatus* Liberibacter solanacearum', the bacterium associated with potato zebra chip disease. PLoS One 6(4):e19135

Lugtenberg BJJ, Dekkers L, Bloemberg GV (2001) Molecular determinants of rhizosphere colonization by *Pseudomonas*. Annu Rev Phytopathol 39:461–490

Lundberg DS, Lebeis SL, Paredes SH (2012) Defining the core *Arabidopsis thaliana* root microbiome. Nature 488:86–90

Martin M, Biver S, Barbeyron T et al (2014) identification and characterization of a halotolerant, cold-active marine endo-ß-1,4-glucanase by using functional metagenomics of seaweed-associated microbiota. Appl Environ Microbiol 80:4958–4967

Mason CJ, Couture JJ, Raffa KF (2014) Plant-associated bacteria degrade defense chemicals and reduce their adverse effects on an insect defoliator. Oecologia 175:901–910

McCormick SP (2013) Microbial detoxification of mycotoxins. J Chem Ecol 39:907–918

Mendes R, Garbeva P, Raaijmakers JM (2013) The rhizosphere microbiome: significance of plant beneficial, plant pathogenic, and human pathogenic microorganisms. FEMS Microbiol Rev 37:634–663

Mitrovic J, Siewert C, Duduk B et al (2014) Generation and analysis of draft sequences of 'Stolbur' *Phytoplasma* from multiple displacement amplification templates. J Mol Microbiol Biotechnol 24:1–11

Mitter B, Petric A, Shin MW et al (2013a) Comparative genome analysis of *Burkholderia phytofirmans* PsJN reveals a wide spectrum of endophytic lifestyles based on interaction strategies with host plants. Front Plant Sci 4:120

Mitter B, Brader G, Afzal M et al (2013b) Advances in elucidating plant-soil-microbe (bacteria) interactions. Adv Agronomy 121:381–445

Morohoshi T, Oikawa M, Sato S et al (2011) Isolation and characterization of novel lipases from a metagenomic library of the microbial community in the pitcher fluid of the carnivorous plant *Nepenthes hybrida*. J Biosci Bioeng 112:315–320

Moubayidin L, Di Mambro R, Sabatini S (2009) Cytokinin-auxin crosstalk. Trends Plant Sci 14:557–562

Neufeld J, Engel K, Cheng J et al (2011) Open resource metagenomics: a model for sharing metagenomic libraries. Stand Genom Sci 5(2):203–210

Nikolic B, Schwab H, Sessitsch A (2011) Metagenomic analysis of the 1-aminocyclopropane-1-carboxylate deaminase gene (*acdS*) operon of an uncultured bacterial endophyte colonizing *Solanum tuberosum* L. Arch Microbiol 193:665–676

Ofek-Lalzar M, Sela N, Goldman-Voronov M et al (2014) Niche and host-associated functional signatures of the root surface microbiome. Nat Commun 5:4950

Osbourn AE, Qi X, Townsend B, Qin B (2003) Dissecting plant secondary metabolism - constitutive chemical defenses in cereals. New Phytol 159:101–108

Parniske M (2008) Arbuscular mycorrhiza: the mother of plant root endosymbioses. Nat Rev Microbiol 6:763–775

Peiffer JA, Spor A, Koren O et al (2013) Diversity and heritability of the maize rhizosphere microbiome under field conditions. Proc Natl Acad Sci U S A 110:6548–6553

Porazinska DL, Morgan MJ, Gaspar JM et al (2014) Discrimination of plant-parasitic nematodes from complex soil communities using ecometagenetics. Phytopathology 104:749–761

Puspita ID, Kamagata Y, Tanaka M, Asano K, Nakatsu CH (2012) Are uncultivated bacteria really uncultivable? Microbes Environ 27:356–366

Raaijmakers JM, Mazzola M (2012) Diversity and natural functions of antibiotics produced by beneficial and plant pathogenic bacteria. Annu Rev Phytopathol 50:403–424
Raaijmakers JM, De Bruijn I, Nybroe O, Ongena M (2010) Natural functions of lipopeptides from *Bacillus* and *Pseudomonas*: more than surfactants and antibiotics. FEMS Microbiol Rev 34:1037
Ramirez-Puebla ST, Servin-Garciduenas LE, Jimenez-Marin B et al (2013) Gut and root microbiota commonalities. Appl Environ Microbiol 79:2–9
Rasche F, Hödl V, Poll C et al (2006a) Rhizosphere bacteria affected by transgenic potatoes with antibacterial activities in comparison to effects of soil, wildtype potatoes, vegetation stage and pathogen exposure. FEMS Microbiol Ecol 56:219–235
Rasche F, Velvis H, Zachow C et al (2006b) Impact of transgenic potatoes expressing antibacterial agents on bacterial endophytes is comparable to effects of wildtype potatoes and changing environmental conditions. J Appl Ecol 43:555–566
Rastogi G, Coaker GL, Leveau JHJ (2013) New insights into the structure and function of phyllosphere microbiota through high-throughput molecular approaches. FEMS Microbiol Lett 348:1–10
Roossinck MJ, Martin DP, Roumagnac P (2015) Plant virus metagenomics: advances in virus discovery. Phytopathology 105:716–727
Rosenberg E, Sharon G, Zilber-Rosenberg I (2009) The hologenome theory of evolution contains Lamarckian aspects within a Darwinian framework. Environ Microbiol 11:2959–2962
Ryan RP, Monchy S, Cardinale M et al (2009) The versatility and adaptation of bacteria from the genus *Stenotrophomonas*. Nat Rev Microbiol 7:514–525
Schardl CL, Leuchtmann A, Spiering MJ (2004) Symbioses of grasses with seedborne fungal endophytes. Annu Rev Plant Biol 55:315–340
Sessitsch A, Hardoim P, Döring J et al (2012) Functional characteristics of an endophyte community colonizing rice roots as revealed by metagenomic analysis. Mol Plant-Microb Interact 25:28–36
Siewert C, Luge T, Duduk B et al (2014) Analysis of expressed genes of the bacterium *Candidatus* Phytoplasma Mali: highlights key features of virulence and metabolism. PLoS One 9:e94391
Singh S (2014) A review on possible elicitor molecules of cyanobacteria: their role in improving plant growth and providing tolerance against biotic or abiotic stress. J Appl Microbiol 117:1221–1244
Singh A, Sarma BK, Harikesh B, Upadhyay RS (2014) *Trichoderma*: a silent worker of plant rhizosphere. In: Gupta VK, Schmoll M, Herrera-Estrella A, Upadhyay RS, Druzhinina I, Ruohy MG (eds) Biotechnology and biology of *Trichoderma*. Elsevier, The Netherlands, pp 533–542
Singh RP, Kumari P, Reddy CRK (2015) Antimicrobial compounds from seaweeds-associated bacteria and fungi. Appl Microbiol Biotechnol 99:1571–1586
Souza RC, Cantão ME, Ribeiro Vasconcelos AT et al (2013) Soil metagenomics reveals differences under conventional and no-tillage with crop rotation or succession. Appl Soil Ecol 72:49–61
Stottmeister U, Wissner A, Kuschk P et al (2003) Effects of plants and microorganisms in constructed wetlands for wastewater treatment. Biotechnol Adv 22:93–117
Suzaki T, Ito M, Kawaguchi M (2013) Genetic basis of cytokinin and auxin functions during root nodule development. Front Plant Science 4:42
Swain SM, Singh DP (2005) Tall tales from sly dwarves: novel functions of gibberellins in plant development. Trends Plant Sci 10:123–129
Toth IK, Bell KS, Holeva MC, Birch PR (2003) Soft rot Erwiniae: from genes to genomes. Mol Plant Pathol 4:17–30
Tsurumaru H, Okubo T, Okazaki K et al (2015) Metagenomic analysis of the bacterial community associated with the taproot of sugar beet. Microbes Environ 30:63–69
Turner TR, Ramakrishnan K, Walshaw J et al (2013) Comparative metatranscriptomics reveals kingdom level changes in the rhizosphere microbiome of plants. ISME J 7:2248–2258
Vacheron J, Desbrosses G, Bouffaud M-L et al (2013) Plant growth-promoting rhizobacteria and root system functioning. Front Plant Sci 4:356

Vadassery J, Ritter C, Venus Y et al (2008) The role of auxins and cytokinins in the mutualistic interaction between *Arabidopsis* and *Piriformospora indica*. Mol Plant-Microbe Interact 21:1371–1383

Verastegui Y, Cheng J, Engel K et al (2014) Multisubstrate isotope labeling and metagenomic analysis of active soil bacterial communities. MBio 5:e01157-14

Vey G, Charles TC (2014). MetaProx: the database of metagenomic proximons. Database 2014:ID bau097

Vorholt JA (2012) Microbial life in the phyllosphere. Nat Rev Microbiol 10:828–840

Vorwerk S, Somerville S, Somerville C (2004) The role of plant cell wall polysaccharide composition in disease resistance. Trends Plant Sci 9:203–209

Weinert N, Piceno Y, Ding GC et al (2011) PhyloChip hybridization uncovered an enormous bacterial diversity in the rhizosphere of different potato cultivars: many common and few cultivar-dependent taxa. FEMS Microbiol Ecol 75:497–506

Yang CH, Crowley DE, Borneman J, Keen NT (2001) Microbial phyllosphere populations are more complex than previously realized. Proc Natl Acad Sci U S A 98:3889–3894

Yilmaz S, Singh AK (2012) Single cell genome sequencing. Curr Opin Biotechnol 23:437–443

Zarraonaindia I, Owens SM, Weisenhorn P et al (2015) The soil microbiome influences grapevine-associated microbiota. MBio 6:e02527-14

Zuccaro A, Schoch CL, Spatafora JW et al (2008) Detection and identification of fungi intimately associated with the brown seaweed *Fucus serratus*. Appl Environ Microbiol 74:931–941

Chapter 12
Metagenome Analyses of Multispecies Microbial Biofilms: First Steps Toward Understanding Diverse Microbial Systems on Surfaces

Christel Schmeisser, Ines Krohn-Molt, and Wolfgang R. Streit

Abstract Microbial biofilms are the dominant form of life on earth. Most naturally occurring microbial biofilms are phylogenetically diverse containing bacteria, archaea, viruses, phages, and smaller eukaryotes such as fungi, which have learned to live together. There are many examples of beneficial biofilms, such as biofilms in the plant rhizosphere and the phyllosphere and as part of the human microbiome. In industries and hospitals, biofilms are often unwanted because they are associated with pathogenicity or they interfere with production processes. On the other hand, especially in industries, biofilms are also used as production systems. Despite their complexity in nature and makeup, there are some common traits of biofilms: they require a surface, either biotic or abiotic; they produce a polymeric matrix (EPS) consisting of different mixtures of polysaccharides, fatty acids, proteins, and DNA; and they are embedded into this structure. The EPS gives a structure and a house to the many microbes, and it allows the exchange of signaling molecules, nutrients, DNAs, RNAs, and other molecules between the cells. Biofilms are not static but rather dynamic systems that are perhaps early forms of multicellular systems. While previous research has mainly focused on research on mono-species biofilms, it was only in the last few years the first examples of polymicrobial and complex biofilms were characterized in detail, using metagenome- and NGS-based technologies. Thus, within this review, we will highlight what we have learned about life in multispecies and complex biofilms through metagenome technologies during the last decade.

C. Schmeisser • I. Krohn-Molt • W.R. Streit (✉)
Division of Microbiology and Biotechnology, Biocenter Klein Flottbek, University of Hamburg, Ohnhorststrasse 18, 22609 Hamburg, Germany
e-mail: wolfgang.streit@uni-hamburg.de

T.C. Charles et al. (eds.), *Functional Metagenomics: Tools and Applications*, DOI 10.1007/978-3-319-61510-3_12

12.1 Introduction

Microbial biofilms are fascinating assemblages of rather diverse microorganisms. In nature they often harbor bacteria, archaea, and eukaryotes, and most biofilms are rather multispecies than mono-species. Well-studied examples are biofilms on catheters and hip implants, periodontal biofilms, or cystic fibrosis-associated biofilms (Singh et al. 2000; Hoiby 2002). While research has often focused on medical biofilms, it is notable that they also play an important role in industries where they interfere with various production processes (Flemming and Wingender 2010; Van Houdt and Michiels 2010; Carpentier and Cerf 1993). While these are examples that highlight microbial biofilms associated with manmade habitats, it is widely accepted that microbial biofilms are a dominant form of life in natural habitats (Costerton et al. 1987; Davey and O'Toole 2000). Prominent examples of microbial biofilms in nature are those observed in microbial mats (Battin et al. 2016; Ward et al. 1998; Treude et al. 2005) or on plant rhizospheres (Danhorn and Fuqua 2007; Bogino et al. 2013).

The diversity within naturally occurring biofilms can range from just a few species as typically expected in cystic fibrosis lung biofilms to not more than 30 species as observed for alga biofilms (Fig. 12.1) to several hundred species within periodontal or drinking water biofilms (Schmeisser et al. 2003; Frias-Lopez and Duran-Pinedo 2012; Krohn-Molt et al. 2013). Biofilms are usually embedded in exopolymeric substances (EPS) that are themselves highly complex and consist of polysaccharides, lipids, proteins, RNA, and DNA (Flemming and Wingender 2010). The EPS helps to adhere to surfaces and protects them against environmental stress and it functions as a nutrient reservoir. Within microbial biofilms, cells show distinct expression patterns that in part make them more resilient against

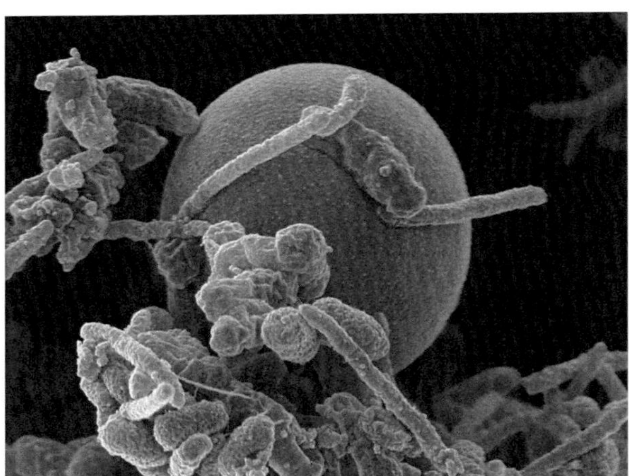

Fig. 12.1 Low diversity microbial biofilm attached to the microalga *Chlorella saccharophila*

biocidal treatment compared to planktonic cells. Cells are also heterogeneous with respect to the expression profiles within the biofilms. Furthermore, biofilms are non-static, and nutrient, pH, temperature, oxygen, and other gradients are continuously changing and thereby affecting the community. It is assumed that biofilm cells communicate, and there is some debate on the extent and at which stages of microbial biofilm formation quorum sensing is involved in biofilm development. It is further accepted that within biofilms, bacteria compete for nutrients and space and they cooperate at different levels (Moons et al. 2009; Burmolle et al. 2014). While in the last decades research has mainly focused on single-species biofilms, nowadays researchers have initiated work with complex multi-species systems. In our view, metagenome technologies can thereby help to better understand the ecology and functions of the polymicrobial biofilms in natural and artificial systems.

Within this review, we will summarize findings achieved through (functional) metagenome analyses of complex multispecies microbial biofilms and highlight a selected number of studies with relevance to biofilm formation in the medical and industrial field but also biofilms occurring in nature. We include only studies that have focused on biofilm microbial communities in a rather narrow sense with respect to the term "biofilm." Thereby this review will mainly focus on such studies that have used sequencing technologies to unravel the genetic content of biofilm DNAs or expression profiles of RNAs and/or that have used proteomic or functional approaches or a mixture of methods.

12.1.1 Biofilm Metagenomes: A Statistical Analysis

Mainly due to the complexity and high diversity of microbial biofilms, the use of metagenomics, metatranscriptomics, or metaproteomics allows an insight into the genetic contents, the regulatory mechanism, and their ecological role. Furthermore the use of these technologies has advanced our understanding in the way that these are perhaps no longer considered to be individual microbes but that they function within a community, share resources, and divide labor. Currently the JGI Genomes OnLine Database (GOLD) set contains 26,080 entries of completed or ongoing projects (July 2016). Surprisingly a search using the keyword biofilm in the field projects turns up less than 50 projects focusing on the metagenome analysis of complex microbial biofilms. Among the projects listed are many incomplete and not-yet-published ones focusing on a diverse set of habitats (e.g., drinking water, concrete pipelines, acid mine drainage) and others. Also using other keywords related to biofilm search only results in the identification of relatively few hits compared to the vast number of other sequencing projects. Similarly, a NCBI PubMed search using the keywords biofilm and metagenome returns not more than 140 publications, while the term "biofilm" alone retrieves over 30,000 entries and the term "metagenome" almost 5000 entries. Thus these simple searches may imply that in-depth metagenome analyses of complex microbial biofilms are underrepresented.

12.1.2 Acid Mine Drainage Biofilms: A Pioneering Study

One of the first metagenome studies ever was the pioneering work published by Phil Hugenholtz and his team on the acid mine drainage biofilm (Tyson et al. 2004). Within this work, the genetic contents and the phylogenetic diversity of acidophilic bacteria and archaea were explored. The metagenome was analyzed by extracting DNA from a community growing on the surface of flowing acid mine drainage of the Richmond mine at Iron Mountain, California. The biofilm analyzed was growing under extreme conditions with respect to the pH (0.83) and the occurrence of some metals as well. The data allowed reconstitution of five genome scaffolds consisting of 2455 contigs and finally resulted in a partial assembly of four bacterial genomes, two *Leptospirillum* and two *Ferroplasma* genomes, and one archaeal genome affiliated with the genus *Thermoplasma*, a facultative anaerobic, thermoacidophilic organism. A more detailed analysis of the gene content of each of the assembled genomes gave first insight into the survival strategies and pathways for energy generation and carbon and nitrogen fixation under these extreme environmental conditions.

Additional proteomic and metabolic studies have increased our understanding of this microbial community and its metabolic and physiological potential. Using a metaproteome assay, 2033 proteins were detected including 48% of the predicted proteins from the dominant biofilm organism, *Leptospirillum* group II. Further transcriptome analyses implied that mainly genes linked to motility and quorum sensing, synthesis of cell wall structures, specific proteases, and stress and genes involved in mixed acid were upregulated in the biofilm cells versus planktonic cells from acidophilic bacteria. Thereby it appears the *Leptospirillum* produces acetate under the microaerophilic conditions observed in the biofilm community. Within this framework, it was speculated that the excretion of acetate may serve as an electron donor for heterotrophic Fe^{3+}-reducing bacteria like *Acidobacterium* spp. and *Sulfobacillus* spp., which were also present in the community (Ram et al. 2005; Tringe et al. 2005; Moreno-Paz et al. 2010).

Only recently, Liljeqvist and colleagues reported on the metagenomic analysis of an acid mine drainage (pH 2.5–2.7) and low-temperature (6–10 °C) stream biofilm collected at the Kristineberg mine, northern Sweden. The metagenome sequence analysis identified the major inhabitants to be affiliated with *Acidithiobacillus ferrivorans* but with a cold-adapted lifestyle.

12.1.3 Drinking Water Biofilm Metagenomes Are Surprisingly Diverse

One of the first ever published metagenome studies focused on the analyses of the metagenome of a drinking water biofilm (Schmeisser et al. 2003). The analyzed biofilm was derived from a rubber-made valve extracted from a drinking water system in Germany. The phylogenetic analyses were still based on the 16S rRNA gene library, and only 650 clones were analyzed at that time. However, the data have largely been reproduced by other studies with respect to the overall community structure (Chao et al. 2015). The study indicated a surprisingly high diversity within the biofilm communities, with the majority of the microbes (86% of all clones) being affiliated with the *Proteobacteria*. Only a smaller fraction of the analyzed 16S rRNA gene sequences were highly similar to rRNA sequences from *Actinobacteria*, low G + C Gram positives, and the *Cytophaga-Flavobacterium-Bacteroides* group. But also not cultivated species were identified. Interestingly, there were a significant number of clones which contained 16S rRNA genes affiliated with the genera *Rhizobium* and *Bradyrhizobium*, indicating a novel function of these microbes under these oligotrophic conditions. Also compared to today's large sequence amounts, this study produced only a relatively small number of DNA sequences. Altogether less than 2 Mb of assembled DNA sequences were analyzed. Despite this rather small number of sequences, the study gave the first insight into the fine structure and the metabolic potential of the analyzed community. The data implied that the biofilm community was able to metabolize a wide variety of different and complex substrates. Thereby, fatty acids, solubilizers, paraffin oils, and other compounds used as the additives in the rubber-made surfaces appeared to play a major role as carbon sources for the bacteria.

Quite recently the study by Chao and colleagues (2015) provided partial insight into the phylogeny and metagenomes of different drinking water biofilms. They analyzed biofilm communities after 60, 120, and 180 days by amplifying the V3-V4 region of the bacterial 16S rRNA gene and then used 454 sequencing for the phylogenetic analyses. In addition, a partial metagenome analysis of chromosomal DNAs was performed. The study is of interest since it used samples from a long-term experiment with respect to the phylogenetic analyses. The metagenome analyses, however, were done only for the 180-day sample. Samples for the long-term studies were harvested from surfaces that were made out of polycarbonate. In addition, they analyzed single biofilms from polyethylene and stainless steel surfaces. Within their samples, they mainly observed bacteria affiliated with the phylum *Proteobacteria*. The *Proteobacteria* appeared to have the highest abundance on plastic and metallic materials that were tested and comprised roughly 90–95% of the overall population. Within the

Proteobacteria, the *Alphaproteobacteria* is the class with the highest number of reads with up to 56% overall reads. Similar to our earlier report (Schmeisser et al. 2003), they observed a significant number of *Bradyrhizobium* in all analyzed samples. Moreover they observed that a minor fraction of the bacteria were affiliated with the phyla of the *Chloroflexi*, *Firmicutes*, *Actinobacteria*, and *Bacteroidetes*. Again in line with our study (Schmeisser et al. 2003), they also observed a few potentially pathogenic microbes within the analyzed samples. Furthermore they generated a larger data set of metagenome data that allowed partial reconstruction of some of the microbial genomes associated within the biofilm. Finally within their study, they attempted to relate the occurrence of certain genes with "life" in microbial biofilms. Among those are the genes involved in glutathione metabolism, the SoxRS system, the OxyR system, RpoS-regulated genes, and the production/degradation of extracellular polymeric substances. The authors indicate that on the stainless steel surface, a greater biodiversity occurred in relation to the other analyzed samples.

Further sequence analyses of single cells from a water distribution system (i.e., biofilms) led to the identification of a hitherto not known bacterial phylum. McLean and colleagues identified through the analysis of a sink biofilm in a hospital the candidate phylum TM6 of which no cultivated isolates existed. Additional data analyses suggested that this phylum is a deep branching phylum and that it is most likely globally distributed (McLean et al. 2013).

12.1.4 Alga-Associated Biofilms Offer Intriguing Insights into the Evolutionary Early Bacteria-Eukaryote Interaction

Recently we reported on the metagenome and functional analysis of a bacterial biofilm associated with microalgae growing in a photobioreactor (Krohn-Molt et al. 2013). Photobioreactors are highly attractive for sunlight-driven production of biofuels and capturing of anthropogenic CO_2. The *Alphaproteobacteria*, *Betaproteobacteria*, and *Bacteroidetes* formed the dominant phyla within the analyzed biofilm (Fig. 12.1). The analyzed biofilm was extracted from a small consortium of microalgae comprising the species *Chlorella vulgaris and Scenedesmus obliquus*. Thereby, the bacteria were tightly associated with the alga surfaces but also showed in part a network of nanofibers on the alga surfaces. While it is assumed that the bacteria associated with the microalgae are essential with respect to vitamin B12 production and perhaps other needed cofactors, the bacteria can interfere with the production efficiency of the photobioreactors. The complex biofilms appear to host between 5 and 30 different microbial species. Assuming that each of the bacteria has a genome of approximately 5 Mb, it is perhaps correct to speculate that the overall metagenome size is around 150 Mb. Within this framework, the study of Krohn-Molt (Krohn-Molt et al.

2013) has produced sufficient sequence data to have an almost complete overview on the metabolic and physiological capabilities of the analyzed biofilm. They generated about 350 Mb of which 165 were assembled into larger contigs and could be used for data mining. One of the most interesting findings was perhaps that the biofilm community is rich in esterases and lipases. Thus it is likely that the bacteria are a sink for the lipids and fatty acids produced by the eukaryotic microalgae.

12.1.5 Metagenomic Sequencing of Marine Periphyton: Taxonomic and Functional Insights into Biofilm Communities

Partial metagenomes of periphytons were described by Sanli and colleagues (2015). Thereby the term periphyton describes the different microbial communities that develop in aquatic and marine habitats on the different surfaces. These communities can be highly diverse including viruses, bacteria, algae, fungi, protozoans, and metazoans. Within the published study, the authors have sampled and analyzed biofilms grown on glass slides in the Gullmar Fjord on the Swedish west coast. They sampled five sites within the fjord and in order to get a broader overview of periphyton composition in this area. Most bacteria in the samples were affiliated with the *Proteobacteria* and the *Bacteroidetes*. For each of the sites, they generated 76,000–103,000 reads with a mean length of 330 bp and analyzed the coding information using functional prediction. From these sequences, they deduced 77,000 protein sequences per sample and analyzed them using the KEGG database. While the overall sequence data generated was perhaps relatively small compared to other studies, the authors concluded that the analyzed biofilms harbor a vast diversity of physiological and metabolic pathways.

12.1.6 Metagenome Analyses of Corroded Concrete Wastewater Pipe Biofilms Reveal a Complex Microbial System

Concrete corrosion affects the life span of almost any wastewater collection system. Not much is known about the complex structure and the metagenomes of microbial biofilms associated with concrete surfaces wastewater collection systems. Gomez-Alvarez and colleagues (2012) provided evidence that roughly 90% of the total diversity was associated with bacteria affiliated with *Actinobacteria*, *Bacteroidetes*, *Firmicutes*, and *Proteobacteria*. As expected, large quantities of sulfide-oxidizing and sulfate-reducing bacteria and human fecal bacteria were observed in these biofilm communities. Most interestingly, the two biofilm samples analyzed were enriched for genes encoding resistance to antibiotics and other toxic compounds. Overall the functional diversity within the concrete biofilms appeared to be quite high.

12.1.7 Integrated Metagenomic and Metaproteomic Analyses of Ship Hull Biofilms

Ship hulls are colonized by different microorganisms forming biofilms. These marine biofilms must resist in general large shear forces, biocidal treatment, and changes in climate due to transit from one climate to the other. Thus the communities growing on ship hulls are certainly highly variable. To give an initial insight in such communities, Leary and colleagues (2014) characterized the metagenomes of ship hull biofilms. The authors gave insight into the phylogenetic, metagenomic, and proteomic structure of two different biofilms growing on the air-water interface from the hulls of Navy destroyers. Interestingly, the phylogenetic analyses indicated largely different compositions of the two samples processed. The metagenome data set that was generated allowed the assembly of 243,146 and 183,173 ORFs for both analyzed biofilms. Of these, 89,504 and 76,123 ORFs could be annotated. Further a metaproteome analysis identified 678 unique proteins, revealing little overlap in species and protein composition between the analyzed ship samples. Interestingly, the majority of the analyzed protein spots originated from eukaryotes, while the DNA sequence analyses indicated that only a minor number of eukaryotes were present within the communities. The eukaryotes were mainly affiliated with the Bacillariophyta, Cnidaria, Chordata, and Arthropoda.

12.1.8 Adaptation of Intertidal Biofilm Communities Is Driven by Metal Ion and Oxidative Stresses

The authors of this study provided convincing evidence indicating that the genes involved into oxidative and metal ion stresses were enriched in intertidal biofilms compared to subtidal biofilms (Zhang et al. 2013). Further they provided evidence that genes responsible for the biosynthesis of EPS were enriched in the intertidal biofilms. The authors postulate that mainly the oxidative stress and metal ions are the key drivers for the selection and shaping of the tidal biofilm communities. Interestingly, largely different profiles of CAZY enzymes exist between the different tidal biofilms analyzed.

12.1.9 Dental Biofilm Microbiomes

Microbial biofilms formed in the oral cavity are highly diverse and linked to periodontal diseases and caries (Pihlstrom et al. 2005; Selwitz et al. 2007; Marsh 2006; Kleinberg 2002). Mainly because of its general importance for human health, the oral microbiome has been subject to intensive characterization over the last decades (Aas et al. 2005; Aas et al. 2008; Zaura et al. 2009; Paster et al. 2001). It is estimated that the oral microbiome hosts several hundred (>600) different microbial species of which roughly 50% will not be cultivable. Parts of these are able to colonize the dental surface and form dense plaques. Plaque formation is associated with gingivitis, periodontics, and caries through the shift in metabolism and community structure in healthy individuals.

One of the first metagenome studies mostly confirmed the data collected by the earlier 16S rRNA gene-based studies and pointed out that the main colonizers are most likely affiliated with the *Firmicutes*, *Proteobacteria*, *Actinobacteria*, *Fusobacteria*, candidate division TM7, and *Bacteroidetes* (Xie et al. 2010). They further reported that many of their metagenome data mapped to streptococci. For the functional prediction, they were able to use slightly over 100,000 sequences. Depending on the pipeline used, they could predict functions of up to 72% of all identified proteins. Based on these findings, these researchers proposed that each individual may have its own community and that additional studies are needed to fully understand the metabolic and physiological potential of dental biofilms. Interestingly, they observed that 2.8% of all predicted genes mapped to antibiotic resistance genes. A similar study (Belda-Ferre et al. 2012) published in 2012 analyzed eight samples and produced 1 Gb of data. Thereby, the dental cavities appeared to be not exclusively colonized by *Streptococcus mutans* but were colonized by tens of different bacteria implying that caries is a polymicrobial disease. Further the functional analysis of this large data set implied that certain functions (gene frequencies) are over-/underrepresented in the oral microbiome of individuals vs. the gut microbiome. Among the underrepresented functions were those linked to glycerate metabolism, some transporters, and sulfatases. Overrepresented appeared to be genes affiliated with type 3 and 4 secretion apparatuses and quorum sensing (Belda-Ferre et al. 2012).

In another study, Liu and colleagues (2012) produced a large data set analyzing 4 individuals with periodontitis and 15 healthy individuals. The phylogenetic analysis of the data set indicated that with the infection, a shift in the community can be observed going from a Gram-positive-dominated community in samples from healthy individuals to a Gram-negative-dominated community in samples derived from individuals with periodontal disease. The functional analysis indicated that a microbiome from an individual suffering from periodontitis is enriched for genes that are involved in a parasitic lifestyle. Among these were genes encoding for functions related to the metabolic use of fatty acids and the degradation of benzoate, tyrosine, and glycerol 3-phosphate. Furthermore they observed that the samples were enriched for genes coding for conjugative systems and genes associated with lipid A of lipopolysaccharide (LPS) biosynthesis. In a similar study, Wang and colleagues (2013) reported genes and metabolic pathways including bacterial chemotaxis and LPS biosynthesis, and the collagenase PrtC was overrepresented in the microbiomes of periodontal disease patients.

These abovementioned metagenome studies have certainly deepened our understanding of the formation and gene content of dental biofilms. Despite this, they have the disadvantage that they will be difficult to reproduce. Furthermore, none of the studies has linked the metagenome data with proteome or RNA expression data.

Thus overall statements on the importance of the enrichment of certain gene categories might be difficult. To overcome these problems, Edlund and colleagues (2013) have recently reported on the development of an in vitro biofilm model system. The system allowed stable cultivation of oral biofilms over time periods of 48 h, and when analyzed, the biofilm communities appeared to be largely similar to oral biofilms reported in earlier studies. Thus developing model systems will allow reproducible gene expression studies giving us better insight into the complex expression patterns of pathogen-associated genes within these highly diverse microbial communities and possible clues for treatment.

Within this framework, recently two reports on combined metagenome and metatranscriptome studies were published, addressing the periodontitis progression and caries formation (Yost et al. 2015; Peterson et al. 2014). Both studies identified a remarkable number of genes differentially regulated in the dental plaque biofilms.

In the study by Yost et al. (Yost et al. 2015), it became evident that periodontal pathogens would upregulate TonB-dependent receptors, peptidases, proteases, aerotolerance genes, iron transport genes, hemolysins, and CRISPR-associated genes. Interestingly, organisms, previously not directly associated with the disease, actively transcribed virulence factors. In a study on dental caries biofilms, the authors showed that 15% of all transcripts were associated with the metabolism of mono- and disaccharides, mostly related to lactose and galactose metabolism (Peterson et al. 2014).

12.1.10 Metagenomes from Microbial Mats

Microbial mats develop on solid/water interfaces and form colored multilayered biofilms of tightly interacting microorganisms. They can be found in different extreme environments, including cold (0.4–3.4 °C) (Bottos et al. 2008; Varin et al. 2012), and high concentrations of iron (Emerson and Revsbech 1994), sulfur (Elshahed et al. 2003), and hydrocarbons (Mills et al. 2004) along coastlines or in hyperthermophilic habitats (Ward et al. 1998).

A recent metagenomic analysis of cyanobacterial mats from Arctic and Antarctic ice shelves revealed that protein-coding genes from *Proteobacteria* and *Cyanobacteria* are dominant in both habitats. However, the Antarctic mats have a higher representation of cyanobacteria, and this is also reflected by a higher percentage of photosynthetic genes. In the Arctic mats, a greater alphaproteobacterial and actinobacterial representation was found, which may reflect the greater access to diasporas from marine sources (Varin et al. 2012). Regarding a functional response to environmental stress, they determined that the genes assigned to copper homeostasis were more represented in the Arctic mats. In contrast, metagenomic sequences of Antarctic mats revealed more reads matching the sigma B genes (Varin et al. 2012).

In a further metagenome study, microbial mats were used as models to get a better understanding of past and present microbial ecosystems. In these studies, the authors compared the metagenomes of two aquatic microbial mats from the oligotrophic Cuatro Cienegas Basin (CCB), a naturally isolated valley in the Chihuahuan Desert (Coahuila, Mexico). One of the two mats, a red one, derived from a desiccation pond and grew under P-limited conditions. The other one was a N-limited green mat collected from a permanent pool at the CCB (Peimbert et al. 2011). In total 150 MB of sequence data could be obtained from both mats. Of these, 105,549 reads were assigned to taxa through the MG-RAST server for the red mat where the *Proteobacteria* formed the dominant phyla with 76.52% followed by *Cyanobacteria* (11.24%) and *Firmicutes* (4.31%). In the green mat, 94,009 reads could be assigned to taxa and most of them to the phyla *Proteobacteria* and *Cyanobacteria*, but no phyla could be identified as the dominant taxon. A functional assignment of the sequenced reads revealed that both mats are metabolically very diverse, but in the red mat, a higher number of transporters and two-component system genes could be found, whereas in the green mat, the gene frequency of carbon fixation pathways seemed to be higher. Based on this data, the authors conclude that microbial communities are able to use many different strategies to survive under environmental pressure.

12.2 Summary

Altogether the abovementioned studies have given us in part a detailed insight into the genetic and genomic structures of the highly diverse microbial biofilm communities. Unfortunately, the majority of these studies are DNA sequence-based, and almost no transcriptome, proteome, or metabolome data exist for these complex communities. Thus the questions remain: which are the dominant microorganisms, what colonizes first, what are the main metabolic pathways turned on in these communities, and which language is spoken in most biofilms? To address these questions, future work has to match the DNA data with solid transcriptome, proteome, and metabolome. While this will certainly be a major challenge, the development of model systems for biofilms with low complexity that can be grown in laboratory systems is also of high relevance to advance this field. It is notable that the studies above mainly relied on natural systems and that these need to be grown under reproducible ways to address many of the abovementioned scientific questions. Therefore obtaining both reproducible and more defined laboratory systems is of high importance to advance this field throughout the next decade (Table 12.1).

Table 12.1 Current examples of metagenome or metatranscriptome analyses of multispecies microbial biofilms

Sample	Key findings	References
Intertidal biofilms	Genes associated with response to oxygen stress and EPS production are enriched in microbes affiliated with these biofilms compared to subtidal biofilms	Zhang et al. (2013)
Marine periphyton	These communities harbor a vast diversity of microorganism and metabolic strategies	Sanli et al. (2015)
Hydrothermal vent biofilms	They observed a relatively high number of transposases within the metagenome of a hydrothermal vent chimney	Brazelton and Baross (2009)
Drinking water biofilm	Unexpectedly diverse community	Schmeisser et al. (2003) and Chao et al. (2015)
Acid mine drainage biofilm	*Leptospirillum*, *Ferroplasma*, and other genomes established in the 2004 report, giving first insight into the function of this community	Tyson et al. (2004), Ram et al. (2005), and Liljeqvist et al. (2015)
Microalga-associated metagenome	Limited diversity associated with microalga; main colonizers were *Alpha*- and *Betaproteobacteria* and various *Bacteroidetes*	Krohn-Molt et al. (2013)
Cellulose-degrading biofilm in marine environment	The authors have identified a number of GH family enzymes mostly matching to nine GH family classes	Edwards et al. (2010)
Dental biofilms	Highly diverse with perhaps 600 different species, 50% most likely not cultivable. Main colonizers affiliated with *Firmicutes*, *Proteobacteria*, *Actinobacteria*, *Fusobacteria*, candidate division TM7, and the *Bacteroidetes*	Edlund et al. (2013), Liu et al. (2012), Yost et al. (2015), Peterson et al. (2014), Xie et al. (2010), Wang et al. (2013), and Belda-Ferre et al. (2012)
Navy ship hulls	Microbial communities are rather variable and contain large amounts of eukaryotes	Leary et al. (2014)
Corroded wastewater pipe	Often composed of *Actinobacteria*, *Bacteroidetes*, *Firmicutes*, and *Proteobacteria*. Large quantities of sulfide-oxidizing and sulfate-reducing bacteria are observed	Gomez-Alvarez et al. (2012)

References

Aas JA, Paster BJ, Stokes LN, Olsen I, Dewhirst FE (2005) Defining the normal bacterial flora of the oral cavity. J Clin Microbiol 43(11):5721–5732. doi:10.1128/jcm.43.11.5721-5732.2005

Aas JA, Griffen AL, Dardis SR, Lee AM, Olsen I, Dewhirst FE, Leys EJ, Paster BJ (2008) Bacteria of dental caries in primary and permanent teeth in children and young adults. J Clin Microbiol 46:1407–1417. doi:10.1128/jcm.01410-07

Battin TJ, Besemer K, Bengtsson MM, Romani AM, Packmann AI (2016) The ecology and biogeochemistry of stream biofilms. Nat Rev Microbiol 14(4):251–263. doi:10.1038/nrmicro.2016.15

Belda-Ferre P, Alcaraz LD, Cabrera-Rubio R, Romero H, Simon-Soro A, Pignatelli M, Mira A (2012) The oral metagenome in health and disease. ISME J 6(1):46–56

Bogino PC, de las Mercedes Oliva M, Sorroche FG, Giordano W (2013) The role of bacterial biofilms and surface components in plant-bacterial associations. Int J Mol Sci 14(8):15838–15859. doi:10.3390/ijms140815838

Bottos EM, Vincent WF, Greer CW, Whyte LG (2008) Prokaryotic diversity of arctic ice shelf microbial mats. Environ Microbiol (4):950–966. doi:10.1111/j1462-2920200701516x. Epub 2008 Jan 22

Brazelton WJ, Baross JA (2009) Abundant transposases encoded by the metagenome of a hydrothermal chimney biofilm. ISME J 3(12):1420–1424

Burmolle M, Ren D, Bjarnsholt T, Sorensen SJ (2014) Interactions in multispecies biofilms: do they actually matter? Trends Microbiol 22(2):84–91. doi:10.1016/jtim201312004

Carpentier B, Cerf O (1993) Biofilms and their consequences, with particular reference to hygiene in the food industry. J Appl Bacteriol 75(6):499–511

Chao Y, Mao Y, Wang Z, Zhang T (2015) Diversity and functions of bacterial community in drinking water biofilms revealed by high-throughput sequencing. Sci Rep 5:10044. doi:10.1038/srep10044

Costerton JW, Cheng KJ, Geesey GG, Ladd TI, Nickel JC, Dasgupta M, Marrie TJ (1987) Bacterial biofilms in nature and disease. Annu Rev Microbiol 41:435–464

Danhorn T, Fuqua C (2007) Biofilm formation by plant-associated bacteria. Annu Rev Microbiol 61:401–422. doi:10.1146/annurev.micro.61.080706.093316

Davey ME, O'Toole GA (2000) Microbial biofilms: from ecology to molecular genetics. Microbiol Mol Biol Rev 64(4):847–867

Edlund A, Yang Y, Hall AP, Guo L, Lux R, He X, Nelson KE, Nealson KH, Yooseph S, Shi W, McLean JS (2013) An in vitro biofilm model system maintaining a highly reproducible species and metabolic diversity approaching that of the human oral microbiome. Microbiome 1(1):1–17. doi:10.1186/2049-2618-1-25

Edwards JL, Smith DL, Connolly J, McDonald JE, Cox MJ, Joint I, Edwards C, McCarthy AJ (2010) Identification of carbohydrate metabolism genes in the metagenome of a marine biofilm community shown to be dominated by gammaproteobacteria and bacteroidetes. Genes (Basel) 1(3):371–384. doi:10.3390/genes1030371

Elshahed MS, Senko JM, Najar FZ, Kenton SM, Roe BA, Dewers TA, Spear JR, Krumholz LR (2003) Bacterial diversity and sulfur cycling in a mesophilic sulfide-rich spring. Appl Environ Microbiol 2003(9):5609–5621

Emerson D, Revsbech NP (1994) Investigation of an iron-oxidizing microbial mat community located near Aarhus, Denmark: field studies. Appl Environ Microbiol 60(11):4022–4031

Flemming HC, Wingender J (2010) The biofilm matrix. Nat Rev Microbiol 8(9):623–633. doi:10.1038/nrmicro2415. Epub 2010 Aug 2

Frias-Lopez J, Duran-Pinedo A (2012) Effect of periodontal pathogens on the metatranscriptome of a healthy multispecies biofilm model. J Bacteriol 194(8):2082–2095. doi:10.1128/JB06328-11

Gomez-Alvarez V, Revetta R, Domingo JW (2012) Metagenome analyses of corroded concrete wastewater pipe biofilms reveal a complex microbial system. BMC Microbiol 12(1):122

Hoiby N (2002) Understanding bacterial biofilms in patients with cystic fibrosis: current and innovative approaches to potential therapies. J Cyst Fibros 1(4):249–254

Kleinberg I (2002) A mixed-bacteria ecological approach to understanding the role of the oral bacteria in dental caries causation: an alternative to streptococcus mutans and the specific-plaque hypothesis. Crit Rev Oral Biol Med 13(2):108–125. doi:10.1177/154411130201300202

Krohn-Molt I, Wemheuer B, Alawi M, Poehlein A, Gullert S, Schmeisser C, Pommerening-Roser A, Grundhoff A, Daniel R, Hanelt D, Streit WR (2013) Metagenome survey of a multispecies and alga-associated biofilm revealed key elements of bacterial-algal interactions in photobioreactors. Appl Environ Microbiol 79(20):6196–6206. doi:10.1128/AEM.01641-13

Leary DH, Li RW, Hamdan LJ, WJt H, Lebedev N, Wang Z, Deschamps JR, Kusterbeck AW, Vora GJ (2014) Integrated metagenomic and metaproteomic analyses of marine biofilm communities. Biofouling 30(10):1211–1223. doi:10.1080/08927014.2014.977267

Liljeqvist M, Ossandon FJ, Gonzalez C, Rajan S, Stell A, Valdes J, Holmes DS, Dopson M (2015) Metagenomic analysis reveals adaptations to a cold-adapted lifestyle in a low-temperature acid mine drainage stream. FEMS Microbiol Ecol 91(4). doi:10.1093/femsec/fiv011

Liu B, Faller LL, Klitgord N, Mazumdar V, Ghodsi M, Sommer DD, Gibbons TR, Treangen TJ, Chang YC, Li S, Stine OC, Hasturk H, Kasif S, Segre D, Pop M, Amar S (2012) Deep sequencing of the oral microbiome reveals signatures of periodontal disease. PLoS One 7(6):e37919. doi:10.1371/journal.pone.0037919

Marsh P (2006) Dental plaque as a biofilm and a microbial community—implications for health and disease. BMC Oral Health 6(Suppl 1):S14

McLean JS, Lombardo MJ, Badger JH, Edlund A, Novotny M, Yee-Greenbaum J, Vyahhi N, Hall AP, Yang Y, Dupont CL, Ziegler MG, Chitsaz H, Allen AE, Yooseph S, Tesler G, Pevzner PA, Friedman RM, Nealson KH, Venter JC, Lasken RS (2013) Candidate phylum TM6 genome recovered from a hospital sink biofilm provides genomic insights into this uncultivated phylum. Proc Natl Acad Sci U S A 110(26):E2390–E2399. doi:10.1073/pnas.1219809110

Mills HJ, Martinez RJ, Story S, Sobecky PA (2004) Identification of members of the metabolically active microbial populations associated with Beggiatoa species mat communities from Gulf of Mexico cold-seep sediments. Appl Environ Microbiol 70(9):5447–5458

Moons P, Michiels CW, Aertsen A (2009) Bacterial interactions in biofilms. Crit Rev Microbiol 35(3):157–168. doi:10.1080/10408410902809431

Moreno-Paz M, Gomez M, Arcas A, Parro V (2010) Environmental transcriptome analysis reveals physiological differences between biofilm and planktonic modes of life of the iron oxidizing bacteria Leptospirillum spp. in their natural microbial community. BMC Genomics 11(1):404

Paster BJ, Boches SK, Galvin JL, Ericson RE, Lau CN, Levanos VA, Sahasrabudhe A, Dewhirst FE (2001) Bacterial diversity in human subgingival plaque. J Bacteriol 183(12):3770–3783. doi:10.1128/JB.183.12.3770-3783.2001

Peimbert M, Alcaraz LD, Bonilla-Rosso G, Olmedo-Alvarez G, Garcia-Oliva F, Segovia L, Eguiarte LE, Souza V (2011) Comparative metagenomics of two microbial mats at Cuatro Cienegas basin I: ancient lessons on how to cope with an environment under severe nutrient stress. Astrobiology 12(7):648–658. doi:10.1089/ast20110694

Peterson SN, Meissner T, AI S, Snesrud E, Ong AC, Schork NJ, Bretz WA (2014) Functional expression of dental plaque microbiota. Front Cell Infect Microbiol 4:108. doi:10.3389/fcimb.2014.00108

Pihlstrom BL, Michalowicz BS, Johnson NW (2005) Periodontal diseases. Lancet (London, England) 366(9499):1809–1820. doi:10.1016/s0140-6736(05)67728-8

Ram RJ, Verberkmoes NC, Thelen MP, Tyson GW, Baker BJ, Blake RC 2nd, Shah M, Hettich RL, Banfield JF (2005) Community proteomics of a natural microbial biofilm. Science 308(5730):1915–1920. doi:10.1126/science.1109070

Sanli K, Bengtsson-Palme J, Nilsson RH, Kristiansson E, Rosenblad MA, Blanck H, Eriksson KM (2015) Metagenomic sequencing of marine Periphyton: taxonomic and functional insights into biofilm communities. Front Microbiol 6:1192. doi:10.3389/fmicb.2015.01192

Schmeisser C, Stockigt C, Raasch C, Wingender J, Timmis KN, Wenderoth DF, Flemming HC, Liesegang H, Schmitz RA, Jaeger KE, Streit WR (2003) Metagenome survey of biofilms in drinking-water networks. Appl Environ Microbiol 69(12):7298–7309

Selwitz RH, Ismail AI, Pitts NB (2007) Dental caries. Lancet (London, England) 369(9555):51–59. doi:10.1016/s0140-6736(07)60031-2

Singh PK, Schaefer AL, Parsek MR, Moninger TO, Welsh MJ, Greenberg EP (2000) Quorum-sensing signals indicate that cystic fibrosis lungs are infected with bacterial biofilms. Nature 407(6805):762–764

Treude T, Knittel K, Blumenberg M, Seifert R, Boetius A (2005) Subsurface microbial methanotrophic mats in the Black Sea. Appl Environ Microbiol 71(10):6375–6378

Tringe SG, von Mering C, Kobayashi A, Salamov AA, Chen K, Chang HW, Podar M, Short JM, Mathur EJ, Detter JC, Bork P, Hugenholtz P, Rubin EM (2005) Comparative metagenomics of microbial communities. Science 308(5721):554–557. doi:10.1126/science.1107851

Tyson GW, Chapman J, Hugenholtz P, Allen EE, Ram RJ, Richardson PM, Solovyev VV, Rubin EM, Rokhsar DS, Banfield JF (2004) Community structure and metabolism through reconstruction of microbial genomes from the environment. Nature 428(6978):37–43. doi:10.1038/nature02340

Van Houdt R, Michiels CW (2010) Biofilm formation and the food industry, a focus on the bacterial outer surface. J Appl Microbiol 109(4):1117–1131. doi:10.1111/j.1365-2672.2010.04756.x
Varin T, Lovejoy C, Jungblut AD, Vincent WF, Corbeil J (2012) Metagenomic analysis of stress genes in microbial mat communities from Antarctica and the high Arctic. Appl Environ Microbiol 78(2):549–559. doi:10.1128/AEM06354-11. Epub 2011 Nov 11
Wang J, Qi J, Zhao H, He S, Zhang Y, Wei S, Zhao F (2013) Metagenomic sequencing reveals microbiota and its functional potential associated with periodontal disease. Sci Rep 3:1843. doi:10.1038/srep01843
Ward DM, Ferris MJ, Nold SC, Bateson MM (1998) A natural view of microbial biodiversity within hot spring cyanobacterial mat communities. Microbiol Mol Biol Rev 62(4):1353–1370
Xie G, Chain PS, Lo CC, Liu KL, Gans J, Merritt J, Qi F (2010) Community and gene composition of a human dental plaque microbiota obtained by metagenomic sequencing. Mol Oral Microbiol 25(6):391–405. doi:10.1111/j.2041-1014.2010.00587.x
Yost S, Duran-Pinedo AE, Teles R, Krishnan K, Frias-Lopez J (2015) Functional signatures of oral dysbiosis during periodontitis progression revealed by microbial metatranscriptome analysis. Genome Med 7(1):27. doi:10.1186/s13073-015-0153-3
Zaura E, Keijser BJF, Huse SM, Crielaard W (2009) Defining the healthy "core microbiome" of oral microbial communities. BMC Microbiol 9:259–259. doi:10.1186/1471-2180-9-259
Zhang W, Wang Y, Lee OO, Tian R, Cao H, Gao Z, Li Y, Yu L, Xu Y, Qian P-Y (2013) Adaptation of intertidal biofilm communities is driven by metal ion and oxidative stresses. Sci Rep 3:3180. doi:10.1038/srep03180

Chapter 13
Functional Metagenomics of a Replicase from a Novel Hyperthermophilic Aquificales Virus

David A. Mead, Scott Monsma, Baigen Mei, Krishne Gowda, Michael Lodes, and Thomas W. Schoenfeld

Abstract Bacteriophage and viral replisomes typically require fewer proteins to replicate their genome compared to their cellular counterparts and are therefore model systems for studying this fundamental and ubiquitous process. Replication elements also tend to be arranged in a cluster or operon, as opposed to the distributed replication genes found in Bacteria and Archaea. A gene encoding a DNA polymerase with innate reverse transcriptase activity was previously isolated from an uncultivated Octopus hot spring viral metagenome sample collected in Yellowstone National Park. This report describes the complete metagenomic sequence of Octopus Spring OS3173 virus, novel structural variants, and new functionally active polymerase derivatives. The 37,256 bp dsDNA circular viral genome contains 48 open reading frames, with numerous genes associated with replication, including a full-length, polyprotein-like variant of the polymerase. OS3173 is predicted to infect an *Aquificales* host, as multiple clustered regularly interspaced short palindromic repeat (CRISPR) sequences matching seven locations in the virus genome are found within a pink filament streamer microbial community metagenome from Octopus Spring. Bioinformatic analysis of the DNA surrounding the CRISPR spacer region matches portions of a cultivated *Thermocrinis ruber* genome from the same location with high sequence identity. Enzymatic screening of large-insert clones yielded numerous polyprotein-containing genes encoding active thermostable variants from this virus, confirming the functional diversity of the polymerase in its native habitat. One variant demonstrated robust PCR capabilities compared to the original "wild-type" enzyme.

D.A. Mead (✉)
Varigen Biosciences, Madison, WI, USA
e-mail: dmead@varigenbio.com

S. Monsma • B. Mei • K. Gowda • M. Lodes
Lucigen Corporation, Middleton, WI, USA

T.W. Schoenfeld
Qiagen, Beverly, MA, USA

T.C. Charles et al. (eds.), *Functional Metagenomics: Tools and Applications*,
DOI 10.1007/978-3-319-61510-3_13

13.1 Introduction

Viruses are the most common biological entities on Earth (Bergh et al. 1989; Ashelford et al. 2003; Clokie et al. 2011; Rosario and Breitbart 2011), and they drive biogeochemical cycles, genetic exchange, and microbial diversity on a global scale (Fuhrman 1999; Suttle 2007; Paul 1999). The 8871 cultivated viruses and phage whose genomes have been sequenced (NCBI RefSeq Virus v80, January 2017) represent a small sample of the estimated 10^{30} viruses found in nature (Chibani-Chennoufi et al. 2004). Cultivated viruses and bacteriophage (collectively referred to herein as viruses) are invaluable model systems for studying genetic and molecular processes. They also provide unique tools for biotechnology applications in the form of novel enzymes, genetic transduction vectors, and therapeutic agents against pathogens (Schoenfeld et al. 2010). However, the study of viruses is limited by the challenges associated with cultivating their hosts.

Classical methods for studying viruses require the isolation of their hosts in axenic cultures before the viruses themselves can be cultured. As the vast majority of Bacteria and Archaea remain uncultured, so too do their viruses. This issue is exacerbated in the case of thermophiles, where many unique viral morphotypes can be observed in the environment (Rice et al. 2001; Rachel et al. 2002), but the production of plaques at high temperature is technically challenging and few hosts can be cultivated. Most of our detailed knowledge about thermophilic viruses stems from enrichment cultures (Geslin et al. 2003b; Snyder et al. 2004; Garrett et al. 2010) or cultured representatives (Snyder et al. 2015; Yu et al. 2006). Currently, over 100 archaeal viruses (Snyder et al. 2015) and a similar number of thermophilic bacteriophage (Yu et al. 2006) have been discovered with 79 sequenced archaeovirus genomes in the NCBI sequence database as of January 2017, 55 of which are from thermophiles (http://www.ncbi.nlm.nih.gov/genomes/GenomesGroup.cgi?taxid=10239&host=archaea). Thermophilic bacteriophages are especially underrepresented in the database; only 17 sequenced genomes can be found (http://www.ncbi.nlm.nih.gov/genomes/GenomesGroup.cgi?taxid=10239&host=bacteria).

Obtaining DNA from uncultured microbes has been relatively easy, and the recent sequence analysis of metagenomes and single-cell genomes has provided an extraordinarily detailed view of thermophilic microbes in their natural habitat (Inskeep et al. 2010, 2013; Eme et al. 2013; Kozubal et al. 2013; Takacs-Vesbach et al. 2013; Hedlund et al. 2014, 2015b; Menzel et al. 2015; Munson-McGee et al. 2015). Similar progress, though, has been more challenging for thermophilic viruses. Even though they are estimated to be tenfold more numerous than prokaryotic cells, recovering viral DNA in amounts sufficient for sequencing was challenging until methods were devised to improve the process (Schoenfeld et al. 2008). The relatively small size of phage genomes compared to their cellular hosts necessitates DNA amplification before metavirome sequencing. Amplification and DNA library construction technologies, as well as sequencing methods, can lead to numerous forms of bias that have an impact on the quality of the resulting data (Bowers et al. 2015; Kim and Bae 2011; Probst et al. 2015).

Despite the high abundance of *Aquificae* in many hot springs (Spear et al. 2005), they have proven difficult to culture, and no cultivated viruses infecting members of this phylum have been described. We have performed metagenomic sequence analysis on viral-enriched samples collected from the same site at Octopus hot spring in 2004 (Schoenfeld et al. 2008) and again in 2007 for this study. The first sample was analyzed using Sanger sequencing chemistry, which required amplification of nanogram amounts of material followed by clone-based sequencing. That study occurred before widespread availability of phi29 DNA polymerase-based multiple displacement amplification (Dean et al. 2002), and there were no published methods for amplifying anonymous DNA at the time. A linker-ligation-based PCR amplification method was developed, which has subsequently been used for numerous viral metagenomic studies (Schoenfeld et al. 2008). The PCR amplicons captured by cloning and long-read Sanger sequencing chemistry from that effort were pivotal in the isolation of a novel thermostable reverse transcriptase with unique biochemical and amplification properties (Moser et al. 2012; Chander et al. 2014). The current study took advantage of phi29 DNA polymerase-based amplification technology (Schoenfeld et al. 2013) followed by direct sequencing using the Roche 454 chemistry platform (Smits et al. 2014) to elucidate the complete viral genome containing the reverse transcriptase and a number of other unique genes. In an attempt to capture an entire replisome, large viral metagenomic DNA inserts were cloned for functional analysis using a novel linear vector designed for this type of application (Godiska et al. 2010). Advantages and disadvantages of the two approaches will be discussed, particularly with regard to capturing functional enzymes.

13.2 Methods and Materials

13.2.1 Isolation of Uncultured Viral Particles from Octopus Hot Spring

Viral particles were isolated from Octopus hot spring in Yellowstone National Park (Permit # YELL-2007-SCI-5240), Wyoming (N 44.5342, W 110.79812) in 2007. Temperature at the sample site ranged from 93°C at the main vent to 87°C at the outflow channel where samples were collected (see Fig. 13.1). Thermal water (between 200 and 630 L) was filtered using a 100 kDa molecular weight cutoff (mwco) tangential flow filter (A/G Technology, Amersham Biosciences, GE Healthcare), and viruses and microbes were concentrated to about 2 L. The resulting concentrates were filtered through a 0.2 μm tangential flow filter to remove microbial cells. The viral fractions were further concentrated to about 100 mL using a 100 kDa tangential flow filter, and 40 mL of viruses were further concentrated to 400 μL and transferred to SM buffer (0.1 M NaCl, 8 mM MgSO4, 50 mM Tris-HCl, pH 7.5) by filtration in a 30 kDa mwco spin filter (Centricon, Millipore).

Fig. 13.1 Octopus hot spring (**a**, *left*) and one of the authors (TS) operating the filtration equipment for collection of virus samples (**b**, *right*)

13.2.2 Isolation of Viral DNA

Serratia marcescens endonuclease (Sigma, 10 U) was added to the viral preparations described above to remove non-encapsidated (nonviral) DNA. The reactions were incubated at 23°C for between 2 and 16 h to degrade non-viral DNA; EDTA (20 mM), sodium dodecyl sulfate (SDS) (0.5%), and proteinase K (100 U) were added and the reactions were incubated at 56°C. Subsequently, sodium chloride (0.7 M) and cetyltrimethylammonium bromide (CTAB) (1%) were added. The DNA was then extracted with chloroform, precipitated with isopropanol and washed with 70% ethanol. Yields of DNA ranged from 20 to 200 ng.

13.2.3 Whole Genome Amplification of Viral Metagenomic DNA

Isolated viral metagenomic DNA was amplified with an Illustra GenomiPhi V2 DNA Amplification Kit (GE Healthcare, Piscataway, NJ) following manufacturer's protocol. Briefly, 9 μL sample buffer and 1 μL sample DNA were mixed and incubated at 95°C for 3 min and then placed on ice. Nine microliters reaction buffer and 1 μL enzyme were then mixed and combined with the 10 μL sample and incubated for 2 h at 30°C and 10 min at 65°C. The amplified DNA was then precipitated with NaCl and ethyl alcohol and resuspended in 40 μL water. The amplified DNA was debranched by adding 10 μL of 5X S1 nuclease buffer and 2 μL S1 nuclease (200 U; Thermo Fisher Scientific Inc., Waltham, MA), mixed and incubated at 25°C for 30 min and then 70°C for 10 min. The sample was again precipitated twice with NaCl and ethyl alcohol and resuspended in 20 μL water.

13.2.4 Construction of Large-Insert Expression Libraries

Amplified DNA was made blunt in preparation for cloning with the DNA Terminator Kit (Lucigen Corp., Middleton, WI). Nineteen microliters debranched DNA was combined with 5 μL of 5X buffer and 1 μL of the enzyme mix and incubated for

30 min at 25°C. End-repaired DNA was separated on a 0.7% agarose gel, and an agar plug containing DNA ranging in size from approximately 8–12 kb was excised. DNA was extracted using a Qiagen QIAquick Gel Extraction Kit (Qiagen Corp., Valencia, CA) following manufacturer's protocol and eluted in 50 μL elution buffer. The DNA was then precipitated with NaCl and ethyl alcohol, resuspended in 8 μL of 10 mM Tris, pH 8, and quantified on a 0.7% agarose gel using a Gel Doc XR+ Molecular Imager with Quantity One densitometry software (version 4.6.9) (Bio-Rad Laboratories) and a mass standard (Thermo Fisher High DNA Mass Ladder).

Approximately 250 ng of 8–12 kb DNA was ligated to 1.5 μL pJAZZ-OC Blunt Vector (Lucigen Corp.) with 1 μL T4 DNA Ligase at 25°C for 2 h followed by 70°C for 15 min. Ligated DNA was transformed into electrocompetent TSA-G5 cells (Lucigen Corp.) with a Gene Pulser Xcell (Bio-Rad Laboratories, Hercules, CA), incubated at 37°C for 1 h in recovery medium, plated on agar plates with chloramphenicol selection, and incubated overnight at 37°C. Single bacterial colonies were cultured overnight in TB medium with chloramphenicol selection, and expression was induced with arabinose as described by the manufacturer.

Full-length 3173 DNA polymerase polyprotein was individually expressed by cloning PCR-amplified DNA from the viral metagenome using sequence-specific primers spanning the start and stop codon of the full-length gene (forward primer, ATGAGTATATCATTTTTTGAGTTATTTTTCAATATAGGTTTG; reverse primer, ACCCATTGTGTACCCTTTTCGGAACT) using Phusion DNA polymerase (New England Biolabs). The amplicon was cloned into the T7 promoter plasmid pETite (Lucigen Corp.), transformed in BL21DE3, and expressed as described by the manufacturer.

13.2.5 Sequencing, Assembly, and Bioinformatics

Octopus hot spring metagenomic DNA was sequenced using Roche 454 chemistry at the Broad Institute (229,553 reads averaging 375 nucleotides each; 86,161,605 bases in total). The data can be accessed at http://data.imicrobe.us/search?query=great+boiling+spring. The full read set was assembled de novo with CLC Genomics Workbench 8.0, using word size of 20 and bubble size of 375. A total of 5143 contigs of length > 500 were assembled with N50 = 1818 bp, average length of 1586 bp, maximum contig length of 35,614 bp (contig _4), and total assembly length of 8,156,404 bp. Of the 229,553 original reads, 66% (152,673 reads) were incorporated into contig assemblies >500 bp.

Of the reads, 56.6% (86,379 reads) mapped to the largest contig (contig_4) at a stringency of 90%, which eventually was closed as Octopus Spring OS3173 virus, resulting in an average coverage of 907-fold. The OS3173 consensus viral sequence was finished by an iterative process of extending the ends of contig_4 with partially mapped reads until the extended consensus ends were found to overlap. This resulted in a 37,256 bp circular genome. A total of 99,924 reads were mapped to the finished genome (also at 90% stringency), and reads were found to map continuously across the joined overlap, consistent with a circular topology. Reads that did not map at

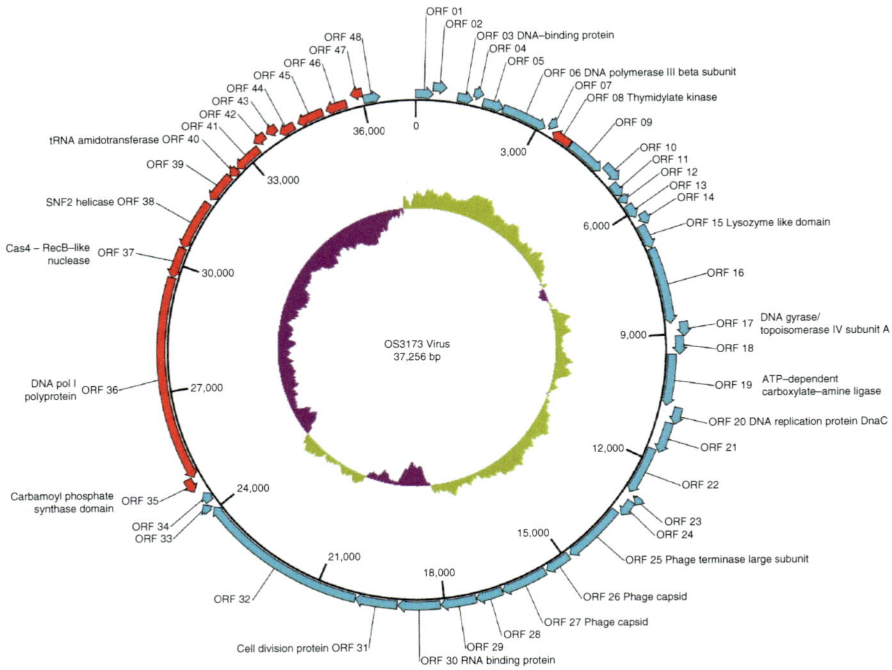

Fig. 13.2 Genome map of Octopus Spring OS3173 virus. *Outer circle* shows numbered ORFs and selected annotation features (clockwise facing ORFs are colored *blue*, counterclockwise *red*). *Second circle* map shows the GC skew (DNAPlotter Release 1.11, window size 1000, step size 15)

90% stringency were saved and remapped at relaxed stringency (80% identity over 80% length). These relaxed stringency reads were found to contain structural variants and polymorphisms as described in the results section. The origin of the reported viral sequence was arbitrarily set to the beginning of the first open reading frame (ORF) clockwise of the negative to positive GC skew transition (Fig. 13.2). ORFs in OS3173 were identified by the GeneMarkS heuristic algorithm (Besemer et al. 2001). Open reading frames (ORFs) identified by GeneMarkS were submitted to NCBI BLASTP (Altschul et al. 1997) using default settings for comparison with proteins in the public database.

13.2.6 *Functional Screening for Thermostable Polymerase Activity*

Four 96-well plates containing overnight cultures (384 recombinant viral DNA clones total) were combined into a single 96-well plate (four clones per well) for thermostable DNA polymerase activity screening. Cell cultures (1.2 mL) were pelleted and resuspended in 200 μL lysis buffer (50 mM Tris pH 8.0, 10 mM EDTA pH 8.0, 0.1 mg/mL RNAse A, 23 U/μL Ready-Lyse enzyme (Epicentre)). The samples were heated at 70°C for 10 min, applied to a Whatman filter plate, and centrifuged at 1500 rpm for 5 min. The eluates were assayed for thermostable polymerase activity (12.5 μL reaction) using a reaction mix containing 1X ThermoPol buffer (20 mM Tris-HCl, 10 mM (NH4)2SO4, 10 mM KCl, 2 mM MgSO4, 0.1% Triton X-100, pH 8.8), 0.2 mM dNTPs (Lucigen Corp.), 2.5 μg activated calf thymus DNA (Sigma-Aldrich, St. Louis, MO), 0.25 uCi [α-33P] dCTP (Perkin-Elmer, Waltham, MA), and 2 μL cell lysate. The plate was incubated at 65°C for 30 min after which 30 μL of water was added to each reaction. The reaction mix was applied to Whatman DE81 filter paper via a vacuum manifold and dried. The filter was washed with 0.5 M sodium phosphate, water, and then ethanol. After drying, the filter was exposed to a K-screen that was imaged on a PharosFX imager (Bio-Rad, Hercules, CA). Seven wells tested positive for thermostable DNA polymerase activity. The same incorporation assay was used to resolve which of the four clones per well expressed the thermostable DNA polymerase activity, except the isotope was measured by scintillation counting rather than autoradiography. A blank reaction without added DNA polymerase was used to determine background activity.

13.2.7 PCR Amplification

DNA polymerases (DNAP) derived from Octopus metaviral clones were assessed in side-by-side PCR reactions using two different-sized amplicons (0.98 and 2.8 Kb). PCR reaction conditions contained 1–20 ng of template DNA, 2.5 U DNAP (Lucigen Corp.), 200 μM dNTPs, and 0.5 μM primers in a 50 μL reaction. DNAP buffer (1X) contained 10 mM Tris-HCl (pH 8.8), 10 mM KCl, 10 mM NH2SO4, 2 mM MgSO4, 0.1% triton X-100, and 15% sucrose. Cycling conditions were 94°C for 2 min and 30 cycles of 94°C for 15 s, 60°C for 30 s, and 72°C for 1 min per kb. The templates and PCR primers are as follows: pUC19 0.9 kb amplicon primers (CCC CTA TTT GTT TAT TTT TCT AAA ATT CAA TAT GTA TCC GCT and TTA CCA ATG CTT AAT CAG TGA GGC ACC TAT CT) and *E. coli* 2.8 kb amplicon primers (TAC TGT CTG CCA TGG TTC AGA TCC CCC AAA ATC CAC TTA TCC TTG TAG A and TTA TCT GTG GTC GAC TTA GTG CGC CTG ATC CCA GTT TTC GCC ACT CCC CA).

13.3 Results and Discussion

13.3.1 Octopus Hot Spring and Virus Sampling

Viral particles were isolated from Octopus hot springs in the White Creek area of Yellowstone National Park as previously described (Schoenfeld et al. 2008). The spring has a pH of 8.0, with low sulfide and high silica content (McCleskey et al. 2004) and a temperature at the source of 93°C (the boiling point at ~2300 m elevation) and 87°C at the site of sampling (Fig. 13.1a, b). Viruses were collected in the proximal outflow channel immediately downstream from rocks containing long adherent pink filament streamers of hyperthermophiles (Fig. 13.1b) previously identified as an *Aquificales* dominated community by 16S rRNA and metagenomic sequence analysis (Reysenbach et al. 1994; Blank et al. 2002; Takacs-Vesbach et al. 2013).

The hyperthermophilic bacterial community of Octopus Spring remains poorly described (Reysenbach et al. 1994; Blank et al. 2002) with one *Thermocrinis* isolate cultivated thus far (Huber et al. 1998). To date, no *Archaeal* isolates have been described from this source. Conserved protein marker analysis of the hyperthermophilic streamer microbial metagenome from Octopus Spring (Takacs-Vesbach et al. 2013) suggests there are eight or nine novel *Thermocrinis*-related populations, none of which are well represented by the sequenced genomes from *Thermocrinis albus* or *T. ruber*. The latter result is incongruent considering *T. ruber* was originally isolated from Octopus hot spring (Huber et al. 1998). Similar observations of high sequence diversity in the Octopus hot spring *Aquificales* population have been documented by phylotyping (Blank et al. 2002), but just the opposite was found for archaeal community members. Octopus Spring contains very little *Archaea* as determined by 16S rRNA sequencing (Reysenbach et al. 1994; Blank et al. 2002); however, ~20% archaeal co-community members were inferred by binning metagenomic sequence reads using functional gene content (Takacs-Vesbach et al. 2013), with Thermoproteaceae populations prevalent.

The nonmicrobial fraction of the hyperthermophilic planktonic community of Octopus hot spring contains viruslike particles identified by electron microscopy (Schoenfeld et al. 2008), with morphological families representing known thermophilic bacterial and archaeal viruses. Tailed morphologies are commonly associated with bacteriophages and euryarchaeotal viruses (Yu et al. 2006, Geslin et al. 2003a); rod-shaped and filamentous morphologies are more commonly associated with crenarchaeal viruses (Prangishvili and Garrett 2004). There are no known cultivated *Aquificales* viruses. However, the PHAST program (Zhou et al. 2011) identified putative prophages present in several species of the 19 sequenced *Aquificales* genomes (Reysenbach et al. 2009), including *Hydrogenobacter thermophilus* TK-6 (Arai et al. 2010), *Hydrogenobaculum* sp. 3684 and SHO (Romano et al. 2013), *Persephonella* sp. KM09-Lau-8 (Reysenbach et al. 2015), *Thermocrinis* sp. GBS (Hedlund et al. 2015a), and *Thermocrinis ruber* strain DSM 23557 (Eisen et al. 2015) (data not shown). All of the sequenced *Aquificales* genomes also contain at least one locus of clustered regularly interspaced short palindromic repeat sequences (CRISPRs) (data not shown), which are involved in the immunity against viruses and plasmids (Horvath and Barrangou 2010).

Metagenomic sequence analysis of the pink filament microbial community of Octopus hot spring (Takacs-Vesbach et al. 2013) has previously identified *Thermocrinis*-like *Aquificales* as the dominant microbe comprising this portion of the thermal spring ecosystem, which is consistent with 16S rRNA analysis results (Blank et al. 2002; Reysenbach et al. 2015). *Aquificales* are microaerophilic gram-negative bacteria that flourish in zones of shallow, high-velocity, turbulent terrestrial springs, as well as marine hydrothermal fissures (Ferrera et al. 2007). *Aquificales* are thermophilic and hyperthermophilic eubacteria believed to comprise the deepest lineage of the domain *Bacteria* (Coenye and Vandamme 2004; Barion et al. 2007; Reysenbach et al. 2005), with three main families: *Thermocrinis* species occupying the higher pH range (6–9), *Hydrogenobaculum* species occupying the lower pH range (<4), and *Sulfurihydrogenibium* species overlapping in the middle range (4–6) (Inskeep et al. 2013).

13.3.2 Viral Amplification, Sequence Analysis, and Genome Assembly

Previous sampling of Octopus hot spring identified numerous viruslike genes by metagenomic sequence analysis (Schoenfeld et al. 2008). In that study, phi29 DNA polymerase and multiple displacement amplification was not commercially available at the time; therefore, we developed a linker-dependent, anonymous method of DNA amplification to access the trace amounts of material present in the virus-enriched samples. DNA samples were physically sheared to 3–8 kb, and short (20 bp) oligonucleotide linkers were ligated to the DNA fragments to serve as priming sites for PCR using a proofreading polymerase. Amplified fragments were cloned into a transcription-free pSMART vector to minimize cloning bias due to cytotoxic sequences (Godiska et al. 2005). This library construction method has been used in the analysis of several cultivated genomes and uncultivated viral metagenomes (Bench et al. 2007; Breitbart et al. 2002, 2003, 2004; Lindell et al. 2004; Paul et al. 2005; Seguritan et al. 2003; Sullivan et al. 2005) as well as the 2004 Octopus viral library (Schoenfeld et al. 2008). That study generated 21,198 paired-end Sanger chemistry sequencing reads from 10,599 clones averaging 981 nucleotides each, or nearly 21 Mb in total, but failed to assemble a complete viral genome. Only four reads were assembled from that project that overlapped the 3173 DNAP gene. Contaminating bacterial DNA was relatively low in the 2004 library; only 4 ribosomal RNA gene sequences were detected in 10.4 "microbial genome equivalents (assuming 2 Mb for hyperthemophilic average genome size)" compared to an expected value of 52 ribosomal genes if 100% of the sample consisted of microbes.

The current work sampled the same Octopus hot spring site in 2007, enriching for the same viral DNA fraction; only this time, DNA was amplified using phi29 DNA polymerase, and 229,553 reads averaging 375 nucleotides each were generated using 454 pyrosequencing chemistry (86 Mb in total). Contaminating bacterial DNA was also relatively low in the 2007 library; only 4 ribosomal RNA gene sequences were detected in 43 "microbial genome equivalents" compared to an expected value of 215 ribosomal genes if 100% of the sample consisted of microbes. Bacterial contamination was measured by binning and inspection of the assembled contig BLASTP results, which suggests a 1.6% level of recognizable bacterial sequences (1,390,256/86,161,605 bp, Table 13.1).

Table 13.1 Octopus hot spring viral-enriched metagenomic reads associated with viruses, bacteria, and Nanoarchaeota

Category	Reads	Base pairs	Contigs assembled
All reads	229,553	86,161,605	5143[a]
OS3173 virus	99,924	36,964,258	1
Bacterial contaminants	3726[b]	1,390,256	–
Remaining reads	124,918	47,807,091	3406[a]
Pyrobaculum spherical virus	985	130,297	2
Hyperthermophilic Archaeal virus	461	54,899	54
Thermoproteus tenax virus	425	34,472	32
Acidianus virus	149	22,776	20
Nanoarchaeota Nst1	189	16,083	11

[a]>500 bp
[b]Includes four rRNA genes

The primary difference between the 2004 versus the 2007 Octopus Spring viral-enriched metagenomic samples was the assembly of several complete viral genomes, as described below. The 229,553 reads from the Roche 454 sequence assembly produced 5143 contigs larger than 500 bp (Table 13.1). Forty-three percent of the nucleotides (36,964,258/86,161,605 bp) assembled into a single contig we are calling Octopus Spring OS3173 virus (Fig. 13.2). The only other nearly complete viral genome to be assembled was a relative of *Pyrobaculum* spherical virus (Häring et al. 2004), whose genome will be described elsewhere. Additional reads were found to associate with hyperthermophilic *Archaeal* virus 1 (Garrett et al. 2010), *Thermoproteus tenax* spherical virus 1 (Ahn et al. 2006), and multiple *Acidianus* viral genomes.

We also noticed a small frequency of contig assembled nucleotides (0.019%) had high similarities to a new terrestrial member of the Nanoarchaeota (*Nanobsidianus stetteri* or Nst1), which is thought to be a cellular parasite associated with a novel crenarchaeal host (Sulfolobales Acd1) (Podar et al. 2013), neither of which have been cultivated. Their closest counterparts are *I. hospitalis* and the only cultivated member of the Nanoarchaeota, *N. equitans* (Jahn et al. 2008). The discovery of Nst1 contigs with sequence similarity E scores up to 2.00E-93 (data not shown) in Octopus Spring is somewhat surprising given that this parasitic *Nanoarchaea* is typically found in sulfidic environments below pH 6.7 (Podar et al. 2013; Clingenpeel et al. 2013) and was not reported in a recent Octopus pink filament metagenomic community analysis (Takacs-Vesbach et al. 2013). The samples collected for the work reported here were filtered planktonic water from the outflow channel of Octopus Spring, so the Nst1 related sequences could be coming from the vent of the spring as opposed to the pink filament streamer community of predominantly bacterial biomass (Takacs-Vesbach et al. 2013). If the finding of *Nanoarchaea* holds up in Octopus and other silica-rich, alkaline hot springs, it will extend the range of this novel branch of life into a new and unexpected geochemical territory.

In this study the hot spring DNA was amplified using phi29 DNA polymerase and random hexamers (Fig. 13.5) to supply the necessary material for metagenomic sequencing as well as functional screening for thermostable polymerases. This approach amplifies single- and double-stranded DNA (ss- or dsDNA), making it difficult to distinguish which nucleic acid form is the native genome for OS3173. The data generated from the 2004 Octopus library utilized cloned PCR amplicons from the dsDNA oligo-adapter ligated material to sequence from both sides of the vector utilizing Sanger chemistry. Because of this technical detail, we are confident that OS3173 has a dsDNA form during part of its replication cycle. A functional clone (3173) containing a thermostable polymerase was eventually discovered from this library (Schoenfeld et al. 2008) and characterized for RT PCR activity (Moser et al. 2012) as well as RT LAMP capabilities (Chander et al. 2014).

13.3.3 OS3173 Viral Genome and Annotations

Octopus Spring metagenomic virus OS3173 assembled into a circular dsDNA contig containing 37,256 bp (37.1% GC). There are 48 predicted protein coding genes greater than 60 amino acids (Fig. 13.2, Table 13.2) with 1915 bp intergenic DNA (94.85% open reading frames (ORFs)). The orientation of the ORFs is neatly divided into a 26 kb clockwise facing set of 35 genes, with the exception of ORF 08, and an 11 kb counterclockwise set of 13 genes, which are separated by 827 and 110 bp intergenic regions, respectively. There is a noticeable GC skew at the junction of the clockwise and counterclockwise facing set of ORFs (Fig. 13.2), suggesting a differential replication module for this part of the virus that begins in the large intergenic region around 37,239 bp and ends in a smaller intergenic space around 24,579 bp. Many bacterial genomes undergo bidirectional replication, which is characterized by a G > C bias in the leading strand of replication and a C > G bias in the lagging strand of replication (Képès et al. 2012). The largest intergenic space, between ORF 48 and ORF 1, contains numerous different categories of repetitive elements capable of forming complex stem-loop structures within its 826 bp. The small intergenic space around 24,605, between ORF 34 and ORF 35 and extending into ORF 33, also contains numerous repetitive elements with the potential for forming multiple stem-loop structures. Based on the GC skew analysis and the presence of putative structure-rich intergenic regions that coincide, the OS3173 origin and termination of replication is hypothesized to lie in these two regions. If correct, this mode of replication does not resemble typical dsDNA phage type modules (Weigel and Seitz 2006), but rather a mini-bacterial genome. Cultivation of the virus and its host will be required to test this hypothesis.

The presence of a polymerase, nuclease/recombinase, and helicase annotated genes in the counterclockwise set of 13 genes also suggests that early gene synthesis of these replication-associated parts of the virus may come before structurally annotated genes found facing the other orientation (Fig. 13.2, Table 13.2). In spite of this arrangement, there are other nucleic acid metabolizing and replication type proteins atypically interspersed among the clockwise facing viral structural genes. Annotated genes for a DNA-binding protein (ORF 3), DNA Pol III beta subunit (ORF 6),

Table 13.2 Open reading frame analysis of OS3173 viral metagenomic assembly

ORF	AA length	pI	Gene annotation	Top BLASTP (cutoff E-value >0.001)
1	138	8.58		
2	104	9.05		
3	119	5.09	DNA-binding protein[A]	Sulfolobus turreted icosahedral virus 2_B116 Expect 9e-35 Identities 41/118 (34%)
4	74	5.80		
5	161	5.42		
6	358	4.72	DNA polymerase III beta subunit	*Thermocrinis ruber* Expect 9e-50 Identities 125/369 (34%)
7	66	10.94	Hypothetical protein	*Hydrogenobacter thermophilus* Expect 3e-17 Identities 39/68 (57%)
8	161	9.34	Thymidylate kinase	Aquifex aeolicus Expect 2e-05 Identities 44/144 (31%)
9	303	5.33	Hypothetical protein	HGMM_F07F09C15 [uncultured *Aquificae* bacterium] Expect 3e-33 Identities 112/316 (31%)
10	143	4.96	Hypothetical protein	HGMM_F07F09C16 [uncultured *Aquificae* bacterium] Expect 2e-67 Identities 101/141 (72%)
11	112	8.4	Hypothetical protein	*Hydrogenobacter thermophilus* Expect 7e-39 Identities 66/109 (61%)
12	68	6.22	Hypothetical protein	*Hydrogenobacter thermophilus* Expect 4e-24 Identities 42/68 (62%)
13	112	4.41		
14	89	9.52	Hypothetical protein	HGMM_F07F09C20 [uncultured *Aquificae* bacterium] Expect 2e-06 Identities 31/89 (35%)
15	186	8.89	Lysozyme-like domain	*Hydrogenobacter thermophilus* Expect 1e-49 Identities 83/156 (53%)
16	593	5.14		
17	106	5.90	DNA gyrase/topoisomerase IV subunit A domain	
18	142	7.75		
19	403	5.83	ATP-dependent carboxylate-amine ligase domain	

Table 13.2 (continued)

ORF	AA length	pI	Gene annotation	Top BLASTP (cutoff E-value >0.001)
20	125	5.24	DNA replication protein DnaC domain	
21	234	8.53		
22	381	5.16		
23	61	9.37		
24	125	4.29	Hypothetical protein	MTPG_00003 [*Methylophilales* phage HIM624-A] Expect 3e-05 Identities 40/102 (39%)
25	486	7.18	Phage terminase large subunit domain	AMJ80_04910 [bacterium SM23_31] Expect 4e-11 Identities 64/216 (30%)
26	204	4.74	Phage capsid domain	
27	361	5.32	Phage capsid domain	
28	211	7.74		
29	276	9.78		
30	328	9.52	RNA-binding protein	*Pyrococcus* sp. ST04 Expect 8e-06 Identities 12/28 (42%)
31	329	8.94	Cell division protein domain	
32	1325	7.92		
33	66	5.24		
34	83	9.25		
35	108	10.31	Carbamoyl phosphate synthase large subunit domain	
36	1606	5.68	OS3173 DNA polymerase I	OS3173 uncultured virus Expect 0.0 Identities 1596/1606 (99%) 3′-5′ exonuclease [*Thermocrinis albus*] Expect 2e-48 Identities 188/623 (30%)
37	238	8.26	Cas4—RecB-like nuclease	*Thermodesulfobium narugense* Expect 4e-09 Identities 57/220 (26%)
38	395	9.05	SNF2 family helicase	OS3173 uncultured virus Expect 6e-52 Identities 83/104 (80%)
39	232	6.21		
40	73	5.45	Glutamyl-tRNA amidotransferase subunit E	*Thermococcus* sp. AM4 Expect 9e-05 Identities 11/46 (24%)

(continued)

Table 13.2 (continued)

ORF	AA length	pI	Gene annotation	Top BLASTP (cutoff E-value >0.001)
41	223	9.32	Hypothetical protein	*Thermocrinis ruber* Expect 1e-50 Identities 93/233 (40%)
42	97	9.79	Hypothetical protein	*Thermocrinis ruber* Expect 4e-18 Identities 47/83 (57%)
43	86	8.53	Hypothetical protein	*Thermocrinis ruber* Expect 2e-11 Identities 27/58 (47%)
44	126	5.73		
45	210	8.80		
46	168	8.30		
47	90	9.75		
48	129	10.59		

Gene annotations of ORFs with BLASTP expect values >0.001 were left blank

thymidylate kinase (ORF 8), DNA gyrase/topoisomerase IV subunit A (ORF 17), DNA replication protein DnaC (ORF 20), and RNA-binding protein (ORF 30) interspersed among the clockwise facing part of the genome suggest a more complex genetic history and replication mode for this virus.

There are 27 ORFs in OS3173 that were annotated based on shared similarity to other genes in the public database at NCBI (Table 13.2) and 21 ORFs that do not resemble anything. Annotated gene products include 10 hypothetical proteins, 11 nucleic acid metabolizing or binding proteins, 4 phage structural or processing proteins, and 2 "other" categories. The hypothetical annotated proteins are all related to *Aquificales* type microbial genes except for ORF 24, which is distantly similar to the undescribed *Methylophilales* phage HIM624-A, a saltwater virus. ORF 9 encodes a thymidylate kinase annotated gene with low similarity to an *Aquifex aeolicus* ortholog.

OS3173 encodes a number of putative nucleic acid metabolizing and replication annotated gene products, with the clockwise open reading frame genes numbering seven and the counterclockwise genes numbering four (Table 13.2). ORF 3 encodes a 119 amino acid protein with high similarity to a *Sulfolobus* virus DNA-binding protein that is highly conserved in diverse crenarchaeal viruses (Larson et al. 2007) and whose structure has been solved (Keller et al. 2007). ORF 6 encodes an annotated sliding clamp beta subunit of DNA polymerase II, but the virus lacks an obvious clamp loader. Whether the viral replicase uses its host clamp loader or has an unrecognized accessory protein that fulfills that function is unknown. Nucleotide kinases are found in numerous phage genomes; ORF 8 encodes a putative thymidylate kinase. ORF 17 encodes a 106 AA DNA gyrase/topoisomerase IV subunit A domain that is exceptionally small compared to orthologous genes in other viruses. It is conceivable that it is functionally linked to the unidentified 519 AA ORF 16.

A DNA replication protein related to DnaC, which is a helicase loader, was annotated for ORF 20; however, the protein is 2–4 times smaller than phage orthologs found in UniProt (www.uniprot.org). A putative RNA-binding protein is encoded within ORF 30, and a cell division protein domain is found in ORF 31 with no significant identities to other known genes.

The counterclockwise operon encodes annotated genes for a tRNA amidotransferase (ORF 40) and three major replicase-associated proteins: an ATP-dependent helicase (ORF 38), a nuclease/recombinase (ORF 37), and a large polyprotein with functionally active polymerase activity (ORF 36). The helicase annotated gene (ORF 38) contains two P-loop containing nucleoside triphosphate hydrolase domains related to the DEAD-like helicase superfamily, but the identity to orthologs is very low. The Cas4-RecB-like nuclease (ORF 37) belongs to the PD-(D/E)XK nuclease superfamily and presumably encodes nuclease activities required for replication or recombination (Zhang et al. 2012; Lemak et al. 2014). Cas4 homologues have been characterized in other viral genomes (Gardner et al. 2011; Guo et al. 2015), implicating their role in replication and/or recombination.

ORF 36 encodes a 1606 amino acid putative polyprotein with multiple annotated domains (Schoenfeld et al. 2013). The amino-terminal region has sequence motifs that imply primase and/or helicase function. These include similarity to DUF927 (conserved domain with carboxy-terminal P-loop NTPase) and COG5519 (superfamily II helicases associated with DNA replication, recombination, and repair [Marchler-Bauer et al. 2011]). A consensus Walker A or P-loop motif and a potential Walker B motif suggest NTP binding and hydrolysis likely associated with helicase activity (Walker et al. 1982). BLASTP analysis showed that the carboxy-terminal 3′exo/pol domains of the viral gene product (585 amino acids) are similar to PolAs of several *Aquificae* species (Schoenfeld et al. 2013), typified by the *Aquifex aeolicus* PolA (Chang et al. 2001), and to 3′exo/pol domains of the nuclear-encoded, apicoplast-targeted DNA polymerases of several *Apicomplexa* species, typified by the PfPrex protein of *Plasmodium falciparum* (Seow et al. 2005). Surprisingly, the PfPrex polymerase is optimally active at 75°C (Seow et al. 2005), much higher than would be encountered during the *Plasmodium* life cycle and much higher than other *Plasmodium* proteins, but remarkably similar to the optimal growth temperature of *Thermocrinis* and to the geothermal springs sampled in this study (Huber et al. 1998), implying lateral gene transfer for this ortholog (Schoenfeld et al. 2013).

The putative polyprotein structure encoded in OS3173 ORF 36 is very common in RNA replicases and reverse transcriptases (Eickbush and Jamburuthugoda 2008), but not in other viral clades. However, a bioinformatics analysis of the 235 kDa *Plasmodium falciparum* PfPrex polyprotein identified potential primase, helicase, exonuclease, and polymerase domains, which were indirectly or directly confirmed experimentally by in vitro assays (Seow et al. 2005). An essential role of the proteolytically mature PfPrex primase in the production of RNA primers for lagging strand DNA synthesis of the apicoplast genome has further implicated its role in replication (Lindner et al. 2011). The reverse transcriptase activity of the 585 amino acid carboxy-terminal domain of OS3173 ORF 36 (Moser et al. 2012) may not be involved in DNA replication as it bears similarity to activities implicated in maintenance of telomeres (Bao and Cohen 2004) and in tropism-switching mechanisms that prevent reinfection by phage (Liu et al. 2004; Medhekar and Miller 2007; Wang et al. 2011). Understanding the biochemical role(s) of the rest of the ORF 36 domains could result in new thermostable accessory proteins for DNA amplification.

Although OS3173 has not been cultivated, indirect evidence that it is a dsDNA tailed virus comes from the presence of ORF 25, which is annotated as a large subunit phage terminase. The small terminase subunit protein is a site-specific endonuclease that restricts the ends of the viral DNA in preparation for packaging and encapsulation by the large subunit terminase (Kala et al. 2014), which work in tandem. The presence of a predicted terminase is indicative of a tailed dsDNA phage belonging to the Caudovirales; however, other hallmark genes of this class of virus, such as the small terminase subunit, tape measure protein, tail formation, and baseplate-related genes, were not annotated. Typically, the large and small subunit terminase genes are tandemly arranged on viral genomes, and ORF 24 is a candidate for the small terminase protein as it is the correct size and pI compared to orthologous enzymes. Another common viral hallmark enzyme is endolysin, which is typically paired with a holin and together serve to lyse the host (Young 2014). ORF 15 is annotated as a lysozyme-like domain. No recognizable holin domain was annotated, but ORF 14 is a likely candidate based on similarity to known holins (Reddy and Saier 2013), i.e., the small size of the protein with three putative transmembrane segments of the appropriate length and an overlapping open reading frame between open reading frames 13, 14, and 15 suggest an anti-holin, holin, lysozyme operon typically found in numerous viruses.

13.3.4 Putative Host for OS3173 Virus Is Thermocrinis ruber

A new tool for the cultivation-independent analysis of virus-microbe coevolution is the study of clustered regularly interspaced short palindromic repeats (CRISPRs) and related Cas (CRISPR associated) genes found in many bacterial and archaeal genomes (Grissa et al. 2007). CRISPR arrays consist of multiple (2–250) direct repeats with each repeat separated by variable spacer sequences (Bolotin et al. 2005; Haft et al. 2005; Kunin et al. 2007). These sequences are often derived from viruses and function as guide sequences to cleave foreign DNA and protect the cell from viral infection (Jansen et al. 2002; Makarova et al. 2006; Barrangou et al. 2007; Beloglazova et al. 2008). Correlation of metaviromic sequences with CRISPR repeats found in microbial genomes and metagenomes can be used to infer virus-host relationships (Heidelberg et al. 2009; Snyder et al. 2010; Anderson et al. 2011). This approach was used to identify a presumptive host for OS3173 virus.

The OS3173 viral genome was used to query the CRISPRFinder spacer database (Grissa et al. 2007), which contains spacer sequences from cellular CRISPR loci from all reported NCBI bacterial and archaeal genomes. Four matches to the *Thermocrinis ruber* genome (NZ_CP007028.1) were found, with E-values ranging from 1e-5 to 1e-6, and one match to the *Thermocrinis albus* genome

Table 13.3 Homologies between Octopus Spring OS3173 virus, Octopus Spring metagenomic contigs, and *Thermocrinis ruber*

OS3173 virus ORF	Octopus Spring microbial CRISPR spacer length/%ID OS3173	Octopus Spring microbial metagenome contig length (bp)	*Thermocrinis ruber* (NZ_CP007028.1) nucleotide identity
3	37 bp—95%	2912	95% ID over 93% coverage
6	47 bp—90%	863	No significant similarity
16	47 bp—90%	17,260	95% ID over 83% coverage
27	35 bp—100%	801	92% ID over 50% coverage
27	36 bp—97%	812	72% ID over 94% coverage
31	34 bp—97%	2060	90% ID over 46% coverage
36	35 bp—97%	738	72% ID over 94% coverage
38	38 bp—97%	718	77% ID over 100% coverage

(NC_013894.1), with an E-value of 0.006 (data not shown). Much higher E scores were obtained when OS3173 was used to blast the DOE-JGI IMG/M (Markowitz 2012) website (http://img.jgi.doe.gov/m) under IMG taxon OID numbers for Octopus Spring (2022920012/2014031007) (Takacs-Vesbach et al. 2013). Table 13.3 shows eight segments of OS3173 (seven different ORFs) with 90–100% homology to Octopus Spring metagenomic spacer sequences within CRISPR repeat arrays. Seven of the eight contigs showed high homology outside of the CRISPR repeats when compared to the published *Thermocrinis ruber* genome, implicating this microbe or a closely related one as the host for Octopus OS3173 virus.

13.3.5 Functional Metagenomics of OS3173 Viral Replicase

OS3173 ORF 36 encodes a PolA type polymerase demonstrated to possess a 3′-5′ exonuclease domain in addition to the RNA-directed DNA polymerase domain which is functional in RT PCR and RT LAMP amplification technologies (Moser et al. 2012; Chander et al. 2014). Most thermostable PolA enzymes lack a functional 3′-5′ exonuclease activity, which is critical for proofreading and enzyme fidelity in biotechnology applications. There are at least two exceptions, *Rhodothermus marinus* PolA (Blöndal et al. 2001) and *Aquifex aeolicus* PolA (Chang et al. 2001), which possess functional 3′-5′ exonuclease domains but are not thermostable enough for thermocycling applications such as PCR. New proofreading thermostable polymerases with RT activity and additional accessory enzyme activities could prove useful in a wide range of amplification and detection applications.

In an effort to capture a full-length OS3173 DNA polymerase clone of ~1606 amino acids, the same Octopus Spring viral DNA used for sequencing was amplified, and an expression library for screening thermostable DNA polymerase activity

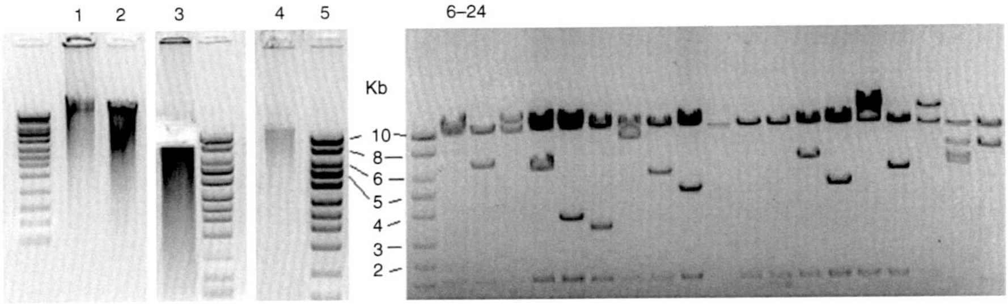

Fig. 13.3 Amplification and cloning of large-insert viral metagenomic DNA. Agarose gel electrophoretic analysis of large-insert DNA manipulations. Lanes: (1) Phi29 DNA polymerase-amplified metagenomic DNA from Octopus hot spring, (2) treatment with S1 nuclease to debranch DNA, (3) void showing plug of >8 kb DNA purified from gel, (4) purified DNA from lane 3, (5) MW marker DNA, (6–23) not I digests of 18 pJAZZ clones containing Octopus viral insert DNA (7.9 kb average insert size), and (24) empty vector

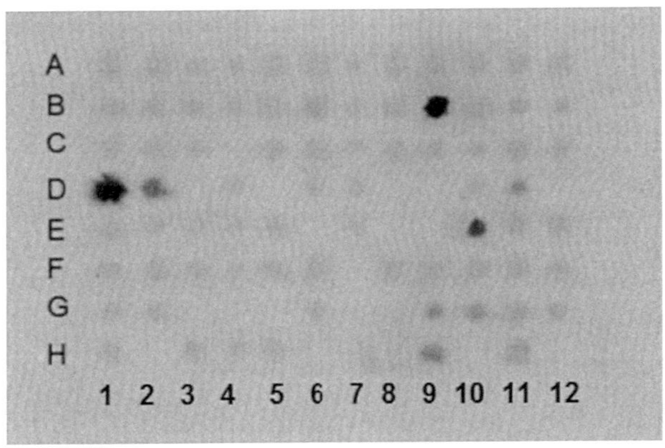

Fig. 13.4 Functional screen for thermostable viral polymerase activity. A 96-well plate containing four recombinant pJAZZ clones each was processed for incorporation of radiolabeled dATP at 65°C. Reactions were filtered onto DEAE paper and exposed to a phosphor imaging screen to visualize putative viral polymerase-containing clones

was constructed using >8 kb size-selected DNA (Figs. 13.3 and 13.4). Cloning and maintenance of larger DNAs is more robust in the linear pJAZZ vector than supercoiled circular plasmids (Godiska et al. 2010; McFarland et al. 2015). Not I restriction digests of 18 pJAZZ clones showed an average insert size of 7.9 kb from Octopus Spring metaviral DNA (Fig. 13.3).

Thermostable DNA polymerase activity from the *E. coli*-cloned Octopus viral DNA was determined by incorporation of radiolabeled nucleotide as described in the methods section. Four 96-well plates containing overnight cultures (384 clones total) were combined into one 96-well plate and processed for thermostable polymerase activity screening. The *E. coli* lysates were incubated at 65°C for 10 min to inactivate host activity, resulting in four strong signals and several weak ones (Fig. 13.4). Seven wells tested positive for thermostable polymerase activity (D1, D2, D11, B9, E10, G9, and H9). To dereplicate which of the four clones/well contained a putative polymerase, the 28 individual clones were assayed by radiolabel incorporation at 65°C (see Table 13.4 for results).

Table 13.4 Thermostable DNA polymerase activity in large-insert clones

Polymerase clone	Insert size (bp)	Incorporation counts	DNA polymerase polyprotein # amino acids	Accession number
Empty vector	0	578		–
PCR 14	4822	28,162	1608	KC440900
				KC440901
POL4B9	7232	36,194	1602	DNAP 85% ID to PCR14
POL3D1	2733	14,020	807	DNAP 99% ID to PCR14
POL3D11	5506	20,004	993	DNAP 99% ID to PCR14
POL4E10	5732	27,467	1206	DNAP 99% ID to PCR14
POL1E10	8488	46,404	1360	DNAP 99% ID to PCR14
POL3G9	3808	17,320	472	DNAP 99% ID to PCR14
POL2H9	6010	18,419	1465	DNAP 99% ID to PCR14

A full-length 1608 amino acid replicase clone for OS3173 ORF 36 (Schoenfeld et al. 2013) was used as a control (PCR 14 clone) for thermostable polymerase activity measurements (Tables 13.4, 28,162 counts). All seven viral metagenomic clones produced for this study yielded significantly more activity than background (578 counts) irrespective of the length of the encoded polyprotein polymerase (472–1602 amino acids). All seven clones were sequenced in their entirety to discovery genetic variants. Six of the seven variants are 99% identical to the original OS3173 polymerase gene (PCR 14 in Table 13.4, GenBank Accession Number KC440900) and OS3173 ORF 36 described here, whereas one clone (POL4B9, NCBI Accession Number KC440901) was significantly divergent at only 85% amino acid identity. Expression of the full-length PCR4 clone (1608 AA) and POL4B9 clone (1602 AA) is predicted to encode a protein of 189 kDa. Sodium dodecyl sulfate-polyacrylamide gel electrophoresis gels of *E. coli* expressing these clones gave rise to a similar size product (~55 kDa) (data not shown), suggesting that the genes are expressed as larger polyproteins and cleaved within the protein, with amino-terminal regions being unstable or insoluble in the expression host. Multiple attempts to express and purify the full-length polymerase genes found in clones PCR14 and POL4B9 were hampered by in vivo proteolysis that generates smaller protein products.

The utility of OS3173 ORF 36 variants encoding the polymerase domains only (585 amino acid carboxy terminus) was evaluated for efficacy in PCR (Fig. 13.5). Viral polymerase gene products derived from four OS3173 variants and the original "wild type" were chosen for comparative analysis of their ability

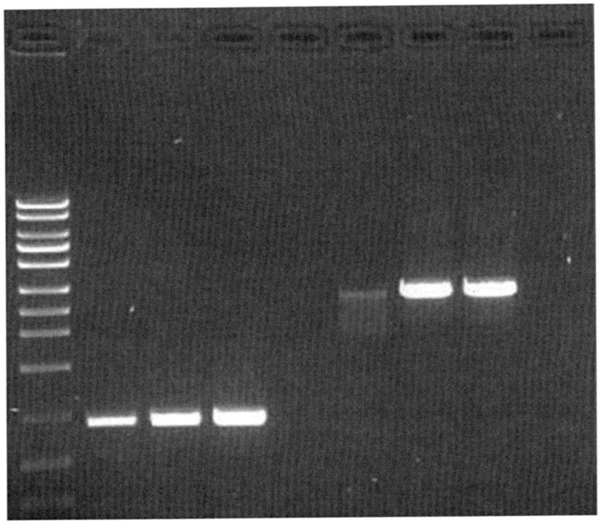

Fig. 13.5 PCR amplification of two genes using four variants of Octopus Spring OS3173 PolA recombinant enzymes. The ampicillin gene product (Amp) is 980 bp and the *E. coli* polymerase A gene product (PolA) is 2.8 kb. Lanes: (1) 3173 original wild-type viral polymerase (exonuclease proficient), (2) OmniAmp DNA polymerase (exonuclease proficient with N-terminal DNA-binding motif), (3) 3173A variant viral polymerase, and (4) 3173B variant viral polymerase

to amplify two marker sequences using standard PCR reaction conditions. The original 3173 wild-type *polA* gene products (Moser et al. 2012), variants 3173A and 3173B, and an engineered version containing an N-terminal DNA-binding domain called OmniAmp DNA polymerase (Chander et al. 2014) were cloned into the pETite T7 or pRham vectors (Lucigen Corp.) for expression and the enzymes purified using standard conditions.

Three of the OS3173 ORF 36 variants were able to amplify the 980 bp ampicillin gene, whereas 3173B failed (Fig. 13.5). The same three variants were also able to amplify the more challenging 2.8 kb genomic DNA. The amplification efficiency of OmniAmp DNAP (lane 2) was significantly improved compared to the original PolA isolate (lane 1) as expected. Surprisingly, the new variant 3173A was equally efficient as the engineered version. Additional research is required to understand the basis for this improved enzyme capability.

13.4 Conclusions

Advances in amplification methods, sequencing instruments, and molecular biology tools have allowed sequencing of the entire genome for Octopus Spring virus OS3173 and determination of its presumptive host, *Thermocrinis ruber*, without cultivating either the virus or its host. The polymerase encoded within this virus has several remarkable properties found in a single enzyme. Further exploration of some of the replication accessory proteins may allow for the development of new capabilities in DNA and RNA amplification.

Acknowledgments We thank the Gordon and Betty Moore Foundation for funding the sequence of the viral metagenome and Matt Henn at the Broad Institute for 454 sequencing.

References

Ahn DG, Kim SI, Rhee JK, Kim KP, Pan JG, JW O (2006) TTSV1, a new virus-like particle isolated from the hyperthermophilic crenarchaeote Thermoproteus tenax. Virology 351:280–290

Altschul SF, Madden TL, Schäffer AA, Zhang J, Zhang Z, Miller W, Lipman DJ (1997) Gapped BLAST and PSI-BLAST: a new generation of protein database search programs. Nucleic Acids Res 25:3389–3402

Anderson RE, Brazelton WJ, Baross JA (2011) Using CRISPRs as a metagenomic tool to identify microbial hosts of a diffuse flow hydrothermal vent viral assemblage. FEMS Microbiol Ecol 77(1):120–133

Arai H, Kanbe H, Ishii M, Igarashi Y (2010) Complete genome sequence of the thermophilic, obligately chemolithoautotrophic hydrogen-oxidizing bacterium Hydrogenobacter thermophilus TK-6. J Bacteriol 192:2651–2652

Ashelford KE, Day MJ, Fry JC (2003) Elevated abundance of bacteriophage infecting bacteria in soil. Appl Environ Microbiol 69:285–289

Bao K, Cohen SN (2004) Reverse transcriptase activity innate to DNA polymerase I and DNA topoisomerase I proteins of Streptomyces telomere complex. Proc Natl Acad Sci U S A 101(40):14361–14366

Barion S, Franchi M, Gallori E, Di Giulio M (2007) The first lines of divergence in the Bacteria domain were the hyperthermophilic organisms, the Thermotogales and the Aquificales, and not the mesophilic Planctomycetales. Biosystems 87:13

Barrangou R, Fremaux C, Deveau H, Richards M, Boyaval P, Moineau S, Romero DA, Horvath P (2007) CRISPR provides acquired resistance against viruses in prokaryotes. Science 315:1709–1712

Beloglazova N, Brown G, Zimmerman MD, Proudfoot M, Makarova KS, Kudritska M, Kochinyan S, Wang S, Chruszcz M, Minor W, Koonin EV, Edwards AM, Savchenko A, Yakunin AF (2008) A novel family of sequence-specific endoribonucleases associated with the clustered regularly interspaced short palindromic repeats. J Biol Chem 283(29):20361–20371

Bench SR, Hanson TE, Williamson KE, Ghosh D, Radosovich M, Wang K, Wommack KE (2007) Metagenomic characterization of Chesapeake Bay virioplankton. Appl Environ Microbiol 73:7629–7641

Bergh O, Borsheim KY, Bratbak G, Heldal M (1989) High abundance of viruses found in aquatic environments. Nature 340:467–468

Besemer J, Lomsadze A, Borodovsky M (2001) GeneMarkS: a self-training method for prediction of gene starts in microbial genomes. Implications for finding sequence motifs in regulatory regions. Nucleic Acids Res 29:2607–2618

Blank CE, Cady SL, Pace NR (2002) Microbial composition of near-boiling silica-depositing thermal springs throughout Yellowstone National Park. Appl Environ Microbiol 68(10):5123–5135

Blöndal T, Thorbjarnardóttir SH, Kieleczawa J, Hjörleifsdóttir S, Kristjánsson JK, Einarsson JM, Eggertsson G (2001) Cloning, sequence analysis and functional characterization of DNA polymerase I from the thermophilic eubacterium Rhodothermus marinus. Biotechnol Appl Biochem 34(1):37–45

Bolotin A, Quinquis B, Sorokin A, Ehrlich SD (2005) Clustered regularly interspaced short palindrome repeats (CRISPRs) have spacers of extrachromosomal origin. Microbiology 151:2551–2561

Bowers RM, Clum A, Tice H, Lim J, Singh K, Ciobanu D, Ngan CY, Cheng JF, Tringe SG, Woyke T (2015) Impact of library preparation protocols and template quantity on the metagenomic reconstruction of a mock microbial community. BMC Genomics 16:856

Breitbart M, Salamon P, Andresen B, Mahaffy JM, Segall AM, Mead D, Azam F, Rohwer F (2002) Genomic analysis of uncultured marine viral communities. Proc Natl Acad Sci U S A 99:14250–14255

Breitbart MI, Hewson B, Felts JM, Mahaffy J, Salamon NP, Rohwer F (2003) Metagenomic analyses of an uncultured viral community from human feces. J Bacteriol 85:6220–6223

Breitbart MB, Felts S, Kelley JM, Mahaffy J, Salamon NP, Rohwer F (2004) Diversity and population structure of a near-shore marine sediment viral community. Proc Biol Sci 271:565–574

Chander Y, Koelbl J, Puckett J, Moser MJ, Klingele AJ, Liles MR, Carrias A, Mead DA, Schoenfeld TW (2014) A novel thermostable polymerase for RNA and DNA loop-mediated isothermal amplification (LAMP). Front Microbiol 5:395

Chang JR, Choi JJ, Kim HK, Kwon ST (2001) Purification and properties of Aquifex aeolicus DNA polymerase expressed in Escherichia coli. FEMS Microbiol Lett 201(1):73–77

Chibani-Chennoufi S, Bruttin A, Dillman ML, Brussow H (2004) Phage-host interaction: an ecological perspective. J Bacteriol 186:3677–3686

Clingenpeel S, Kan J, Macur RE, Woyke T, Lovalvo D, Varley J, Inskeep WP, Nealson K, McDermott TR (2013) Yellowstone Lake Nanoarchaeota. Front Microbiol 4:274

Clokie MR, Millard AD, Letarov AV, Heaphy S (2011) Phages in nature. Bacteriophage 1(1):31–45

Coenye T, Vandamme P (2004) A genomic perspective on the relationship between the Aquificales and the epsilon-Proteobacteria. Syst Appl Microbiol 27:313

Dean F, Hosono S, Fang L, Wu X, Faruqi AF, Bray-Ward P, Sun Z, Zong Q, Du Y, Du J, Driscoll M, Song W, Kingsmore SF, Egholm M, Lasken RS (2002) Comprehensive human genome amplification using multiple displacement amplification. Proc Natl Acad Sci U S A 99(8):5261–5266

Eickbush TH, Jamburuthugoda VK (2008) The diversity of retrotransposons and the properties of their reverse transcriptases. Virus Res 134(1–2):221–234

Eisen J, Huntemann M, Han J, Chen A, Kyrpides N, Mavromatis K, Markowitz V, Palaniappan K, Ivanova N, Schaumberg A, Pati A, Liolios K, Nordberg HP, Cantor MN, Hua SX, Woyke T (2015) Thermocrinis ruber strain DSM 23557, complete genome. http://www.ncbi.nlm.nih.gov/nuccore/NZ_CP007028.1

Eme L, Reigstad LJ, Spang A, Lanzén A, Weinmaier T, Rattei T, Schleper C, Brochier-Armanet C (2013) Metagenomics of Kamchatkan hot spring filaments reveal two new major (hyper) thermophilic lineages related to Thaumarchaeota. Res Microbiol 164:425–438

Ferrera I, Longhorn S, Banta AB, Liu Y, Preston D, Reysenbach AL (2007) Diversity of 16S rRNA gene, ITS region and aclB gene of the Aquificales. Extremophiles 11:57–64

Fuhrman JA (1999) Marine viruses and their biogeochemical and ecological effects. Nature 399:541–548

Gardner AF, Prangishvili D, Jack WE (2011) Characterization of *Sulfolobus islandicus* rod-shaped virus 2 gp19, a single-strand specific endonuclease. Extremophiles 15:619–624

Garrett RA, Prangishvili D, Shah SA, Reuter M, Stetter KO, Peng X (2010) Metagenomic analyses of novel viruses and plasmids from a cultured environmental sample of hyperthermophilic neutrophils. Environ Microbiol 12:2918–2930

Geslin C, Le Romancer M, Erauso G, Gaillard M, Perrot G, Prieur D (2003a) PAV1, the first virus-like particle isolated from a hyperthermophilic euryarchaeote, "Pyrococcus abyssi". J Bacteriol 185:3888–3894

Geslin C, Le Romancer M, Gaillard M, Erauso G, Prieur D (2003b) Observation of virus-like particles in high temperature enrichment cultures from deep-sea hydrothermal vents. Res Microbiol 154:303–307

Godiska R, Patterson M, Schoenfeld T, Mead D (2005) Beyond pUC: vectors for cloning unstable DNA. In: Kieleczawa J (ed) DNA sequencing: optimizing the process and analysis. Jones and Bartlett, Sudbury, MA, pp 55–75

Godiska R, Mead DA, Dhodda V, Hochstein R, Karsi A, Usdin K, Entezam A, Ravin N (2010) Linear plasmid for cloning large or repetitive sequences in *E. coli*. Nucleic Acids Res 38:e88

Grissa I, Vergnaud G, Pourcel C (2007) The CRISPRdb database and tools to display CRISPRs and to generate dictionaries of spacers and repeats. BMC Bioinformatics 8:172

Guo Y, Kragelund BB, White MF, Peng X (2015) Functional characterization of a conserved archaeal viral operon revealing single-stranded DNA binding, annealing and nuclease activities. J Mol Biol 427:2179–2191

Haft DH, Selengut J, Mongodin EF, Nelson KE (2005) A guild of 45 CRISPR-associated (Cas) protein families and multiple CRISPR/Cas subtypes exist in prokaryotic genomes. PLoS Comput Biol 1:e60

Häring M, Peng X, Brügger K, Rachel R, Stetter KO, Garrett RA, Prangishvili D (2004) Morphology and genome organization of the virus PSV of the hyperthermophilic archaeal genera Pyrobaculum and Thermoproteus: a novel virus family, the Globuloviridae. Virology 323:233–242
Hedlund BP, Dodsworth JA, Murugapiran SK, Rinke C, Woyke T (2014) Impact of single-cell genomics and metagenomics on the emerging view of extremophile "microbial dark matter". Extremophiles 18:865–875
Hedlund B, Huntemann M, Han J, Chen A, Kyrpides N, Mavromatis K, Markowitz V, Palaniappan K, Ivanova N, Schaumberg A, Pati A, Liolios K, Nordberg HP, Cantor MN, Hua SX, Woyke T (2015a) Thermocrinis sp. GBS K217DRAFT_scaffold00001.1_C, whole genome shotgun sequence. http://www.ncbi.nlm.nih.gov/nuccore/NZ_JNIE01000002.1
Hedlund BP, Murugapiran SK, Alba TW, Levy A, Dodsworth JA, Goertz GB, Ivanova N, Woyke T (2015b) Uncultivated thermophiles: current status and spotlight on 'Aigarchaeota'. Curr Opin Microbiol 25:136–145
Heidelberg JF, Nelson WC, Schoenfeld T, Bhaya D (2009) Germ warfare in a microbial mat community: CRISPRs provide insights into the co-evolution of host and viral genomes. PLoS One 4(1):e4169
Horvath P, Barrangou R (2010) CRISPR/Cas, the immune system of bacteria and archaea. Science 327(5962):167–170
Huber R, Eder W, Heldwein S, Wanner G, Huber H, Rachel R, Stetter KO (1998) Thermocrinis ruber gen. nov., sp. nov., A pink-filament-forming hyperthermophilic bacterium isolated from Yellowstone National Park. Appl Environ Microbiol 64(10):3576–3583
Inskeep WP, Rusch DB, Jay ZJ, Herrgard MJ, Kozubal MA, Richardson TH, Macur RE, Hamamura N, Rd J, Fouke BW, Reysenbach A-L, Roberto F, Young M, Schwartz A, Boyd ES, Badger JH, Mathur EJ, Ortmann AC, Bateson M, Geesey G, Frazier M (2010) Metagenomes from high-temperature chemotrophic systems reveal geochemical controls on microbial community structure and function. PLoS One 5:e9773
Inskeep WP, Jay ZJ, Herrgard MJ, Kozubal MA, Rusch DB, Tringe SG, Macur RE, Rd J, Boyd ES, Spear JR, Roberto FF (2013) Phylogenetic and functional analysis of metagenome sequence from high-temperature archaeal habitats demonstrate linkages between metabolic potential and geochemistry. Front Microbiol 4:95
Jahn U, Gallenberger M, Paper W, Junglas B, Eisenreich W, Stetter KO, Rachel R, Huber H (2008) *Nanoarchaeum equitans* and *Ignicoccus hospitalis*: new insights into a unique, intimate association of two archaea. J Bacteriol 190:1743–1750
Jansen R, Embden JD, Gaastra W, Schouls LM (2002) Identification of genes that are associated with DNA repeats in prokaryotes. Mol Microbiol 43:1565–1575
Kala S, Cumby N, Sadowski PD, Hyder BZ, Kanelis V, Davidson AR, Maxwell KL (2014) HNH proteins are a widespread component of phage DNA packaging machines. Proc Natl Acad Sci U S A 111(16):6022–6027
Keller J, Leulliot N, Cambillau C, Campanacci V, Porciero S, Prangishvili D, Forterre P, Cortez D, Quevillon-Cheruel S, van Tilbeurgh H (2007) Crystal structure of AFV3-109, a highly conserved protein from crenarchaeal viruses. Virol J 4:12
Képès F, Jester BC, Lepage T, Rafiei N, Rosu B, Junier I (2012) The layout of a bacterial genome. FEBS Lett 586:2043–2048
Kim KH, Bae JW (2011) Amplification methods bias metagenomic libraries of uncultured single-stranded and double-stranded DNA viruses. Appl Environ Microbiol 77:7663–7668
Kozubal MA, Romine M, Rd J, Jay ZJ, Tringe SG, Rusch DB, Beam JP, McCue LA, Inskeep WP (2013) Geoarchaeota: a new candidate phylum in the Archaea from high-temperature acidic iron mats in Yellowstone National Park. ISME J 7:622–634
Kunin V, Sorek R, Hugenholtz P (2007) Evolutionary conservation of sequence and secondary structures in CRISPR repeats. Genome Biol 8:R61
Larson ET, Eilers BJ, Reiter D, Ortmann AC, Young MJ, Lawrence CM (2007) A new DNA binding protein highly conserved in diverse crenarchaeal viruses. Virology 363:387–396
Lemak S, Nocek B, Beloglazova N, Skarina T, Flick R, Brown G, Joachimiak A, Savchenko A, Yakunin AF (2014) The CRISPR-associated Cas4 protein Pcal_0546 from Pyrobaculum

calidifontis contains a [2Fe-2S] cluster: crystal structure and nuclease activity. Nucleic Acids Res 42(17):11144–11155
Lindell D, Sullivan MB, Johnson ZI, Tolonen AC, Rohwer F, Chisholm SW (2004) Transfer of photosynthesis genes to and from *Prochlorococcus* viruses. Proc Natl Acad Sci U S A 101:11013–11018
Lindner SE, Llinás M, Keck JL, Kappe SH (2011) The primase domain of PfPrex is a proteolytically matured, essential enzyme of the apicoplast. Mol Biochem Parasitol 180(2):69–75
Liu M, Gingery M, Doulatov SR et al (2004) Genomic and genetic analysis of Bordetella bacteriophages encoding reverse transcriptase-mediated tropism-switching cassettes. J Bacteriol 186(5):1503–1517
Makarova KS, Grishin NV, Shabalina SA, Wolf YI, Koonin EV (2006) A putative RNA-interference-based immune system in prokaryotes: computational analysis of the predicted enzymatic machinery, functional analogies with eukaryotic RNAi, and hypothetical mechanisms of action. Biol Direct 1:7
Markowitz VM, Chen IM, Chu K, Szeto E, Palaniappan K, Grechkin Y et al (2012) IMG/M: the integrated metagenome data management and comparative analysis system. Nucleic Acids Res. 40. D123–D129. doi:10.1093/nar/gkr975
Marchler-Bauer A, Lu S, Anderson JB, Chitsaz F et al (2011) CDD: a conserved domain database for the functional annotation of proteins. Nucleic Acids Res 39(Database issue):D225–D229
McCleskey RB, Ball JW, Nordstrom DK, Holloway JM, Taylor HE (2004) Water-chemistry data for selected hot springs, geysers, and streams in Yellowstone National Park, Wyoming, 2001–2002. U.S. Geological Survey Open-File Report 2004-1316, U.S. Geological Survey, Boulder, CO
McFarland KN, Liu J, Landrian I, Godiska R, Shanker S, Yu F, Farmerie WG, Ashizawa T (2015) SMRT sequencing of long tandem nucleotide repeats in SCA10 reveals unique insight of repeat expansion structure. PLoS One 10(8):e0135906
Medhekar B, Miller JF (2007) Diversity-generating retroelements. Curr Opin Microbiol 10(4):388–395
Menzel P, Gudbergsdóttir SR, Rike AG, Lin L, Zhang Q, Contursi P, Moracci M, Kristjansson JK, Bolduc B, Gavrilov S, Ravin N, Mardanov A, Bonch-Osmolovskaya E, Young M, Krogh A, Peng X (2015) Comparative metagenomics of eight geographically remote terrestrial hot springs. Microb Ecol 70:411–424
Moser MJ, DiFrancesco RA, Gowda K, Klingele AJ, Sugar DR, Stocki S, Mead DA, Schoenfeld TW (2012) Thermostable DNA polymerase from a viral metagenome is a potent RT-PCR enzyme. PLoS One 7(6):e38371. doi:10.1371/journal.pone.0038371
Munson-McGee JH, Field EK, Bateson M, Rooney C, Stepanauskas R, Young MJ (2015) Nanoarchaeota, their Sulfolobales host, and Nanoarchaeota virus distribution across Yellowstone National Park Hot Springs. Appl Environ Microbiol 81:7860–7868
Paul JH (1999) Microbial gene transfer: an ecological perspective. J Mol Microbiol Biotechnol 1:45–50
Paul JH, Williamson SJ, Long A, John D, Segall A, Rohwer F (2005) Complete genome sequence of phiHSIC, a pseudotemperate marine phage of *Listonella pelagia*. Appl Environ Microbiol 71:3311–3320
Podar M, Makarova KS, Graham DE, Wolf YI, Koonin EV, Reysenbach AL (2013) Insights into archaeal evolution and symbiosis from the genomes of a nanoarchaeon and its inferred crenarchaeal host from Obsidian Pool, Yellowstone National Park. Biol Direct 8:9
Prangishvili D, Garrett RA (2004) Exceptionally diverse morphotypes and genomes of crenarchaeal hyperthermophilic viruses. Biochem Soc Trans 32:204–208
Probst AJ, Weinmaier T, DeSantis TZ, Santo Domingo JW, Ashbolt N (2015) New perspectives on microbial community distortion after whole-genome amplification. PLoS One 10(5):e0124158
Rachel R, Bettstetter M, Hedlund BP, Häring M, Kessler A, Stetter KO, Prangishvili D (2002) Remarkable morphological diversity of viruses and virus-like particles in hot terrestrial environments. Arch Virol 147(12):2419–2429
Reddy BL, Saier MH Jr (2013) Topological and phylogenetic analyses of bacterial holin families and superfamilies. Biochim Biophys Acta 1828(11):2654–2671

Reysenbach AL, Wickham GS, Pace NR (1994) Phylogenetic analysis of the hyperthermophilic pink filament community in Octopus Spring, Yellowstone National Park. Appl Environ Microbiol 60:2113–2119

Reysenbach AL, Banta A, Civello S, Daly J, Mitchell K, Lalonde S et al (2005) The aquificales of Yellowstone National Park. In: Inskeep WP, McDermott TR (eds) Geothermal biology and geochemistry in Yellowstone National Park. Montana State University/Thermal Biology Institute, Bozeman, pp 129–142

Reysenbach AL, Hamamura N, Podar M, Griffiths E, Ferreira S, Hochstein R, Heidelberg J, Johnson J, Mead D, Pohorille A, Sarmiento M, Schweighofer K, Seshadri R, Voytek MA (2009) Complete and draft genome sequences of six members of the Aquificales. J Bacteriol 191:1992–1993

Reysenbach AL, Huntemann M, Han J, Chen A, Kyrpides N, Mavromatis K, Markowitz V, Palaniappan K, Ivanova N, Schaumberg A, Pati A, Liolios K, Nordberg HP, Cantor MN, Hua SX, Woyke T (2015) Persephonella sp. KM09-Lau-8 strain KM09_Lau8, whole genome shotgun sequencing project. http://www.ncbi.nlm.nih.gov/nuccore/657727105

Rice G, Stedman K, Snyder J, Wiedenheft B, Willits D, Brumfield S, McDermott T, Young MJ (2001) Viruses from extreme thermal environments. Proc Natl Acad Sci U S A 98:13341–13345

Romano C, D'Imperio S, Woyke T, Mavromatis K, Lasken R, Shock EL, McDermott TR (2013) Comparative genomic analysis of phylogenetically closely related Hydrogenobaculum sp. isolates from Yellowstone National Park. Appl Environ Microbiol 79:2932–2943

Rosario K, Breitbart M (2011) Exploring the viral world through metagenomics. Curr Opin Virol 1:289–297

Schoenfeld T, Patterson M, Richardson P, Wommac E, Young M, Mead DA (2008) Assembly of viral metagenomes from Yellowstone hot Springs. Appl Environ Microbiol 74:4164–4174

Schoenfeld T, Liles M, Wommack EK, Polson SW, Godiska R, Mead D (2010) Functional viral metagenomics and the next generation of molecular tools. Trends Microbiol 18:20–29

Schoenfeld TW, Murugapiran SK, Dodsworth JA, Floyd S, Lodes M, Mead DA, Hedlund BP (2013) Lateral gene transfer of family A DNA polymerases between thermophilic viruses, aquificae, and apicomplexa. Mol Biol Evol 30(7):1653–1664

Seguritan V, Feng I, Rohwer F, Swift M, Segall AM (2003) Genome sequences of two closely related *Vibrio parahaemolyticus* phages, VP16T and VP16C. J Bacteriol 185:6434–6447

Seow F, Sato S, Janssen CS, Riehle MO, Mukhopadhyay A, Phillips RS, Wilson RJ, Barrett MP (2005) The plastidic DNA replication enzyme complex of Plasmodium falciparum. Mol Biochem Parasitol 141(2):145–153

Smits SL, Bodewes R, Ruiz-Gonzalez A, Baumgärtner W, Koopmans MP, Osterhaus ADME, Schürch AC (2014) Assembly of viral genomes from metagenomes. Front Microbiol 5:714

Smits SL, Bodewes R, Ruiz-González R, Baumgärtner W, Koopmans MP, Osterhaus AD, Schurch AC (2015) Recovering full-length viral genomes from metagenomes. Front Microbiol 6:1069. doi:10.3389/fmicb.2015.01069

Snyder JC, Spuhler J, Wiedenheft B, Roberto FF, Douglas T, Young MJ (2004) Effects of culturing on the population structure of a hyperthermophilic virus. Microb Ecol 48:561–566

Snyder JC, Bateson MM, Lavin M, Young MJ (2010) Use of cellular CRISPR (clusters of regularly interspaced short palindromic repeats) spacer-based microarrays for detection of viruses in environmental samples. Appl Environ Microbiol 76(21):7251–7258

Snyder JC, Bolduc B, Young MJ (2015) 40 Years of archaeal virology: expanding viral diversity. Virology 479-480:369–378. doi:10.1016/j.virol.2015.03.031

Spear JR, Walker JJ, McCollom TM, Pace NR (2005) Hydrogen and bioenergetics in the Yellowstone geothermal ecosystem. Proc Natl Acad Sci U S A 102(7):2555–2560

Sullivan MB, Coleman M, Weigele P, Rohwer F, Chisholm SW (2005) Three *Prochlorococcus* cyanophage genomes: signature features and ecological interpretations. PLoS Biol 3:1–17

Suttle CA (2007) Marine viruses—major players in the global ecosystem. Nat Rev Microbiol 5:801–812

Takacs-Vesbach C, Inskeep WP, Jay ZJ, Herrgard MJ, Rusch DB, Tringe SG, Kozubal MA, Hamamura N, Macur RE, Fouke BW, Reysenbach AL, McDermott TR, Jennings RD, Hengartner NW, Xie G (2013) Metagenome sequence analysis of filamentous microbial

communities obtained from geochemically distinct geothermal channels reveals specialization of three aquificales lineages. Front Microbiol 4:84
Walker JE, Saraste M, Runswick MJ, Gay NJ (1982) Distantly related sequences in the alpha- and beta-subunits of ATP synthase, myosin, kinases and other ATP-requiring enzymes and a common nucleotide binding fold. EMBO J 1(8):945–951
Wang C, Villion M, Semper C, Coros C, Moineau S, Zimmerly S (2011) A reverse transcriptase-related protein mediates phage resistance and polymerizes untemplated DNA in vitro. Nucleic Acids Res 39(17):7620–7629
Weigel C, Seitz H (2006) Bacteriophage replication modules. FEMS Microbiol Rev 30:321–381
Young R (2014) Phage lysis: three steps, three choices, one outcome. J Microbiol 52(3):243–258
Yu MX, Slater MR, Ackermann HW (2006) Isolation and characterization of *Thermus* bacteriophages. Arch Virol 151:663–679
Zhang J, Kasciukovic T, White MF (2012) The crispr associated protein Cas4 is a 5′ to 3′ DNA exonuclease with an iron-sulfur cluster. PLoS One 7(10):e47232
Zhou Y, Liang Y, Lynch KH, Dennis JJ, Wishart DS (2011) PHAST: a fast phage search tool. Nucleic Acids Res 39(Web Server issue):W347–W352. PMCID: PMC3125810

Chapter 14
Functional Metagenomics and Antimicrobial Resistance

Fiona Walsh

Abstract The currently available arsenal of antibiotics is diminishing. This has indirectly led to the expansion of research in antibiotic resistance into identifying the reservoirs of antibiotic resistance outside the pathogen and understanding how antibiotic resistance has emerged from their original sources to pathogenic bacteria. Functional metagenomics has enabled researchers to investigate the unculturable bacteria and nonclinical microbiomes to enhance these new areas of research. This chapter will outline how functional metagenomics has enabled the discovery of new antibiotic resistance mechanisms and exploration of the total microbiome to understand the source of antibiotic resistance. The functions and origins of many genes and DNA fragments, which were discovered, remain a mystery. This chapter will also identify the next frontiers that this discovery science needs to explore.

What has functional metagenomics ever done for antimicrobial resistance research?

14.1 The Pre-functional Metagenomics Era and Antibiotic Resistance

Since the discovery of antibiotics, we have been aware of the presence of antibiotic resistance. However, it is only in relatively recent years that the medical community has faced a realistic end to the efficacy of our currently available arsenal of antibiotics. For decades antimicrobial resistance has been the realm of medical microbiologists, a few veterinary or agriculture microbiologists and even fewer microbiologists interested in ecology and environmental microbiology. The perspective that we viewed antibiotics prior to the 2000s was almost exclusively in terms of antibiotics as anti-infective agents used in human and animal medicine. This led to the idea that

F. Walsh
Department of Biology, Maynooth University, Maynooth, Co. Kildare, Ireland
e-mail: Fiona.walsh@mu.ie

T.C. Charles et al. (eds.), *Functional Metagenomics: Tools and Applications*,
DOI 10.1007/978-3-319-61510-3_14

we only need to investigate antibiotic resistance in pathogenic bacteria to identify the resistance mechanisms present and identify how they transfer within a species or between pathogenic species.

The questions relating to the origins of antibiotic resistance or the emergence of new antibiotic resistance genes from a source not exposed to antibiotics were rarely considered. There was no need to look for new antibiotic resistance genes or understand where antibiotic resistance originated as we had sufficient antibiotics. The driving need was to understand how we minimize the spread of the resistance genes that are currently in pathogens. There were also other factors that limited the investigation of antibiotic resistance outside the pathogen, such as the limitation of laboratory techniques, which were widely used to identify the resistant bacteria. These techniques were designed and created in order to selectively isolate pathogenic bacteria and genes known to cause resistance in pathogens. Therefore, it was not possible to use them to culture all known bacteria from, e.g. the gut microbiome or a soil sample.

With the advent of molecular tools, we could use PCR and DNA sequencing to identify resistance genes present in a total DNA sample. However, as our view was still very focused on the medical aspects of resistance, we investigated the total DNA for known resistance genes of clinical importance. We had no other choice in terms of techniques, as in order to amplify a gene by PCR, we needed to know the sequence. So to date we can see that we could only detect a resistance mechanism once it was identified in a clinical isolate, which was identified as resistant using culture techniques such as agar dilution minimum inhibitory concentration (MIC) or disk diffusion techniques. Once the novel resistance gene was identified, we could then investigate other environments using PCR and sequencing. This resulted in our current dilemma: we could only identify the problem of a novel resistance mechanism after it has reached the pathogen. As we know, prevention is better than cure, so it would have been much better for the continued use of antibiotics if we could have identified where antibiotic resistance genes come from and how they move from that source into a pathogen, before it reached the pathogenic bacteria. However, we did not have the tools to investigate all bacteria in all locations or potential non-pathogen-associated sources for as-yet-unknown antibiotic resistance mechanisms. Thus, we could not source antibiotic resistance genes prior to their arrival in the pathogenic bacteria that we wished to inhibit or kill with our ever-decreasing armoury of antibiotics.

14.2 Using Functional Metagenomics to Understand Antibiotic Resistance

The advent of functional metagenomics enabled scientists to expand their areas of investigation and to dream of possibly tackling some fundamental questions concerning antibiotic resistance such as where do reservoirs of antibiotic resistance exist outside the pathogen, or to what extent are the source environments of antibiotics also the source environments of antibiotic resistance genes, or how have the anthropogenic activities of farming practices, antibiotic use or pollution changed the environments of soil and water to enable them to become reservoirs of antibiotic resistance? These are some of the questions that I will discuss during this chapter.

14.3 Soil: Where It All Began

Functional gene analysis from uncultured bacteria first evolved by cloning small fragments into host bacteria (Stein et al. 1996; Healy et al. 1995). This provided an excellent method to identify specific genes or small fragments required to confer a specific trait or function. However, many bacterial functions, including the production of antibiotics, their corresponding resistance genes and the regulatory sequences, are carried in large operons. Such large DNA fragments could not be transferred by the earlier cloning methods. The first large-fragment functional metagenome was cloned from soil DNA (Rondon et al. 2000) and identified a set of genes encoding an antibacterial agent from uncultured bacteria. The aim of this study was to develop methods to fully explore the as-yet-unidentified bacteria from soil. This study provided the method required to analyse bacterial traits and products without the need to culture the bacteria harbouring these genes. This was the birth of metagenomic libraries created by cloning large fragments of total DNA extracted from soil into the bacterial artificial chromosome (BAC) and using *Escherichia coli* to express these genes and produce the phenotypes associated with them. Sequence analysis identified that the pieces of DNA being expressed using this technique were novel. This new tool was a great leap towards the discovery of truly novel sets of genes and their functions.

Functional metagenomics has been used to detect novel antibiotic resistance genes and known antibiotic resistance genes in novel bacterial hosts predominantly from soil microbiomes but also water, animal, insect and human microbiomes. The pioneers of using functional metagenomics to study the antibiotic resistome of total sample DNA were professors and researchers at the University of Wisconsin-Madison and Cornell University (Handelsman et al. 1998). Their first study investigated soil for the presence of β-lactamases using nitrocefin as an overlay on the clones (Rondon et al. 2000). However, no clones were identified from the selection investigated. In a follow-up study (Riesenfeld et al. 2004) of the same libraries and additional libraries from the same sampling site, the researchers identified nine clones expressing resistance to aminoglycosides, six of which encoded 6′-*N*-acetyltransferases. The predicted amino acid sequences differed considerably from previously identified 6′-*N*-acetyltransferases. One tetracycline-resistant clone was identified, but no nalidixic acid-resistant clones were identified.

Additional investigations of soil continued from these initial experiments. A remote Alaskan soil was investigated for the presence of β-lactam resistance genes using functional metagenomics (Allen et al. 2008). The sampling site was as isolated as possible as it was an island in a fast-flowing river in central Alaska. A total of 714,000 clones were identified, of these 14 contained metagenomic DNA that conferred resistance to β-lactams. Transposon mutagenesis was used to identify the gene associated with the antibiotic resistance phenotype; this technique was frequently used following this study to identify the specific gene responsible for a resistance phenotype. The genes from 13 clones were identified as novel β-lactamases (classes A, B and C). The single representative of the class D β-lactamases detected was linked to a class C β-lactamase as a single open reading frame. This was the first description of a bifunctional β-lactamase. Evolutionary analysis of the β-lactamases identified that the β-lactamases diverge deeply from previously described β-lactamases, thus suggesting that these β-lactamases are more closely related to ancestral β-lactamases than clinical β-lactamases. A later study of the same functional metagenomic library generated from the Alaskan soil identified a novel resistance gene to florfenicol and chloramphenicol (Lang et al. 2010). The new resistance protein (PexA) was aligned with known phenicol exporter proteins of the major facilitator superfamily (MFS) using the ClustalW method. The closest sequence similarity was to drug-resistant transporters from *Wolbachia* spp., but the highest level of gene similarity was only 33%. Further novel resistance enzymes and bifunctional proteins were discovered by this group in a functional metagenomic library from apple orchard soil (Donato et al. 2010). The bifunctional enzyme, which conferred resistance to β-lactams, was a fusion of a gene with sequence similarity (35% amino acid identity) to an extracytoplasmic function (ECF)-type sigma factor from *Rhodoferax ferrireducens* T118 and sequence similarity (49% identity) to class A β-lactamase isolated from the previous study of Alaskan soil. The other bifunctional enzyme conferred resistance to kanamycin and was composed of an aminoglycoside acetyltransferases (*aac6'*) fused to an acetyltransferase. Each of the two genes was separately cloned. Only the aminoglycoside acetyltransferases section of this open reading frame conferred resistance to kanamycin.

The same scientists returned to the functional metagenome of the Alaskan soil to understand one specific clone, which contained a two-component response regulator gene that conferred resistance to the β-lactam carbenicillin in *E. coli* (βLR16) (Allen et al. 2015). Their experimental observations were consistent with their gene expression analysis, which showed negligible changes in expression to the *acrA* and *acrB* genes in the presence of the βLR16 response regulator gene yet increases in expression of both the *acrD* gene and the *mdtA* and *mdtB* genes. The βLR16 response regulator gene, therefore, was responsible for both repressing porin expression and activating multidrug efflux pump expression. The introduction of a foreign response regulator gene into *E. coli* directly impacted gene expression, leading to an antibiotic resistance phenotype, and showed that antibiotic resistance can be achieved by the modulation of gene regulation by heterologous DNA.

The initial studies in the USA led to further studies around the world, including the investigation of Spanish soils for the presence of novel antibiotic resistance genes. DNA fragments (9–10 kb fragments) extracted from three different Spanish soils were inserted into vectors, expressed in *E. coli* and screened for resistance to ampicillin, gentamicin, chloramphenicol, erythromycin, streptomycin and trimethoprim (Torres-Cortés et al. 2011). Eleven novel resistance genes were identified from three soil samples, three conferring resistance to ampicillin, two to gentamicin, two to chloramphenicol and four to trimethoprim. To date, this is the only European soil (Spain) to be used to identify novel antibiotic resistance mechanisms using functional metagenomics. Phylogenetic analysis identified that two of the β-lactamases were closely related to chromosomal β-lactamases of *Phenylobacterium*, *Rhizobium* and a third to a member of the order *Solibacter*. The two open reading frames conferring resistance to chloramphenicol were similar to each other and to other drug transporters present in proteobacteria. The sequence of one of the trimethoprim resistance genes was not similar to any known dihydrofolate reductase genes sequenced to date.

A further study of agricultural soil from China used three different agricultural amended soils to generate metagenomic libraries of short insert lengths (1–3 kb) (Su et al. 2014). As this study used pUC19, they could not screen for β-lactamases but could investigate cephalosporin resistance. This study identified a large number of novel resistance genes, with low sequence similarity to known genes of any function. These included putative efflux pump genes and a wide variety of enzymes. A total of 45 unique clones with an antibiotic resistance phenotype to tetracyclines, aminoglycosides, chloramphenicol and rifampin were detected. Most (67%) had <60% sequence identity to known amino acids. Antibiotic resistance genes conferring resistance to cefotaxime, ciprofloxacin, trimethoprim, nitrofurantoin and erythromycin were not detected. This however does not suggest that such genes are not present in these soils but rather that there are limitations to using small fragment inserts or *E. coli* as the sole host and that the libraries represent only a small fraction of the total microbial content of soil. This study, however, has identified a large reservoir of functionally active resistance genes with low sequence similarity to any known genes in GenBank, therefore identifying the untapped reservoir of potential resistance genes capable of expression in *E. coli* present in soil. When such a large reservoir exists, this then leads to the question, why are there relatively few resistance genes in pathogenic bacteria?

An urban soil functional metagenomic library (McGarvey et al. 2012) was screened for resistance to trimethoprim, chloramphenicol, kanamycin, gentamicin, rifampin or tetracycline. The average insert size was 2 kb. Fragments containing 39 unique resistance genes were identified. The majority ($n = 20$) of which conferred resistance to trimethoprim and were almost all dihydrofolate reductase enzymes of varying sequence similarity to known dihydrofolate reductase enzymes. Chloramphenicol resistance was mediated by a putative major facilitator family (MFS) transporter and two novel genes. Aminoglycoside acetyltransferases, with sequence similarity to known acetyltransferases between 38 and 71%, conferred resistance to kanamycin and gentamicin. Rifampin resistance was most frequently mediated by rifampin ADP-ribosyltransferase with sequence identity of between

69 and 77% to *Oscillatoria* sp. PCC 6506. Tetracycline resistance was also mediated by MFS transporters and an ABC transporter.

The addition of animal manure to soils has been shown to increase the diversity and abundance of antibiotic resistance genes using PCR and qPCR (Heuer and Smalla 2007; Binh et al. 2008). The hypothesis of Udikovic-Kolic et al. was that unique β-lactamases present in soil after manure application could be traced back to the manure. Functional metagenomics was used to identify the potential origins of β-lactamases identified in bacteria cultured from soil after manure or inorganic fertilization and the cow manure (Udikovic-Kolic et al. 2014). They constructed five metagenomic fosmid libraries: four libraries from cultured β-lactam-resistant bacteria isolated from soil after manure or inorganic fertilizer treatment and one library from the manure that had been used for fertilization. From the libraries, they identified seven unique β-lactamases, two from the manure library and five from the cultured β-lactam-resistant bacterial community. The β-lactamases were from Ambler classes A, B and C.

The data generated from soil functional metagenomics studies are limited by geographical area, antibiotic selection and the number of studies performed. The majority of studies have investigated resistance to the β-lactams and have therefore identified novel β-lactamases. However, the carbapenems, imipenem or meropenem, were not included in these β-lactamases to date. Aminoglycoside, chloramphenicol, tetracycline, trimethoprim and rifampin resistance have also been investigated and novel resistance genes identified. Two studies investigated soil as a potential reservoir of quinolone resistance but did not generate resistant clones, even though the *qnr* genes are thought to have originated from environmental sources. Transferable colistin resistance (*mcr-1*) was recently identified in commensal *E. coli* in pigs, but no functional metagenomics study to date has investigated the environment or animals for colistin resistance (Liu et al. 2015).

Although the number of studies using functional metagenomics to identify resistance genes in soil is limited, this section comprises the greatest amount of knowledge. In comparison there have been few studies on water as a reservoir of novel resistance genes or resistance genes in novel bacteria using functional metagenomics. One river sediment study has been performed upstream and downstream of an urban wastewater treatment plant (Amos et al. 2014). Ten clones were investigated from each of the upstream and downstream sediment samples. Aminoglycoside resistance was conferred by acetyltransferases, phosphotransferases and novel mechanisms. Ampicillin resistance was conferred by the β-lactamase bla_{TEM}. This study resulted in the identification of five clones with elevated ciprofloxacin MIC (1 mg/L) in comparison to the strain without a fosmid (0.25 mg/L). The reported mechanism for ciprofloxacin resistance was the presence of genes with 74% identity to *recA* (*Geobacter lovleyi*) and 33% identity to *recX* (*Listeria seeligeri*). These proteins could possibly be involved in repair of the damage inflicted by ciprofloxacin on the DNA gyrase (Cardenas et al. 2012).

14.4 Water: A Potential Antibiotic Resistance Reservoir

Only one study has utilized functional metagenomics to investigate antibiotic resistance in marine water (Hatosy and Martiny 2015). Ampicillin, tetracycline, sulfadimethoxine and nitrofurantoin were chosen for screening. A total of 466 clones were sequenced, which contained genes not previously associated with antibiotic resistance or with low sequence similarity to any known genes. Sequences matching previously characterized antibiotic resistance genes included β-lactamases, efflux pumps, bacitracin and vancomycin resistance genes. These results have clearly identified the marine environment as reservoirs of known and novel resistance mechanisms. In the only freshwater functional metagenomics study to date, clone libraries were screened for resistance to ampicillin (Vercammen et al. 2013). The water was from a highly polluted river in Belgium. The resulting β-lactamase had 74% sequence identity to the bla_{TEM} gene.

14.5 The Functional Metagenome of Human Microbiomes

Functional metagenomics has been used to investigate the antibiotic resistome of the oral microflora, the adult human gut as well as infant and child gut (Diaz-Torres et al. 2003; Sommer et al. 2009; Moore et al. 2013; Fouhy et al. 2014). The functional metagenomic resistome analysis of the oral microflora identified that human microbiomes are themselves reservoirs of novel resistance mechanisms. They identified the previously characterized tetracycline resistance genes *tetM*, *tetO* and *tetQ* and a novel tetracycline resistance gene *tet*37. The functional metagenome resistomes of saliva and faeces from two adult humans were investigated for resistance to (Sommer et al. 2009) five different classes of antibiotics (amino acid derivatives, aminoglycosides, amphenicols, β-lactams and tetracyclines). Only 22% of the clones contained gene sequences with high sequence similarity (>90% amino acid identity) to known antibiotic resistance genes. Sommer et al. identified 78 genes with low homology (<90% amino acid identity) to known resistance genes. This study raises the question of whether the presence of a novel resistance gene poses a risk to the treatment of human pathogens. The 78 genes conferring a resistance phenotype in the gut microbiome bacteria have not to date been identified in human pathogens. However, they are capable of being expressed in pathogenic bacteria, as they conferred a resistance phenotype when expressed in the host *E. coli*. Two independent studies investigated infant and child gut microbiomes for the presence of known and novel antibiotic resistance genes (Moore et al. 2013; Fouhy et al. 2014). In total, the faecal microbiomes of 44 children were analysed for resistance. Both studies showed the presence of known β-lactamases and aminoglycoside resistance genes. The functional metagenomic clones created by

Moore et al. described resistance to seven classes of antibiotics conferred by known and novel resistance genes (Moore et al. 2013). These genes were confirmed to be present on mobile fragments. These results indicate that the antibiotic resistome is already established in the infant gut at 6 months old in the absence of antibiotic selective pressure.

14.6 The Birds and the Bees Are Reservoirs of Antibiotic Resistance Genes

Three studies have used functional metagenomics to identify the antibiotic resistomes in animal or bird microbiomes and one in an insect (Allen et al. 2009; Kazimierczak et al. 2009; Martiny et al. 2011; Wichmann et al. 2014). The functional metagenome of the organic pig gut resulted in 9000 bacterial artificial chromosome clones. When screened for tetracycline resistance, ten contained known resistance genes (*tetC*, *tetW* and *tet40*). Novel resistance genes, *galE1* and *galE2*, that confer resistance to minocycline and doxycycline were identified. Martiny et al. surveyed the diversity and genomic composition of antibiotic resistance genes in gulls in order to investigate wildlife as a potential method of spreading antibiotic resistance (Martiny et al. 2011). The BAC clones were screened for β-lactam and tetracycline resistance and resulted in the identification of a wide variety of antibiotic resistance genes, which had not previously been identified in wildlife. The β-lactam resistance mechanisms comprised of class A and class C β-lactamases mediated by genes similar to the bla_{TEM}, bl1_ec and RAHN-2 as well as efflux-mediated resistance. Over a third (38%) of the clones contained genes with no known sequence similarity. However, the tetracycline-resistant clones contained only well-characterized resistance mechanisms such as *tet* C, *tet* J and *tet*L efflux pumps and *tet*M, *tet*O and *tet*W ribosomal modification proteins.

A study of dairy cows in the USA screened the functional metagenomic libraries generated from the cow faeces for the presence of β-lactam, aminoglycoside, tetracycline and chloramphenicol resistances (Wichmann et al. 2014). Of 87 clones with genes conferring resistance to chloramphenicol, kanamycin, tetracycline or β-lactam antibiotics, 80 carried unique antibiotic resistance genes. Similar to the gull study, the tetracycline resistance genes shared high sequence similarity to known resistance genes. The only functional metagenomic study of an insect to date was performed on the midgut of a gypsy moth (Allen et al. 2009). This was an early functional metagenomic study, which investigated the presence of β-lactams, erythromycin and chloramphenicol. The resistance mechanisms included efflux pumps, a transcriptional regulator of an efflux pump protein and an extended spectrum class A β-lactamase.

14.7 Where to Next?

The use of functional metagenomics to search non-pathogenic bacteria and unculturable bacteria has opened the door to discover the untapped resources of the microbiomes around us. Many gigabases of DNA are now present in *E. coli* clones, which prior to functional metagenomics were as remote as another galaxy to microbiologists. My final section will address this topic, the untapped resource of the potential antibiotic resistance genes in the functional metagenomes already created. Others will have addressed the complexities of generating the functional metagenomic libraries, but I think that the final frontier is in understanding the many gigabases of DNA contained in the already-generated functional metagenomes and those which are still to be generated. We can create these libraries and interrogate them for a phenotype, but we are still limited in how we can identify specific gene clusters, operons or linked genes. Many of the DNA sequences in libraries, as well as in many of the genomes of well-characterized bacteria, are simply described as unknown function. How do we unlock the functions of these genes and proteins? These are novel resistance genes, for which no relative has yet jumped the barrier into pathogenic bacteria. These I believe are the ones for which we must use functional metagenomics. The reasons for this are to enable us to understand the emergence and spread of resistance genes from their origins in a time when all antibiotic resistance genes were completely novel in all pathogens, i.e. in the pre-antibiotic era.

The identification of the specific genes associated with a phenotype on a DNA insert that may be 50 kb in length is particularly challenging. While DNA sequencing can provide the solution in some cases, if the insert sequence does not resemble known resistance genes, then further investigations are required. The majority of these are low throughput, thereby limiting the efficiency at which novel genes will be detected. One option is transposon mutagenesis. However, this usually functions by inactivating single genes and may not reveal other genes of relevance. Subcloning of the individual genes comprising the 50 kb insert can potentially also provide a solution but can be time-consuming and doesn't guarantee success, especially if more than one gene is required to create the phenotype. A recent study returned to a functional metagenomic library generated from soil collected in 2003, 2004 and 2005 (Allen et al. 2015). They had identified an unusual clone, which did not contain a β-lactamase gene or resistance gene with similarity to a β-lactamase gene. The actual mechanism of resistance was too complex to be directed by a single gene. The resistance phenotype was generated by manipulating the chromosomal DNA of the *E. coli* host to activate efflux pump genes and suppress an outer membrane porin. I think that the second phase of functional metagenomics has now begun. This will be to identify the truly unusual and unexpected mechanisms that the inserted pieces of DNA on the fosmids use to control the host chromosome and physiology. With high-throughput screening, one may identify hundreds or thousands of clones of interest. Therefore, individually cloning each section of each putative gene to test for the phenotype is neither efficient nor feasible. If several genes are required for the resistance phenotype, they too would not be detected by functional metagenomics alone. Therefore, we need to develop techniques, which can identify the functions of the unknown genes and operons. The identification and analysis of novel resistance genes are time-consuming and/or costly, but if we can identify resistance genes before they enter pathogenic bacteria or if we can understand the transfer from reservoirs to the pathogen, then it could potentially save not only money but many lives.

References

Allen HK, Moe LA, Rodbumrer J, Gaarder A, Handelsman J (2008) Functional metagenomics reveals diverse β-lactamases in a remote Alaskan soil. ISME J 3:243–251

Allen HK, Cloud-Hansen KA, Wolinski JM, Guan C, Greene S, Lu S, Boeyink M, Broderick NA, Raffa KF, Handelsman J (2009) Resident microbiota of the gypsy moth midgut harbors antibiotic resistance determinants. DNA Cell Biol 28(3):109–117

Allen HK, An R, Handelsman J, Moe LA (2015) A response regulator from a soil metagenome enhances resistance to the β-lactam antibiotic carbenicillin in *Escherichia coli*. PLoS One 10(3):e0120094

Amos GC, Zhang L, Hawkey PM, Gaze WH, Wellington EM (2014) Functional metagenomic analysis reveals rivers are a reservoir for diverse antibiotic resistance genes. Vet Microbiol 171:441–447

Binh CT, Heuer H, Kaupenjohann M, Smalla K (2008) Piggery manure used for soil fertilization is a reservoir for transferable antibiotic resistance plasmids. FEMS Microbiol Ecol 66:25–37

Cardenas PP, Carrasco B, Defeu Soufo C, Cesar CE, Herr K, Kaufenstein M, Graumann PL, Alonso JC (2012) RecX facilitates homologous recombination by modulating RecA activities. PLoS Genet 8:e1003126

Diaz-Torres ML, McNab R, Spratt DA, Villedieu A, Hunt N, Wilson M, Mullany P (2003) Novel tetracycline resistance determinant from the oral metagenome. Antimicrob Agents Chemother 47:1430–1432

Donato JJ, Moe LA, Converse BJ, Smart KD, Berklein FC, McManus PS et al (2010) Metagenomic analysis of apple orchard soil reveals antibiotic resistance genes encoding predicted bifunctional proteins. Appl Environ Microbiol 76:4396–4401

Fouhy F, Ogilvie LA, Jones BV, Ross RP, Ryan AC, Dempsey EM et al (2014) Identification of aminoglycoside and β-lactam resistance genes from within an Infant gut functional metagenomic library. PLoS One 9:e108016

Handelsman J, Rondon MR, Brady SF, Clardy J, Goodman RM (1998) Molecular biological access to the chemistry of unknown soil microbes: a new frontier for natural products. Chem Biol 5:R245–R249

Hatosy SM, Martiny AC (2015) The ocean as a global reservoir of antibiotic resistance genes. Appl Environ Microbiol 81(21):7593–7599

Healy FG, Ray RM, Aldrich HC, Wilkie AC, Ingram LO, Shanmugam KT (1995) Direct isolation of functional genes encoding cellulases from the microbial consortia in a thermophilic, anaerobic digester maintained on lignocellulose. Appl Microbiol Biotechnol 43:667–674

Heuer H, Smalla K (2007) Manure and sulfadiazine synergistically increased bacterial antibiotic resistance in soil over at least two months. Environ Microbiol 9:657–666

Kazimierczak KA, Scott KP, Kelly D, Aminov RI (2009) Tetracycline resistome of the organic pig gut. Appl Environ Microbiol 75(6):1717–1722

Lang KS, Anderson JM, Schwarz S, Williamson L, Handelsman J, Singer RS (2010) Novel florfenicol and chloramphenicol resistance gene discovered in Alaskan soil by using functional metagenomics. Appl Environ Microbiol 76(15):5321–5326

Liu Y-Y, Wang Y, Walsh TR, Yi L-X, Zhang R, Spencer J et al. (2015) Emergence of plasmid-mediated colistin resistance mechanism MCR-1 in animals and human beings in China: a microbiological and molecular biological study. The Lancet Infectious Diseases, Available online 19 November 2015

Martiny AC, Martiny JB, Weihe C, Field A, Ellis JC (2011) Functional metagenomics reveals previously unrecognized diversity of antibiotic resistance genes in gulls. Front Microbiol 2:238

McGarvey KM, Queitsch K, Fields S (2012) Wide variation in antibiotic resistance proteins identified by functional metagenomic screening of a soil DNA library. Appl Environ Microbiol 78:1708–1714

Moore AM, Patel S, Forsberg KJ, Wang B, Bentley G, Razia Y et al (2013) Pediatric fecal microbiota harbor diverse and novel antibiotic resistance genes. PLoS One 8:e78822

Riesenfeld CS, Goodman RM, Handelsman J (2004) Uncultured soil bacteria are a reservoir of new antibiotic resistance genes. Environ Microbiol 6(9):981–989

Rondon MR, August PR, Bettermann AD, Brady SF, Grossman TH, Liles MR, Loiacono KA, Lynch BA, MacNeil IA, Minor C, Tiong CL, Gilman M, Osburne MS, Clardy J, Handelsman J, Goodman RM (2000) Cloning the soil metagenome: a strategy for accessing the genetic and functional diversity of uncultured microorganisms. Appl Environ Microbiol 66(6):2541–2547

Sommer MOA, Dantas G, Church GM (2009) Functional characterization of the antibiotic resistance reservoir in the human microflora. Science 325:1128–1131

Stein JL, Marsh TL, KY W, Shizuya H, DeLong EF (1996) Characterization of uncultivated prokaryotes: isolation and analysis of a 40-kilobase-pair genome fragment from a planktonic marine archaeon. J Bacteriol 178:591–599

Su JQ, Wei B, Xu CY, Qiao M, Zhu YG (2014) Functional metagenomic characterization of antibiotic resistance genes in agricultural soils from China. Environ Int 65:9–15

Torres-Cortés G, Millán V, Ramírez-Saad HC, Nisa-Martínez R, Toro N, Martínez-Abarca F (2011) Characterization of novel antibiotic resistance genes identified by functional metagenomics on soil samples. Environ Microbiol 13:1101–1114

Udikovic-Kolic N, Wichmann F, Broderick NA, Handelsman J (2014) Bloom of resident antibiotic-resistant bacteria in soil following manure fertilization. Proc Natl Acad Sci U S A 111:15202–15207

Vercammen K, Garcia-Armisen T, Goeders N, Van Melderen L, Bodilis J, Cornelis P (2013) Identification of a metagenomic gene cluster containing a new class A beta-lactamase and toxin-antitoxin systems. Microbiology 2:674–683

Wichmann F, Udikovic-Kolic N, Andrew S, Handelsman J (2014) Diverse antibiotic resistance genes in dairy cow manure. MBio 5(2):e01017

FSC
www.fsc.org
MIX
Papier | Fördert
gute Waldnutzung
FSC® C083411

Kolibri 360 GmbH
Ferdinand-Jühlke-Straße 7
99095 Erfurt, Deutschland
produktsicherheit@kolibri360.de